Optimized Energy Management Strategies for Electric Vehicles

Parita Jain
KIET Group of Institutions, India

Puneet Kumar Aggarwal
ABES Engineering College, India

Mandeep Singh
Bennett University, India

Sushil Kumar Singh
Marwadi University, India

Shailendra Rathore
Division of Cyber Security, School of Design and Informatics, Abertay University, UK

IGI Global
Scientific Publishing
Publishing Tomorrow's Research Today

Vice President of Editorial	Melissa Wagner
Managing Editor of Acquisitions	Mikaela Felty
Managing Editor of Book Development	Jocelynn Hessler
Production Manager	Mike Brehm
Cover Design	Phillip Shickler

Published in the United States of America by
IGI Global Scientific Publishing
701 East Chocolate Avenue
Hershey, PA, 17033, USA
Tel: 717-533-8845
Fax: 717-533-8661
E-mail: cust@igi-global.com
Website: https://www.igi-global.com

Library of Congress Cataloging-in-Publication Data

CIP Data Pending
ISBN:979-8-3693-6844-2
eISBN:979-8-3693-6846-6

British Cataloguing in Publication Data
A Cataloguing in Publication record for this book is available from the British Library.

Table of Contents

Detailed Table of Contents

Urban transport systems in major cities generally use diesel buses and thus, they contribute to greenhouse gas (GHG) emissions. Several cities across the world are now migrating from diesel-powered buses to battery electric buses. Such a migration offers excellent opportunities in terms of both financial and environmental benefits, if the electricity for charging the buses comes from renewable energy resources. This chapter analyzes various issues related to transformation of urban transport systems into futuristic and climate friendly eco-systems. A novel computational framework for comparative analysis is proposed. City of Windhoek, Namibia is considered for the study and the framework is applied to obtain the results.

Electric vehicles, often known as EVs, are often touted as more environmentally friendly and less noisy alternatives to gasoline-powered automobiles. EVs are gaining popularity due to several factors, including declining prices and increased awareness of climate change and environmental issues. This chapter examines the recent developments in electric vehicles (EVs) concerning the trends in battery technology, performance matrix, charging techniques, new challenges in research, and unexplored possibilities. To be more specific, there is an analysis of the current market for electric vehicles across the globe and the potential for the market's growth in the following years. This article presents an in-depth review of the various battery technologies now available, bearing in mind that the battery is one of the most essential components of electric vehicles. In addition, we investigate the several standards presently accessible for charging electric automobiles and the concepts of power regulation and battery energy management.

Renewable energy sources (RES) show great promise in providing sustainable power for electric vehicle (EV) charging infrastructure, offering cleaner alternatives to traditional fossil fuels. Solar-powered and wind-powered EV charging stations tap into the abundant energy from the sun and wind to replenish EV batteries, significantly reducing carbon emissions and promoting environment-friendly transportation. However, integrating RES into EV charging infrastructure presents certain technical challenges. The main issue lies in the variability of renewable energy generation, as it heavily relies on weather conditions. To address this, advanced energy storage solutions and cutting-edge Vehicle-to-Grid (V2G) technology have been explored to ensure a stable and reliable energy supply for EV charging. Additionally, effectively integrating RES with existing power grids requires sophisticated power management and grid balancing strategies to optimize energy usage and prevent overloading. Renewable energy sources' roles in EV charging station infrastructure will discuss in this chapter.

The Energy Management Strategy (EMS) to reduce comprehensive cost of EVs can be approached based on the efficiency improvement using a model, considering various parameters in co-ordination. The main parameters considered are vehicle mass with various resistances such as rolling resistance and average resistance. The energy desired from primary source of battery and emergency source of supercapacitor is consider in the model. Further the model considers energy conversion via converter, bus to motor. The mechanical output of motor is managed by power train/drive train, Hence, overall efficiency improvements is achieved by reducing losses at various stages, such as switching losses of converter, transmission loss of bus, frictional and magnetic loss of motor and energy loss of drive train. The internal considerations for the above loss reduction are discussed in detail along with supercapacitor support in case of emergency and also reverse generation by means of regenerative braking.

Nowadays, Hybrid Electric Vehicles (HEVs) have become the most demanded technology for reducing pollution emissions and fuel saving in the transport sector. The demand growth rate for electric vehicles is rapidly increasing due to zero emissions of carbon dioxide from electric vehicles. The main challenges with HEVs are their limited battery capacity and difficulty in deciding when and where to charge EVs. The Concerned EVs charging problem and range anxieties still make customers uneasy about purchasing an electric vehicle but new advancements in EVs try to resolve the above problems and improve EVs range. The effectiveness of HEVs relies on energy management strategies (EMSs). A big challenge for the researcher is to develop suitable EMSs for HEVs, to enhance the range of EVs and increase their performance and battery life.

The transition to electric vehicles (EVs) is a pivotal step towards achieving sustainability in transportation. However, maximizing the environmental benefits of EVs relies not only on their adoption but also on efficient energy management strategies. This chapter reviewed the different approach to optimize sustainability by leveraging artificial intelligence (AI) for personalized electric vehicle energy management. Traditional energy management systems for EVs often employ generic, one-size-fits-all approaches, neglecting the diverse needs and behaviours of individual drivers. In contrast, our study harnesses the power of AI to tailor energy management strategies to the specific preferences, driving patterns, and charging behaviours of each EV owner. The core of our approach lies in the utilization of machine learning algorithms to analyse vast amounts of data collected from individual EVs, including driving habits, charging patterns, environmental conditions, and energy prices.

Battery electric vehicles are an evolution in the automobile industry as they also contribute to reducing the harmful effects of vehicles on the environment because of fossil fuels and the release of carbon dioxide. The purchasing decision of the consumer is highly dependent on the specific criteria when it comes to the adoption of Battery electric vehicles. This work helps to analyze those specific criteria that are important from a consumer point of view before purchasing a BEV. The selected criteria analyzed in this study are the range of the vehicle, its full charging time, the purchasing price of the vehicle, the battery capacity of the vehicle, and the energy consumption of the vehicle. The analysis to find out the most prominent criteria that affect the purchasing decision of the consumer is done with the help of the Fuzzy AHP approach which provides the weights of the selected criteria and the other technique used is the Technique for Order Preference by Similarity to Ideal Solution TOPSIS, which provides the rank of the selected alternative electric vehicles.

Self-driving cars or also known as Autonomous Vehicles have incorporate with sensors, actuators, sophisticate algorithms, machine learning systems, and high-level processors to run the software. These are vehicles that are programmed to move and be run without the need of a human being. They are of different categories depending on the low-level to high-level automation. The focus of this research is on the ability of SVM to forecast the effectiveness of self-driving cars. Especially in the context of safety and efficiency of autonomous driving considerable importance is given to the prediction of vehicle behaviour. SVM is used in this study to analyse and forecast the performance out the different sensors and environmental measures. The conclusion drawn based on the analysis of results show that the application of SVM enhances the improvement of the level of reliability of the autonomous systems. Thus, the study is demonstrated on various datasets, namely, the KITTI datasets with up to 95.64% accuracy. Finally, K-folds cross validation is performed for the robustness of the Model.

Electric cars (EVs) are becoming more popular, leading to cleaner transportation and energy use expand more However, as EVs connect more with power grids and other digital systems, they become at risk of complex cyberattacks expand more This research explores ways to protect EV energy systems from hacking. It examines current threats targeting EVs, such as weaknesses in charging stations, communication between cars and the grid, and car software. The study emphasizes the need for layered security that includes physical and digital defences. Some key strategies discussed are encryption, intrusion detection, secure software updates, and strong authentication. It also looks at how artificial intelligence, machine learning both can be used to improve real-time threat detection and response. This research aims to provide a comprehensive view of cybersecurity measures needed to protect future EV energy systems from hacking.

Preface

Electric vehicles (EVs) have become a focal point in the global shift towards sustainable transportation. As environmental concerns intensify, governments and industries rapidly transition from fossil-fuel-powered transportation to electric alternatives. However, the effective operation of EVs requires more than just battery innovation—it demands optimized energy management strategies. These strategies include advancements in charging infrastructure, energy-efficient powertrains, battery management systems, renewable energy integration, and AI-driven models for managing power consumption. This book takes a deep dive into the strategies that optimize energy management for EVs. It's about more than improving battery life or building charging stations—it's about developing more innovative ways to manage energy flow using AI, machine learning, and renewable energy integration. We'll explore topics like vehicle-to-grid (V2G) technology, hybrid energy storage systems, and AI-powered traffic management that enhance EV performance and support sustainable energy ecosystems. Every chapter is designed to help us understand the future of transportation and energy through practical solutions that tackle today's challenges. We're living in a time when climate change is forcing us to rethink how we power our lives, and transportation is one of the most significant contributors to global emissions. EVs have become a beacon of hope, offering a cleaner alternative. Governments worldwide are setting deadlines to phase out fossil-fueled vehicles, and cities are racing to build the infrastructure needed to support electric mobility. At the same time, grids are being pushed to the limit—demand is rising, renewable energy sources are being integrated, and energy management has never been more critical. Electric vehicles aren't just passengers in this transformation but active players. With technologies like V2G, EVs can store excess energy when supply is high and return it to the grid when demand peaks. To make this transition work smoothly, we need innovative strategies that ensure energy flows seamlessly between vehicles, homes, and grids—and that's precisely where this book comes in. This book speaks to a broad audience because electric mobility touches all of us—whether you're designing the next generation of EVs or just curious about the changes happening on the roads. Researchers and engineers working to solve the technical challenges behind EVs, and energy management will find the latest insights and innovations here. Energy and automotive industry professionals can explore practical strategies to make electric mobility more efficient, scalable, and sustainable. Policymakers and environmental advocates will benefit from chapters aligning technology with policy, offering ways to drive meaningful transportation and climate action changes. Finally, whether studying electric mobility or interested in how EVs are shaping the future, students and enthusiasts will gain a comprehensive understanding of the evolving transportation and energy landscape. We hope this book will inspire and support further research and innovation in electric mobility, encouraging new strategies for sustainable energy management. We believe this work will be vital in advancing electric vehicle technologies and promoting cleaner, more efficient transportation systems by addressing key challenges such as energy efficiency, grid integration, and renewable energy use. Our goal is to contribute to a future where electric vehicles are not only practical and accessible but also integrated into more intelligent energy ecosystems, driving progress toward a more sustainable world for all.

The Chapter “Electric Buses for Urban Transport Systems” examines electric buses' critical role in reducing urban air pollution and promoting sustainable mass transit. It explores how cities can adopt electric buses to lower operational costs and meet environmental regulations. The chapter discusses charging infrastructure challenges and proposes innovative scheduling solutions to optimize battery usage. It offers insights into how governments and municipalities can scale electric public transportation while benefiting from renewable energy integration, making it essential for urban planners, policymakers, and environmental researchers.

The Chapter “Communication Requirements and Performance Metrics for Electric Vehicle Charging” addresses the need for reliable communication protocols to ensure efficient energy transfer between vehicles and charging stations. It analyzes existing standards such as ISO 15118 and OCPP, emphasizing the importance of interoperability and secure data transmission. The chapter is relevant to researchers and industry professionals focused on building robust EV ecosystems, as it presents strategies for minimizing latency and maximizing system reliability. The chapter contributes to scalable and secure electric vehicle infrastructure globallyby ensuring seamless communication.

The Chapter “Advanced Battery Cooling Systems for Enhanced Efficiency and Longevity” focuses on thermal management technologies essential for maintaining battery performance and longevity. With EV batteries susceptible to degradation under extreme conditions, this chapter offers solutions such as liquid cooling systems and phase change materials. It discusses the impact of these systems on vehicle efficiency, safety, and total cost of ownership, making it relevant to manufacturers, engineers, and researchers in the automotive industry.

The Chapter "Life Cycle Assessment of Emerging Battery Technologies for Electric Vehicles" provides a comprehensive environmental assessment of various battery technologies, including lithium-ion and solid-state batteries. It evaluates the ecological impact of battery production, use, and disposal, offering recommendations for sustainable manufacturing practices. The chapter targets environmental scientists, policymakers, and energy experts, aiming to align battery development with circular economy principles to minimize the carbon footprint of EVs.

The Chapter "Advancements in Renewable Energy Harvesting for EV Charging Infrastructure" discusses how renewable energy sources like solar and wind can be integrated into EV charging networks. It highlights the technological challenges of managing intermittent energy sources and offers solutions such as smart grids and energy storage systems. The chapter is particularly relevant for energy professionals and policymakers looking to reduce reliance on fossil fuels while meeting the rising demand for electric mobility.

The Chapter "Sustainable Energy Solutions for Charging Electric Vehicles" explores innovative energy management solutions that combine renewable sources with EV charging infrastructure. It presents models for energy distribution that enhance grid stability while reducing costs. The chapter is essential for energy companies and urban planners focused on integrating green technologies into public and private transportation networks, advancing the global shift toward sustainability.

The Chapter, "An Efficient Energy Management Strategy to Reduce the Comprehensive Cost of Electric Vehicles", introduces hybrid energy storage systems (HESS) that combine batteries and supercapacitors to optimize power delivery. It explains how real-time energy management can extend battery life and lower operating costs. The chapter is significant for automotive engineers, researchers, and policymakers focused on reducing the total cost of ownership for EVs, making electric mobility more accessible to consumers.

The Chapter, "A Smart Traffic Management System for Improving EVs' Range Using Machine Learning Algorithms", demonstrates how machine learning algorithms can be applied to traffic management systems to extend EV range. It discusses energy consumption challenges in congested areas and offers predictive models for optimizing traffic flow. This chapter is highly relevant to researchers, city planners, and transportation authorities, providing innovative solutions for managing urban traffic while improving vehicle performance.

The Chapter "Optimizing Sustainability: Harnessing AI for Personalized Electric Vehicle Energy Management" explores how artificial intelligence can create personalized energy management strategies for individual EV users. It presents AI-driven models for optimizing charging schedules, reducing energy costs, and enhancing user convenience. The chapter is designed for researchers, automotive companies, and smart grid operators seeking to leverage AI for sustainable transportation solutions.

The Chapter "A Soft Computing-Based MCDM Approach: Selection of the Best Electric Vehicle toward SDG-7" introduces a multi-criteria decision-making (MCDM) model that helps consumers select the most suitable electric vehicle. It aligns with the United Nations Sustainable Development Goal 7 (SDG-7) by promoting access to affordable, sustainable energy. The chapter is relevant to policymakers, manufacturers, and environmentalists, offering a framework for evaluating EVs based on environmental, economic, and performance criteria.

The Chapter "Autonomous Vehicle Performance Prediction System Using Support Vector Machine" applies support vector machine (SVM) models to predict the performance of autonomous vehicles (AVs). It emphasizes the importance of accurate predictions for enhancing AV safety and reliability under various traffic and environmental conditions. The chapter is geared toward researchers, data scientists, and automotive companies developing autonomous driving technologies that rely on robust machine learning frameworks.

The Chapter "Securing Vehicle-to-Grid Communications: A Cyber-Physical Approach" addresses the cybersecurity challenges of integrating electric vehicles with the power grid through vehicle-to-grid (V2G) communication. It discusses advanced encryption methods, intrusion detection systems, and real-time threat mitigation. This chapter is essential for cybersecurity professionals, energy providers, and policymakers focused on building secure and resilient innovative grid systems.

The Chapter "Innovations in Energy Management for Enhanced Efficiency and Sustainability: Advancing Electric Mobility", explores innovative solutions such as bidirectional charging, blockchain technology, and advanced battery management systems. It discusses the economic and environmental benefits of integrating these technologies into EV networks. The chapter is relevant to energy experts, automotive engineers, and researchers working toward sustainable, efficient electric transportation.

This book aims to bring together technology, policy, and sustainability in a way that's accessible and actionable. Each chapter addresses a specific challenge in EV energy management—from optimizing batteries and traffic flow to securing vehicle-to-grid communications. The ideas shared here don't just look good on paper; they have the potential to shape real-world systems that are more efficient, resilient, and environmentally friendly. This book encourages collaboration among researchers, engineers, industry leaders, and policymakers by exploring the intersections between energy, transportation, and technology. It shows how electric vehicles can go beyond reducing emissions to becoming integral to a more brilliant, sustainable energy future. Ultimately, this book invites us to learn, innovate, and work together toward a world where clean energy powers our journeys and progress.

Parita Jain

KIET Group of Institutions, Ghaziabad, India

Puneet Kumar
Aggarwal, ABES Engineering College, Ghaziabad, India

Mandeep Singh
Bennett University, Greater Noida, India.

Sushil Kumar Singh
Rajkot, Gujarat, India

Shailendra Rathore
Abertay University, UK.

Section 1
Electric Mobility and Urban Transportation

Chapter 1
Electric Buses for Urban Transport Systems

Taimi K. Uutoni
https://orcid.org/0009-0008-4391-4319
Namibia University of Science and Technology, Namibia

K. S. Sastry Musti
https://orcid.org/0000-0003-4384-7933
Namibia University of Science and Technology, Nambia

Hilde Amushembe
https://orcid.org/0009-0006-9557-3136
Namibia University of Science and Technology, Namibia

ABSTRACT

Urban transport systems in major cities generally use diesel buses and thus, they contribute to greenhouse gas (GHG) emissions. Several cities across the world are now migrating from diesel-powered buses to battery electric buses. Such a migration offers excellent opportunities in terms of both financial and environmental benefits, if the electricity for charging the buses comes from renewable energy resources. This chapter analyzes various issues related to transformation of urban transport systems into futuristic and climate friendly eco-systems. A novel computational framework for comparative analysis is proposed. City of Windhoek, Namibia is considered for the study and the framework is applied to obtain the results.

DOI: 10.4018/979-8-3693-6844-2.ch001

INTRODUCTION

Fossil fuel-powered vehicles, particularly public transit, are causing serious environmental problems, including air pollution and intensifying the greenhouse effect. In China, emissions reached 43.6 million tons in 2017, and the transportation sector accounted for 70% of domestic petroleum consumption in 2016. Public transit contributes to 34% of total vehicle emissions and consumes significant amounts of fossil fuel, posing a challenge to energy security(Lin et al., 2019). The country's growing population and increasing economic development are driving up the number of vehicles. As a result, these vehicles use billions of litters of fuel and pollute the air quality index of the cities at the global level. Due to an increase in global warming worldwide and the target of reaching net zero carbon emissions by 2050 (Colombo et al., 2023), various regions are moving away from internal combustion engine vehicles to electric vehicles (John & G., 2021).

Over the last decade, several cities worldwide have been adopting electric vehicles as a green and clean mobility solution to address emerging issues (Colombo et al., 2023; John & G., 2021). Replacing public transport, such as internal combustion engine buses, with electric buses will make a better choice, when it comes to reducing carbon emissions and energy savings. A study was carried out to develop an integrated planning model for a heterogeneous fleet consisting of electric buses and electric taxis, along with the associated charging stations to accommodate travel requirements (Tian et al., 2022). Trends over the recent years clearly indicate that electric buses are preferred, due to their energy-efficiency when compared to diesel buses for city transportation. A few recent electric buses do come with solar panels on top, which means the bus will be charging while in operation. This might reduce the cost of charging infrastructure to an extent. At the same time, a few important aspects should be taken into account – such as heating, ventilation, recharging schedule, energy storage, and air conditioning besides physical attributes of the bus itself (Tian et al., 2022).

Several studies have indicated about the transition to electric buses for urban transport. Majority of the studies are about planning and/or operation related to specific cities due to renewable energy resource availability locally. This chapter presents the case of the capital city of Windhoek, Namibia. As of today, Windhoek is experiencing increase in the usage of private vehicles and, an increase in travel demand, leading to travel congestion with the increase in level of local air pollution. Due to the rapid increase in diesel-based private vehicles and decreased use of public transport, the country will experience higher greenhouse gas emissions. Then, pressure to reduce greenhouse gas emissions in the transport sector will increase. On the other hand, if the transition to electric vehicles happens abruptly, electricity demand increases and this also results in shortage of electricity for other purposes in general. As of now, Namibia imports 60 to 70% of electricity from neighbouring countries and thus it is recommended that external dependency in energy sector be reduced to the extent possible.

Therefore, to protect the environment and air quality of the cities, and also to reduce external dependency for electricity, adoption of electric vehicles in Namibia is highly desirable. Given the challenges associated with electric vehicles, such as the availability of charging stations, the ability to maintain and repair vehicles, the introduction of Battery Electric Buses (BEBs) presents significant challenges. Although Namibia has included a target for 10,000 EVs by 2030 in its Nationally Determined Contribution (NDC), the adoption of EVs is happening slowly and as of 2024, only five light vehicle charging stations exist throughout the country. Overall, Namibia contributes relatively lesser carbon emissions when compared to other countries, due to various measures have been taken to offset the carbon emissions. At the same time, it is desirable to increase the use of renewable energy sources to meet the future electricity demands and further reduce the emissions from the energy and transport sectors (Matheus and Musti, 2023; Musti, 2023a.).

This chapter reviews many research publications on the energy consumption of electric vehicles utilizing diverse methodologies. This chapter examines several methodologies for designing charging stations for electric buses and analyses the associated costs using a life cycle cost analysis. This chapter concludes with a review of carbon emissions published by different researchers and recommendations.

Literature Review

Several studies worldwide were conducted to do energy consumption calculations for EVs to transform internal combustion engine vehicles (An et al., 2023; Bracco et al., 2021; Colombo et al., 2023; Preetha Yesheswini et al., 2020). A study in Malanga, Spain, promoted the decarbonization of public transportation, focused on a diesel-powered line, and analyzed its benefits (Colombo et al., 2023). The 2030 Carbon Peak Action Plan emphasizes the need for the country to actively encourage the use of new energy vehicles, with a particular focus on pure electric vehicles (PEVs) as the primary means of transportation at this stage (An et al., 2023). Nevertheless, the adoption and implementation of PEVs in Northeast China continue to face challenges. The primary factor is that the real-world energy consumption and driving range of vehicles on the road in low-temperature conditions are somewhat worse than the officially certified values (An et al., 2023). This discrepancy surpasses the expectations of the general public (An et al., 2023). Estimating the energy consumption of electric buses is necessary for efficient energy management of electric vehicles. This includes tasks such as determining the best routes, coordinating charging, and implementing vehicle-to-grid (V2G) technology (Jiang et al., 2022). Various studies of solar PV-powered charging were undertaken, charging stations were designed using the MPPT algorithms, and the performance simulations were done using software solutions like Matlab etc., (John & G., 2021; Kaur & Bindal, 2023; Preetha Yesheswini et al., 2020; Shariff et al., 2020). At the same time, it is interesting to see various authors have reported on migration to EVs in specific cities around the world. This section provides a brief overview of such contributions.

Case study in Bufallo, Newyork City

Electrifying public transportation is being implemented in several US cities and globally, with 13% of all U.S. transit agencies having electric buses. The number of transit agencies committing to purchasing electric buses is increasing, with California committing to having all electric buses statewide by 2040. Other countries have already begun their transition from diesel buses to electric buses, with approximately 385,000 electric buses in operation worldwide by the end of 2017. However, 99% of these electric buses are located in China (Meyers, 2021). Analysis of energy consumption between electric buses and internal combustion engine buses was conducted in Buffalo where the data for the annual maintenance cost of a diesel bus was borrowed from the NYC Metro Transportation Authority (MTA). The data for the mix of buses is assumed to be the approximate fleet size at the year end of 2019. The Niagara Frontier Transportation Authority (NFTA) is both a job creator and supporter, creating jobs and an income source for families in large metropolitan areas. By investing in zero emission electric buses, the NFTA would help further reduce their emissions and contribute to a more sustainable future for public transportation. The study examined the use of electric buses in other U.S. cities and their experiences with the new public transit bus technology to learn from and apply lessons to the local community.

A Case Study in Malanga city Spain

The study aims to transition from ICE vehicles to electric vehicles (EVs) using a mathematical model, considering Newton's second law and four resistances such as rolling, aerodynamic, grade, and curvature resistance, to evaluate performance, fuel consumption, and emissions. The study presents a case study in Spain promoting the decarbonization of urban public transport, focusing on a diesel-powered line, to analyze its benefits. A case study is performed in which the vehicle performs one cycle, equivalent to one trip. The model does not consider the possibility of the E-bus speed changing. The study examined the carbon footprint reduction of a bus line in Malaga, Spain, transitioning from diesel to electric energy. The E-bus produced significantly less CO_2 emissions than the Euro5 standard engine, reducing emissions by nearly one-third.

Therefore, the following formula (Newton's law) was used:

$$F_i - \sum R_i = m_{eq} \times a \qquad (1)$$

Variables in the equations are defined as follows;

F_i Is the force in the direction of the motion;

$\sum R_i$ Is the sum of the resistive force applied to the vehicle;

m_{eq} is the equivalent mass of the vehicle;

a is the acceleration to which the vehicle of equivalent mass is subjected.

The case study considered a less than 13 km route that required 45 minutes to complete with 33 stops. The route also took into account the direction with the largest slope, which may affect energy usage. There are three stages to a driving cycle: braking, constant speed, and acceleration.

The performance of the electric bus, together with the formulas used in this case study, are presented in Figure 1. The graph presented the speed versus the speed to cover for one trip. At the start, the vehicle is stationary, resulting in zero speed (v) and traction effort (ft) (as shown in the formula for when the bus is at the station). Electric power consumption (P) is not considered zero to maximize E-bus consumption. The vehicle begins motion, with speed (v_i) and power (P) following a discrete integration, involving traction effort (F_t), speed (v_i), efficiency (μ), and auxiliary power consumption (P).

Figure 1. Bus performance for the selected route in Malanga City, Spain ((Colombo et al., 2023)

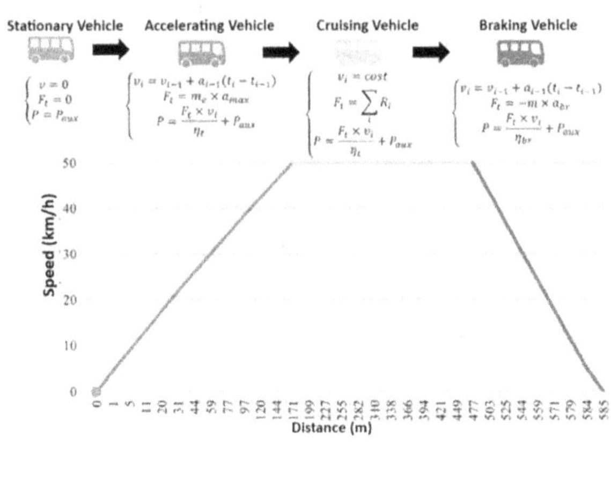

The E-bus speed changes its trend once the vehicle reaches its base speed, but this model does not consider this possibility, as the vehicle maintains its speed constant after reaching its local speed limit which is 50km/h. This is formulated under cruising vehicles. Finally, during braking vehicle, the vehicle's tractive effort and power decrease as it approaches a stop due to a braking comfort acceleration of -0.9 [m/s2]. The study presents a graphic analysis of e-bus behaviours in cities, focusing on their power and energy needs. This study reports that the bus has a maximum power output of 260 kW, improving traction force and speed. The energy diagram tracks the vehicle's usage along the journey, with a maximum State of Charge (SoC) of 80% and a Depth of Discharge (DoD) of 13%. Carbon emission is also analyzed based on the data collected from Spanish national energy mix. The case study focused on the performance of e-buses, not diesel buses.

Case Study of North Italy

The paper presents a methodology to estimate the energy consumption of a fully electric bus in different traffic and operating conditions. Starting from driving cycles evaluated through a simulation model developed in Aimsun, the Matlab/Simulink energy model calculates the main electrical and mechanical performance indicators such as the state of charge of the battery, power, and torque provided by the electric motor. The methodology is applied to a real case study in an Italian town in the North of Italy, for which different traffic conditions and drivers' behaviours are analyzed (Bracco et al., 2021). The methodology proposed to evaluate the energy consumption of an electric vehicle is based on two different cascading simulation models: one in Aimsun (to derive driving cycles) and the other in Simulink (to calculate energy consumption and other technical indicators). Aimsun is a widely adopted software package for traffic modelling, planning, and simulation at macroscopic levels in both urban and extra-urban contexts. It can reproduce actual traffic conditions and perform detailed assessments of the state of the network, such as flow analysis, congestion levels, queue propagation dynamics estimation, and emission evaluation.

The main inputs of the model can be divided into two categories: vehicle-rated data and time-dependent operating data. The main outputs of the Simulink model are the real speed profile of the vehicle and its acceleration, current and voltage of the battery during the charging and discharging phases, the power exchanged between the battery and the motor, the battery state of charge (SOC) profile, and the electric consumption measured in Wh/km. The data collected by 14 cameras and 12 sensors between 7:00 a.m. and 9:00 a.m. is used to estimate and calibrate the origin-destination matrix representing the mobility demand of users. The traffic conditions are used to define traffic scenarios, such as Sc (standard traffic condition), Cc (congested case), and CPc (parking manoeuvres). The simulation model evaluates the final SOC value of a battery and the mean energy consumption of a bus. Results show a strong dependency on factors like passenger count, tire pressure, external temperature, driver behaviour, and traffic conditions. Energy consumption increases with passengers, while external temperature significantly impacts thermal comfort. Driver behaviour affects the state of charge slightly more than tire pressure.

Case Study, Beijing City, China

A case study in Beijing city China, by (Jiang et al., 2022) introduced two probabilistic models, utilizing Bayesian and quantile regression, to estimate the probabilistic distribution of energy consumption in electric buses. The models are trained and validated using real-world driving data from a fleet of 10 electric buses over a year. The study validates proposed methods on electric bus driving data from the National Big Data Alliance of New Energy Vehicles (NDANEV). The data was collected from May 2020 to May 2021 and consisted of 1 year of driving data from 10 buses with a 15-second sampling period. The study uses a preprocessing method to separate the original data into trip segments and charging segments based on vehicle state and latitude and longitude information. The model's convergence is evaluated using the Mean Square Relative Error (MAPE) and the parameters used were reported as well. The results show that the Bayesian model's PIT distributions are more skewed for each temperature range, indicating that the Bayesian model's estimated intervals underestimate high energy consumption at low temperatures, overestimate low energy consumption at high temperatures, and overestimate the entire width of the probabilistic interval at normal temperatures. In contrast, the PIT distribution of quantile regression is substantially less skewed, as it estimates intervals based on quantiles, which are sensitive to the deviated samples and better capture the distribution of energy consumption.

Case Study in India

This case study proposes the use of solar-based sustainable power sources for charging stations, as EVs require renewable energy. The research presents findings from the design of an Indian-specific solar-powered EV charging station, which was tested using MATLAB-Simulink (Kaur & Bindal, 2023). Power demand has surged due to industrialization and population growth, leading to an increase in the use of sustainable energy sources and EVs. Solar PV panels through converter MPPT algorithm systems are implemented to generate energy while the battery of EVs fulfils the demand of power during deficiency of solar radiation. The system has been implemented in MATLAB to verify the system's design and performance analysis.

Several studies have concluded that plug-in hybrid electric cars (PHEVs) have little effect on the power system if they are charged during the nightly off-peak hours. However, the biggest displacement of petroleum will occur if cars are permitted to charge throughout the day, increasing the range of all-electric vehicles. Solar photovoltaics (PV) provide a potential solution for providing peaking capacity in the middle of the day. Plug-in electric vehicle (PEV) grid integration is a major innovation in the electric power system, and this study provides a PEV charging strategy that responds to market signals across time scales and systems while taking transmission and distribution integration difficulties into account. The research investigates the feasibility of developing an infrastructure for charging electric vehicles (EVs) at workplaces. The goal is to use PV energy as much as possible for EV charging while utilizing the grid as little as possible. The EV battery serves as a PV energy storage system, which lessens the adverse effects of integrating large amounts of PV into the distribution network.

Case Study of Haikou City, China

According to the case presented in Haikou City, China, the challenge to transforming to electric vehicles is the longer charging time of electric vehicles (Tian et al., 2022). To address this issue, the case study focused on investing in electric vehicle fleets such as taxi and bus charging stations in Haikou City. The road map was obtained using the built-in function in MATLAB, and the planning model was validated using the actual data of Haikou City. Four typical bus routes were selected for this case study, and one electric bus and one taxi were considered in this research to compose the mixed fleet. The mathematical formulation was completed using mixed-integer linear programming (MILP). The proposed model achieves minimal investment and operation costs, optimal deployment of a mixed EB and ET fleet, and charging stations, according to simulation results.

Migration to alternate fuels, cleaner fuels is not an easy task and can take long time due to several factors. One of the major challenges is to secure financing. However, due to the advent of renewable energy technologies and green financing models securing the funding seems easier than earlier times. Since transition to BEBs happens over the time and involves several stakeholders, the role of government in terms of setting appropriate policies is critical (Musti, 2023b; Shirunda et.al, 2023).

From the above, it can be seen that different researchers have used different methodologies to do energy consumption analysis, but in most cases, they did not incorporate internal combustion energy consumption analysis and mostly focused on electric buses only. Various approaches have been taken to design charging stations, with many focusing solely on small electric vehicles and not on buses. Despite reviewing more than ten articles, none have compiled a full report of the investigation that includes all four components, such as energy consumption analysis for electric buses and internal combustion engine buses, charging station design, carbon emission analysis, and life cycle cost analysis, in one document. In view of these identified research gaps, this chapter makes an effort to address the issues related to environmental, physical, and economic implications of migrating to BEBs for the public transportation and specifically the case of Windhoek City, Namibia is considered. The objectives of this chapter include:

- Energy consumption analysis of individual bus categories (ICEB and Electric Bus)
- Designing a solar-powered charging station for electric buses
- Calculating the charging station cost estimate and doing a life cycle cost analysis
- Carbon emission analysis

Methodology

To meet the set objectives as stated in the earlier section, different computations will have to be undertaken for different stages of engineering analysis. Therefore, this chapter proposes a comprehensive computational framework for the end-to-end engineering analysis as presented in figure 2. Various stages of the proposed computational framework are explained with appropriate mathematical equations and sequence of computations.

Figure 2. Proposed Computational Framework

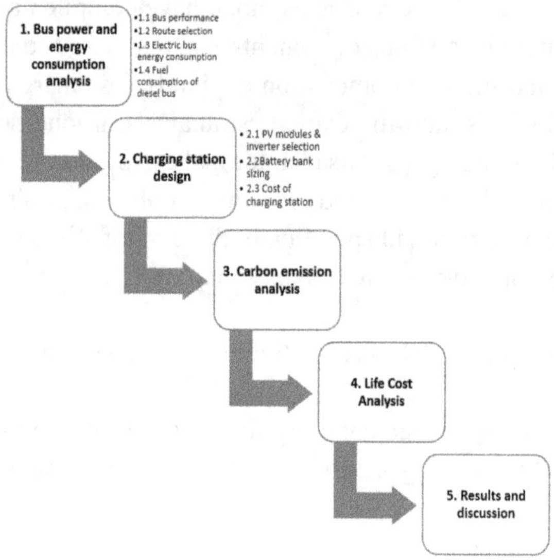

Different sets of calculations need to be performed at each stage, as shown in the framework and the figure 2. Since this chapter considers solar energy for charging the BEBs, location needs to be fixed due to obvious reasons. For this work, City of Windhoek is considered and the same is explained in the next section.

Case of City of Windhoek, Namibia

For the bus to move the total force applied to a bus to move has been calculated using simple physics equations. In stage one, the energy and fuel consumption of a public transport bus are computed and influenced by numerous factors. The fuel consumption of a city bus is affected by factors such as different speeds, modes of acceleration and deceleration, uncertain durations at bus stops, congestion, and the load which depends on the number of passengers being transported (Ji et al., 2022; Jiang et al., 2022). The capital city of Windhoek is considered for this study, just as several authors have conducted their investigations for a specific location. In stage two, the total force and the selected route power required are calculated. Special emphasis was given to investigating the influence of rush hour traffic and congestion on energy use, therefore the acceleration was assumed to be constant. To support the growing EV charging station market in the context of global warming concerns caused by the conventional Internal Combustion Engines(John & G., 2021). Solar energy powers the charging station, representing a possible source of pollution-free.

To understand further that the replacement of existing buses with electric buses is economically feasible, the life cycle cost analysis is to be determined. The LCC analysis consists of initial costs, annual fuel expenses, and maintenance expenditures. Additionally, it will encompass the infrastructure and the expense of battery replacement for the electric buses. The LCC calculation for each bus alternative involved determining the present value of each cost component and subsequently aggregating all cost components for each year throughout the bus's lifespan of 25 years.

Given that the case study further aims to reduce CO_2 emissions produced by the outdated fleet, it is vital to examine the extent of carbon footprint reduction when substituting diesel with electric energy. Although energy generation significantly influences global CO_2 emissions; so, this case study examines the emissions generated to provide the service. All the results and discussions are presented in the form of graphs and tables for better understanding.

Data Collection, Parameters and Assumptions

Various sources were consulted to gain a comprehensive understanding of the bus fleet. Data on the average yearly maintenance cost was incomplete, necessitating certain assumptions to ensure a comprehensive study. The data about the existing diesel buses, including the timetable and route map are obtained from the City of Windhoek. The number of diesel buses and the current bus schedule in the COW fleet serve as crucial data points in this analysis. The precise number of buses fluctuates annually for numerous reasons; therefore, this analysis utilizes a constant of 10 buses.

Applying the Computational Framework

Optimizing the powertrain of a battery-electric bus (BEB) is crucial for reliable operation and cost-effectiveness. Bus operators need to understand energy consumption under various conditions, such as weather, driving cycles, and passenger occupancy, to intelligently integrate BEBs in cities and identify suitable bus types. The characteristics of the bus are based on the specifications presented in Table 1. The initial step involves selecting the electric bus and determining the principal load forces that it will encounter, namely aerodynamic drag F_d, rolling resistance F_{rr}, climbing resistance F_{cc}, and acceleration force F_a.

Table 1. Specifications of the buses (Source: Volvo 9700, 2024)

Description	Electric bus (Volvo 9700)	ICE-bus(MAN LION'S CITY 12)
Bus length (m)	12	12.18
Width (m)	2.55	2.55
Height (m)	3.3	3.06
Permitted GVW (kg)	19 000	19 000
Number of passengers	90 persons	90 persons
Emission standard	electric	diesel
Storage (l or kWh)	350	200
Charging	Electric charger	Oil station

The following equations illustrate the determination of various forces associated with a moving bus.

$$gravitational = g = 9.806 \text{ m/s}^2$$

$$Airdensity@27°C = \rho = 1.2kg/m^3$$

$$Frontbusarea = A = 2.55x3.3 = 8.415m^2$$

$$Dragcoeffient = C_d = 0.645$$

$$Rollingeffiecient = \mu = 0.007$$

Considering the city traffic jam, the bus will reach a maximum speed of 22 km/h in 10 minutes, which is equivalent to 600 seconds. For four trips per day, buses will travel 11 km one way and 22 km two ways. The route consists of nine stops. In our calculations, the stops will be ignored and the speed will be set constant at 22 km/h and velocity calculate in equation 1.

$$\mathbf{Velocity} = \mathbf{v} = \frac{22 * 1000}{3600} = 6.111\mathbf{m/s} \quad (1)$$

Therefore, the forces applied to the bus are calculated and shown in figure 3.

$$\mathbf{Dragingforce} = \mathbf{F_d} = \frac{1}{2} * \boldsymbol{\rho} * \mathbf{C_d} * \mathbf{A} * \mathbf{V}^2 = 121.575\mathbf{N}$$

Considering the weight (w) of the bus in Table 1 and incline angle $\theta° is 5°$, the rooling force calculated as follows;

$$\textbf{Rollingforce} = \mathbf{F_{rr}} = \mathbf{w * g * \mu * cos\theta} = 1299.23\mathbf{N}$$

$$\textbf{Climbingforce} = \mathbf{F_{cc}} = \mathbf{w * g * \mu * sin\theta} = 113.668\mathbf{N}$$

$$\textbf{Accelerationforce} = \mathbf{F_{a}} = \mathbf{wa} = 193.483\mathbf{N}$$

Where;

$$Acceleration = a = \frac{v}{t} = 0.01 m/s^2$$

$$\textbf{Totalforce} = \mathbf{F_{t}} = \mathbf{F_{d} + F_{rr} + F_{cc} + F_{a}} = 1727.956\mathbf{N}$$

Figure 3. Applied forces for a bus to move

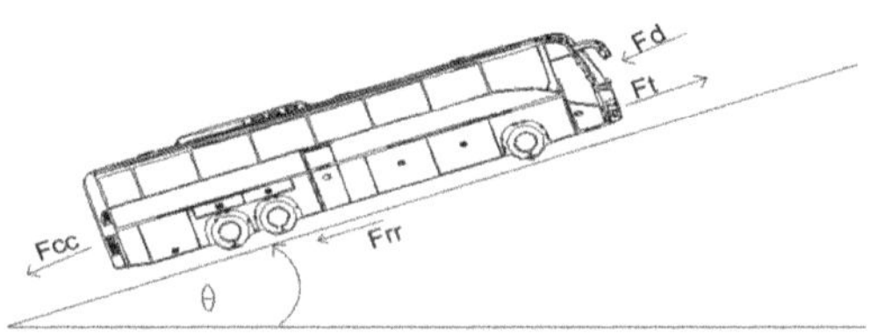

Now, route selection computations are taken up. Stops in the route have been identified, distances between these stops, and the average times for the travel are shown in table 2. For simplifying the discussion, this research assumes a fully loaded bus at stop No. 2, which maintains a constant acceleration until it offloads at stop No. 9. The $Constantacceleration = a = 0.01 m/s^2$.

Table 2. Distance between different stops

Stop No.	Stop name	Distance (Km)	Time
1	Siyanda	1	05:25
2	Justine	2	06:00
3	Werhnil	8	06:20
Total distance		11	

The electric bus trip started at 05h25 from Siyanda and arrived at Werhnil at 06h20. The total distance for one trip is 11 km, and it takes 40 minutes (2400 seconds) to complete it. The selected bus route consists of nine stops with maximum slope of 7.7% as shown in figure 4.

Figure 4. The section bus route with stops, elevation profile, and high traffic volume

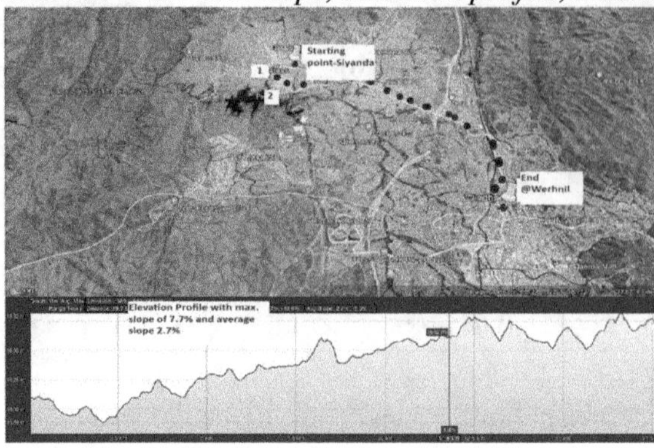

For better analysis, further the route simplified in figure 5.

Figure 5. Route further simplified

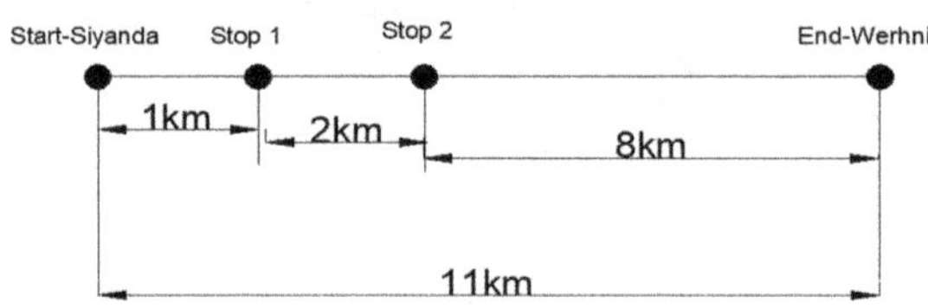

Considering the physical dimensions and technical specifications of the bus for the computations, the wheel radius is 0.5 m. The torque required is calculated in equation 2;

$$\textbf{Torque} = \textbf{F}_t * \textbf{r} * \textbf{d} \tag{2}$$

= 19 007 516 *Nm*, Where *d* is the total distance for one trip which is 22km.

The power required is $\textbf{P} = \textbf{Torque} * \textbf{a} = 190075.16\ \textbf{W}$

As far the energy consumption of a typical electric bus, the researchers have found that BEBs typically consume between 1–2 kWh/km for standard buses, 1.5–3 kWh/km for double-decker buses, and 2–3.5 kWh/km for articulated buses (Verbrugge et al., 2021). In heavy traffic and extreme weather conditions, BEBs consume more energy, with propulsion systems and auxiliary systems potentially increasing energy consumption by up to 35%. Thus, our work considers bus energy consumption to be 2 kWh/km.

The energy required per trip= $\mathbf{E} = \mathbf{E_c * d}$

$= 2 * 22 = 44 \ kWh/trip$

Where;
E_c is energy consumption per km
d is the total distance per trip.

Therefore, the overall energy consumption for one round trip, leave and back to Siyanda, will consumes 44 kWh. Further, one bus requires four trips per day, therefore the energy required daily is $= 44 * 4 = 176 \ kWh/bus$

Assume electricity costs N$4 /kWh (which is about 0.23 USD). Therefore, the total electricity cost per day per bus is $= 176 * 4 = N\$704.00$.

$Electricity\ cost\ per\ year = N\$704 * 365\ days$

=N$256, 960.00 (USD 14,631.41)

For comparing the energy consumptions, the fuel consumptions of both BEBs and diesel buses will be considered. Fuel consumption in public transport buses is influenced by various factors such as speed, acceleration, topping time, congestion and passenger load same applies as in the previous section of energy consumption for electric buses. Śmieszek & Mateichyk (2021) used the Vehicle Specified Power (VSP) software, since it is very difficult to determine the fuel consumption through manual calculations. The fuel consumption model, based on actual city bus mileage measurements, uses various parameters to determine its relevant parameters. A model based on this method was used in computer simulations to determine bus fuel consumption in urban traffic, focusing on the impact of rush hours and congestion. The results reveal that the number is unrealistic for a city bus with passengers, therefore, for this research, a manual methodology is used to get better results of bus with passengers.

According to fuel eco driving monitoring case of city and intercity buses in Sweden (AMTFT, 2024), bus consumes 25 to 30 litters per 100 km

Therefore, in one km a bus consumes= $\frac{25}{100} = 0.25 \ litres$

The maximum speed of a city bus is 22km/h, taking into account traffic jams and number of stops, bus consume 0.25 litres. Given the same daily distance of 88 km, total diesel consumption per day is $=0.25 * 88 \ km = 22 \ liters$

Cost of fuel determination: A cost of N$23.00 =1.31 USD is considered for a litter of diesel based on the prevailing prices. Therefore, the total cost of diesel fuel per day is equal to 22 x 23 = N$506.00 (USD $ 28.00). The annual fuel cost per bus is equal to = 506.00*365 days=N$ 184 690.00 = USD 10, 516.32

Electricity Requirement Analysis

Now, to provide the required electricity to charge the BEBs, calculations related to charging station are taken up. Electric bus utilization in specific areas can further decrease carbon dioxide emissions by utilizing renewable energy sources like photovoltaic (PV) charging. Additionally, this could yield advantages such as mitigating the risk of local grid overloading, promoting self-sufficiency, and enhancing energy self-consumption. Integrating BESS (battery energy storage system) into these charging stations will enhance the proportion of solar energy provided to electric vehicles (EVs) while bolstering the grid's ability to withstand the inherent fluctuations in PV (photovoltaic) electricity. The charging station with main components is presented in figure 6.

Figure 6. Charging station scheme

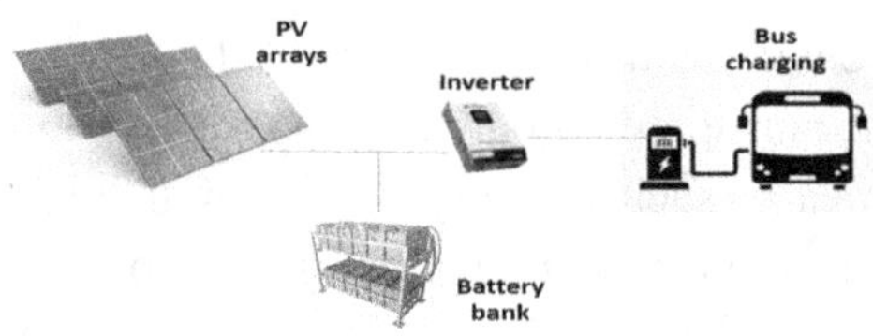

Assume the city transport system has 10 electric buses each with a battery capacity of 350kWh and energy consumption per bus per day is 176 kWh.

The energy required for 10 buses is TE $=10*176 = 1760$ **kWh**

Peak Sun Hours: $PSH = 6hours$. It needs to be noted that peak sunshine hours are different from location to location. This key aspect can affect the overall results dramatically based on the location. In general, most places in Namibia specifically Windhoek City, enjoy more than 7hours of sunshine per day. So, a moderate value of 6 hours is considered in this study.

Solar power required to charge 10 buses $\mathbf{SE} = \frac{\mathbf{TE}}{\mathbf{PSH}} = \frac{1760\mathbf{kWh}}{6}$

$$= 293.33 \textbf{ kW}$$

Considering the system efficiency=90% and panel degradation 85%

Then solar power (SE) is required to charge 10 buses = $\frac{293.33}{0.9*0.85}$ =383.44 kW

Sizing the Solar PV Plant

To meet the electricity / energy requirements as determined earlier, further computations are required to design the solar PV plant with appropriate number of PV modules and inverters. Using the specifications presented in Table 3, the number of panels, series connections, and parallel strings were calculated.

Table 3. Solar panels specifications (Ningbo, 2024)

Parameters	Values
Panel power W_P	570 W
V_{oc}	51 V
V_{mp}	36 V
I_{mp}	8.6 A
I_{SC}	14.23 A
PSH	6 hours

Using the data in Table 3, the number of panels to power the station is calculated as follows;

$$\textbf{Solarpanelrequired} = \frac{\textbf{SE}}{\textbf{W}_\textbf{p}} = \frac{383.44\textbf{kW}}{0.57\textbf{kW}} = 672.7 \textbf{ panels}$$

Therefore, the station requires 673 panels of 570 W.

An inverter of the specifications presented in Table 4 is selected for these charging stations;

Table 4. Inverters' specifications (Ningbo, 2024)

Parameters	Values
Max. Power (W)	18200
Max. Voltage (V)	500
Short circuit voltage (V)	125
PV input current (A)	26+26+26
Max PV Short circuit current (A)	44+44+44
No. of MPP trackers	3
No. of strings	2+2+2

$$\textbf{Number of inverters required} = \frac{\textbf{SystemPower}}{\textbf{Inverterpower}} = \frac{383.44\textbf{kW}}{18.2\textbf{kW}} = 21.06$$

The above computation results in a total of 22 inverters. A combination of series and parallel of the panel requires the system to operate effectively. Using the information from the solar panel and inverter specification above series connection is calculated as follows:

Series connections:

Number of panels to be connected on one inverter

$$= \frac{\textbf{InverterMax.Voltage}}{\textbf{V}_{\textbf{oc}}} = \frac{500}{51} = 9.8 \textbf{ panels}$$

total number of panels connected in series=10

Parallel Connections

Max. PV current to an inverter:$I_{SC} = 44A$

PV short circuit current is $I_{PV} = 14.7A$

$$\text{Number of strings} = \frac{\mathbf{I}_{\mathbf{SC}}}{\mathbf{I}_{\mathbf{PV}}} = \frac{44}{14.7} = 2.993$$

Therefore 10 panels in series with 3 strings and 22 inverters are required for this charging station. Figure 7 shows the overall series and parallel connections of the solar panels.

Figure 7. Series and Parallel connections of the panels

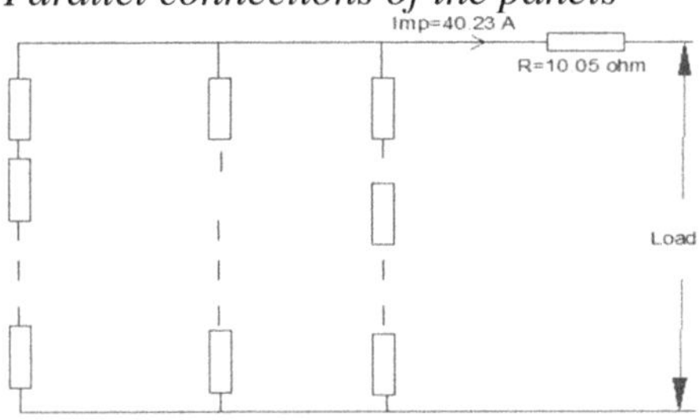

The full load voltage is 500V, therefore the voltage regulation is calculated as per equation (3);

$$\mathbf{VR} = \frac{\mathbf{V}_F - \mathbf{V}_{NL}}{\mathbf{V}_{NL}} \tag{3}$$

$$= \frac{500 - 422.3}{422.3} * 100\%$$

$$= 18.39\%$$

Where;

VR=Voltage Regulation
V_F= Full load Voltage
V_{NL}=No load Voltage

Sizing the Battery Storage System

Since, solar power cannot be produced during the non-sunshine hours, it is a common practice to set up a battery energy storage system. A battery fundamentally stores electrical energy into chemical energy, which can be released as electricity when required. The battery charging process entails the conversion of electrical energy into chemical energy while discharging reverses this process. Battery energy storage systems regulate energy charging and discharging, frequently employing advanced control systems to deliver power when necessary or most economically advantageous. Battery storage is essential in the energy balance for multiple reasons. It facilitates the integration and expansion of renewable energy sources, hence diminishing dependence on fossil fuels. Storing surplus energy generated during periods of high renewable output (sunny or windy conditions) alleviates the intermittency challenges linked to renewable resources. To avoid energy not being served, the excess energy produced when the buses are in operation and to charge buses when in the absence of sunshine; battery storage is designed as illustrated below:

Let us consider the number of days of autonomy as 3 days. Energy required for 10 buses is 1760 kWh.

State of Charge (SOC) is 50%; Depth of Discharge is 85% battery efficiency is 85% and the system voltage is 400 V.

$$\text{Therefore, battery bank capacity} = \frac{1760 * 3days}{1-(0.5*0.85*0.85)}$$

$$= 19411.76\ kWh$$

$$\text{Amp-hour} = \frac{1000 * \textbf{Batterybankcapcity}}{\textbf{System voltage}}$$

$$\text{Therefore, the Amp-hour of the battery bank} = \frac{1000 * 1760}{400}$$

$$36666.66\ AH@48Volts$$

Lithium battery of 48 V, 200AH is selected, therefore, the number of batteries required is

$$No.ofbatteries = \frac{36666.66}{200} = 183.33$$

Total number of batteries for this charging station is 184.

Sizing the Charger System

One of the key aspects in the BEBs is the selection and sizing of the electric charger. Electric vehicle (EV) chargers are categorized into several levels based on their charging speed, which is denoted by the number of kilowatts (kW) they can deliver. The pace at which the bus battery recharges increases proportionally with the power of the charger, up to the maximum charging rate permitted for the vehicle. Currently, three types of chargers can be found in the market: Level 1 (L1), Level 2 (L2), and DC Fast Charger (DCFC), which is sometimes referred to as Level 3. For this chapter level 3 and Table 5 present the evaluation of the charging standards as per IEC (International Electro-Technical Commission) 61851-1.

Table 5. EV charger and standards as per IEC 61851-1(Mohamed & Mohamed, 2020)

Types of EV charger	Ordinary charging 1-phase	Ordinary Charging 3-phase	Fast and extreme fast charging	
	V=230 V AC, I=16 A, P=3.7 kW	V=400 V AC, I=32 A, P=43 KW	V=500 V AC, I =250 A, P=220 kW	V=600 V AC, I= 400 A, P= 240 kW
Battery (kWh)	**Charging Time**			
100	27 h	2.5-4.5 h	30 min	25 minutes
40	11 h	60-120 minutes	10 min	10 minutes
20	5.5 h	30-60 minutes	5 min	5 minutes
10	3 h	15-30 minutes	<5 min	<5 minutes

To minimize the charging period of an electric vehicle battery, sufficiently powerful chargers are necessary, adhering to fundamental physical rules:

$$P_{ch} = \frac{E}{t_{ch}}$$

where:
P_{ch} is charger power in watts,
E is bus battery energy in watt-hours, and
t_{ch} is charging time in hours.

Although ABB recently introduced an electric bus fast charger that can charge an electric city bus within 4-6 minutes, this type of charger is not currently available in African countries. Therefore, to fully charge a 350kWh battery (Volvo 9700 electric bus) in 2 hours, a charger with a power rating of 220 kW is considered.

Cost Estimation

After determining the parameters and sizing the key components, cost estimation is necessary to determine the practicality of installing a charging station. Rather than installing panels on top of an existing roof, it would be better to initiate the carport PV system where in shaded parking can be provided to the vehicles during charging. Even roof-tops of offices owned by city transport departments can be used. In fact, the recent trends show that solar based electricity is being produced in rivers and canals with other benefits as well (Matheus and Musti, 2023; Musti, 2023b). Therefore, the system cost estimation must include the supporting structure for the carport PV system, in addition to the PV array, charge controller, batteries, and inverter. Any electrical system requires protection, necessitating the determination of fuses and circuit breaker sizes.

PV panels connected in parallel require protection therefore DC fuse rating is calculated as follows:

$$\mathbf{DC_{fuse} = I_{SC} * SF}$$

Where SF is the safety factor.

$= 14.23 * 1.25 = 17.78A$, this points to the selection of the closest fuse, which is of 20 A rating. The system requires 97 fuses since the panels connected in parallel are 97. The panels are divided into groups of 10, 9 groups have 10 panels each, and 1 group has 7 panels and spare for future planning.

$$\mathbf{CB_{group} = I_{SC} * 10 = 142.3\ A}$$

A circuit breaker rating of 150 A is selected, therefore the system will have 13 of 150 A circuit breaker. From this, the main circuit breaker for the entire PV panel system was calculated.

$$\mathbf{CB_{PV} = I_{SC} N_P}$$

$= 14.23 * 97 = 1380.31\ A$, this points to the selection of a circuit breaker of 1500 A rating.

According to the literature review, an approximate cost estimate of a charging station runs between N$700,000 to N$ 5,500,000. In this work, a detailed cost analysis is undertaken, however the details are omitted for the sake of simplicity and only the final results are presented in Table 6.

Table 6. Cost estimate of charging station for 10 buses

Description	Units	Unit price	Total price
PV module 570W each	673	200	134,600
MPPT (1000V 80A)	7	4,000	28,000
Inverters(18200 W)	22	20,000	44,0000
Storage battery 200Ah 48 V	183	2500	457,500
Electric bus charger (DC-450kW fast charger)	10	2,500	25,000
DC fuse (20 A)	13	100	1,300
PV circuit breaker 1500A	1	800	800

continued on following page

Table 6. Continued

Description	Units	Unit price	Total price
Group circuit breaker 150 A	13	500	6,500
DC cabling (10 m^2 x 2C) (m)	10000	300	3million
Carport structure (Area=1628 m^2)	1	200 000	200 000
Total			4milion
Labour 15%			644 055
Contingencies 15%			644 055
VAT 15%			644 055
Grand Total			$6 million

Carbon Emission Analysis

One of the primary goals of the transition to BEBs is to offset the carbon emissions. For this, Carbon dioxide emission analysis needs to be done, given the size of solar energy produced as above. Carbon dioxide emissions are classified into two categories; activity-based method and energy-based method. In this chapter, the energy-based method is considered and carbon emission is calculated in Equation 4.

$$CO_2 \text{ emission} = V \times D \times CF \times FF \qquad (4)$$

where;
V= Number of buses
D= Distance travelled during the day
CF= Average fuel consumption factor per km
FF= Average carbon dioxide factor per km

Consider 10 buses' travel distance of 88 km each per day, carbon emission is 1150.1 g/km for each bus (Euro 6 standard), and each bus consumes 0.25 L/km of fuel. Therefore, the average fuel consumption factor is calculated in equation 5:

$$CF = \frac{1.150\,\mathbf{Kg/Km}}{0.25\ \mathbf{L/km}} = 4.6\ \text{Kg/L} \qquad (5)$$

Total Carbon emission for 10 buses= 10 x 88 x 1.150 x 4.6 =4655.2 kg/day

Life Cycle Cost Analysis

One of the standard expectations of migrating to a different energy class or an alternative option, is the Life cycle cost analysis. Life Cycle Costing (LCC) is a management tool that estimates investment support for alternatives, aiding managers in decision-making by distributing expenses over the product life cycle, focusing on inflation rate, time value of money, and price fluctuations. The study aims to decide between electric buses and diesel buses in terms of environment and economic perspective by calculating the life cycle cost.

The required distance per annum has been set to 16 060km per bus on the bus line of Siyanda to Wernil for a total distance of 11km per trip. According to (El-agouz et al., 2021) the following assumptions have been set to determine the life cycle cost of the buses;

- Lifetime for 15, 20 and 25 years
- The estimated purchase cost charging station cost of the electric bus is based on the values presented in Table 4.
- The inflation rate is set at 7.5%
- The interest rate is set to 4.6%

As per (Potkány et al., 2018) the LCC was calculated using equation (6).

$$LCC = Inital\ cost + operating\ cost * NPF + \left(Resedual\ price * \frac{1}{(1+r)^n}\right) \quad (6)$$

Where;

The initial cost includes the cost of the bus and the charging station for the electric bus. Operating expenses for the specified cost categories (fuel expenses, electricity expenses, maintenance expenses, oil change expenses, tire change expenses).

Cost of a typical electric bus: N$ 336, 966.21; Cost of the charging station: N$ 6, 225, 865.00

Initial cost= N$ 336, 966.21 + N$ 6, 225, 865.00= N$ 6, 562, 831.21

The overall operating cost of an electric bus is electricity consumption is N$ 4 per km (includes expenses towards tire change and general maintenance and services etc., for every 10,000 km) on an average. Thus, the operating cost= 256, 960.00 + (4 * 32120) = N$ 385, 440.00

The initial cost of the diesel bus is N$500 000.00 no charging station is required because of the existing fuel station. Operating costs of diesel buses include fuel consumption, replacement of tires, and annual bus service.

Table 7. Structure of cost for LCC calculations

Type of Buses	Electric bus	Diesel bus
Initial cost (includes charging station)	6million	5million
Distance (per year)	32120	32120
Power required to move a bus (kW)	190	190
Average electricity consumption (kWh/year)/fuel consumption (Liters/year)	65240	8030
Cost of electricity consumption/ fuel consumption (N$/year)	23milion	29milions
Maintenance cost (N$/ year)	128480	200000
Residual value	0	0
Discount rate %	1.63	1.63

NPF is Net Present Factor= $\frac{(1+\mathbf{d})^{n}-1}{(1+\mathbf{d})^{n}*\mathbf{d}}$

d is the discount rate
n is the lifetime of the bus that set to 25 years.

$$d = \frac{interest\ rate * 100}{inflation\ rate\ 9 * 100}$$

$$= \frac{7.5*100}{4.6*100} = 1.63\%$$

$$.: NPF = \frac{(1+0.0163)^{25}-1}{(1+0.0163)^{25}*0.0163} = 20.40$$

Assume residual price is Zero, LLC for 25 years calculated:

Electric Bus LCC= 6, 225, 865 + 385, 440.00 * 20.4 + 0 = N$ 492, 204, 289.6

Diesel Bus LCC=500, 000+ 3,130, 950 * 20.4+ 0 = N$ 64, 367, 364.01

$$\text{Payback period} = \frac{\textbf{The grand total cost of Electric bus}}{\textbf{Annual Savings}}$$

Where,

Annual savings = Total Number of PV modules x Module Peak Power x PSH x 365 days' x Electricity tariffs =673 x 0.57 x 6 x 365 x 4= N$ 3, 360, 423.60.

The total cost of an electric bus (including charging infrastructure) as calculated earlier is N$ 7, 525, 865.00 and this results in a payback period of 2.2 years.

Comparison of Results and Discussion

From the previous sections, it can be seen that the proposed computational framework for determination of benefits and overall costs is applied for BEBs and diesel engine buses. The benefits can be different in the sense that BEBs enjoy the environment friendly aspect, due to their zero carbon emissions. On the other hand, diesel buses have the advantage of matured technology and know-how in both maintenance and operation. A precise comparison of results obtained in this work to those reported by other researchers is impossible due to the use of different systems, different bus model specifications, and different locations. However, overall benefits and costing estimates obtained through the use of our proposed framework more or less agree with the results reported in the literature thus far. At the same time, it should be noted that the use of BEBs in urban areas is still in early development stages.

On the other hand, the life cycle cost analysis in this chapter, compares various scenarios over 15, 20, and 25 years. This is in contrast to previous studies that focused solely on 10 years (Potkány et al., 2018) and 12 years (Meyers, 2021). Design of charging stations and sizing the circuit breaker and other protective apparatus are not considered by others. Additionally, many researchers did not do off-grid systems (Lin et al., 2019) but focused on grid-tied systems, which are not 100% carbon emission-free.

Our proposed framework like other researchers, considers the behaviour of the route, power, and energy requirements. By understanding the power and energy needs, it is feasible to analyse other factors regarding the service and ultimately, the reduction of carbon footprint associated with replacing the bus. Based on this aspect, further comparisons have been made and the results are presented in figure 8.

Figure 8. Comparison of BEB with ICE bus

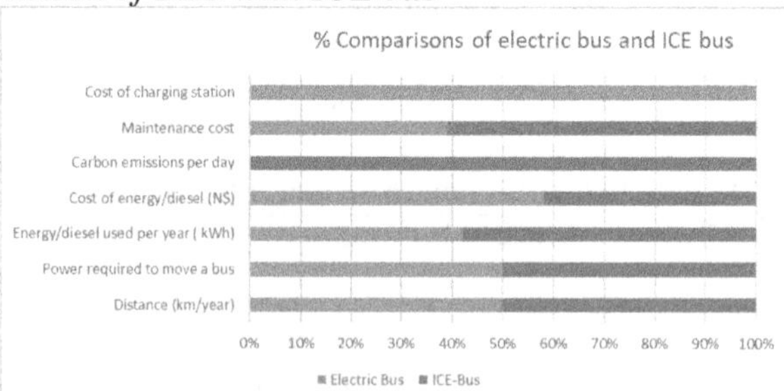

Based on the results presented in Figure 8, and with approximate costs of electricity and diesel; [N$ 4.00 (or USD 0.23) for electricity and N$23.00 (USD 1.30) for diesel], it is evident that the electric buses can be economical in the long run, when compared to the internal combustion engine buses. BEBs in general tend to do better during operation due to their ability to restore energy at stops and when sloping. Further, the results show that the maintenance cost of an electric bus is less compared to an ICE bus. However, such a benefit can be offset to an extent as BEBs require charging stations to be established, which comes with a certain initial capital investment. On the other hand, it is evident that migration to BEBs results in the reduction of carbon emissions, while the diesel bus releases a total of 4655.2 kg CO_2 per day.

RECOMMENDATIONS

Migration to BEBs in urban transport systems will help in reducing the carbon emissions, if the electricity supply comes from clean sources such as the solar. In the case of Windhoek, Namibia, this is achievable since the sunshine hours are higher when compared to most other places. One of the challenges can be to source the land and finances for migration to BEBs. Gradually integrating electric buses into the fleet would provide the transport authorities with the opportunity to acquire knowledge and comprehension about the operations of electric buses on a limited scale, before determining the optimal strategy for expanding the deployment to the entire fleet. To realize tangible benefits, a prudent approach would be to prioritize the replacement of the oldest diesel buses with the new electric buses. By participating in the transition to electric buses, it is possible for Windhoek city, and by extension, any city in general, to significantly advance towards cleaner, futuristic and sustainable urban transport systems.

CONCLUSION

This chapter has considered the problem of replacing legacy urban transport buses by electrical buses to reduce overall greenhouse gas emissions. A new, computational framework is proposed for determining the energy requirements for a typical electric bus. Based on the energy requirements, a solar PV based charging station is designed using standard computations. Then a detailed life cycle cost analysis along with the comparative analysis of legacy and electrical buses is presented. City of Windhoek in Namibia is considered for this study and the proposed computational framework is utilized to estimate the reduction of greenhouse gas emissions. Results and analysis point to the fact that migrating to electric buses will result in economic and environmental benefits

REFERENCES

Abd Elsamii Mahfouz, A. E., & Idris, A. (2023). Toward sustainable transportation in developing areas. *Journal of Al-Azhar University Engineering Sector*, 18(66), 74–82. DOI: 10.21608/auej.2023.283022

AMTFT. (2024), Advanced machinery telematics and fuel monitoring by Technoton, available online, Retrieved October 2024, from https://jv-technoton.com/

An, X., Liu, Y., Li, J., Yu, H., & Hu, X. (2023). Travel Scenario Characteristics Effects on the Energy Consumption Deterioration of Electric Vehicles under Low Temperature. 2023 4th International Conference on Clean and Green Energy Engineering (CGEE), 128–133. DOI: 10.1109/CGEE59468.2023.10351963

Barragán-Escandón, E. A., Zalamea-León, E. F., Terrados-Cepeda, J., & Vanegas-Peralta, P. F. (2020). Energy self-supply estimation in intermediate cities. *Renewable & Sustainable Energy Reviews*, 129, 109913. DOI: 10.1016/j.rser.2020.109913

Bohlourihajar, K., & Harper, M. (2024). Catalyzing Urban Health and Sustainability: The Impact of Electric Bus Fleet Electrification in Justice 40 Communities of Salt Lake County. 2024 Forum for Innovative Sustainable Transportation Systems (FISTS), 1–6. DOI: 10.1109/FISTS60717.2024.10485534

Bracco, S., Bianco, G., Siri, S., Barbagelata, C., Casati, C., & Siri, E. (2021). Simulation Models for the Evaluation of Energy Consumptions of Electric Buses in Different Urban Traffic Scenarios. *2021 Sixteenth International Conference on Ecological Vehicles and Renewable Energies (EVER)*, 1–6. DOI: 10.1109/EVER52347.2021.9456661

Colombo, C. G., Longo, M., Foiadelli, F., Garcia, M. D., & Yaici, W. (2023). Green Mobility Application in Malaga: Analysis of the Transition from ICE to EV. IEEE Int. Conference on Environment and Electrical Engineering (EEEIC / I&CPS Europe), 1–5. DOI: 10.1109/EEEIC/ICPSEurope57605.2023.10194877

Colombo, C. G., Miraftabzade, S., Aimar, M., Zaninelli, D., Longo, M., & Yaici, W. (2022). A Comprehensive Study on Electrification of Old Bus Fleets: A Real Case Study in Ottawa. *2022 International Symposium on Power Electronics, Electrical Drives, Automation and Motion (SPEEDAM)*, 844–849. DOI: 10.1109/SPEEDAM53979.2022.9842033

Elagouz, N., Onat, N., & Kucukvar, M. (2021). Environmental, Economic, and Social Life Cycle Impacts of Alternative Fuel Buses: The Case for Qatar. 2021 10th International Conference on Power Science and Engineering (ICPSE), 92–97. DOI: 10.1109/ICPSE53473.2021.9656863

Holland, S. P., Mansur, E. T., Muller, N. Z., & Yates, A. J. (2021). The environmental benefits of transportation electrification: Urban buses. *Energy Policy*, 148, 111921. DOI: 10.1016/j.enpol.2020.111921

Ji, J., Bie, Y., Zeng, Z., & Wang, L. (2022). Trip energy consumption estimation for electric buses. *Communications in Transportation Research*, 2, 100069. DOI: 10.1016/j.commtr.2022.100069

Jiang, J., Bao, B., Meng, F., Ma, Y., Zhang, H., Jin, Y., Pan, F., & Yuan, X. (2022). Probabilistic Energy Consumption Estimation for Electric Buses. 2022 4th International Conference on Smart Power & Internet Energy Systems (SPIES), 1767–1771. DOI: 10.1109/SPIES55999.2022.10082616

John, S., & G., V. (2021). PV fed Electric Vehicle Charging Station With Power Backup. 2021 13th IEEE PES Asia Pacific Power & Energy Engineering Conference (APPEEC), 1–6. DOI: 10.1109/APPEEC50844.2021.9687739

Kaur, N., & Bindal, R. K. (2023). Modeling and Simulation Analysis of Solar Charging Station for Electric Vehicle. 2023 5th International Conference on Smart Systems and Inventive Technology (ICSSIT), 344–349. DOI: 10.1109/ICSSIT55814.2023.10061147

Lin, Y., Zhang, K., Shen, Z.-J. M., Ye, B., & Miao, L. (2019). Multistage large-scale charging station planning for electric buses considering transportation network and power grid. *Transportation Research Part C, Emerging Technologies*, 107, 423–443. DOI: 10.1016/j.trc.2019.08.009

Matheus, M. N., & Musti, K. S. (2023). Design and Simulation of a Floating Solar Power Plant for Goreagab Dam, Namibia. In Human Agro-Energy Optimization for Business and Industry (pp. 1-26). IGI Global. DOI: 10.4018/978-1-6684-4118-3.ch001

Meyers, J. C. (2021) "The Great Transition: A Cost-Benefit Analysis of Transitioning from Diesel Fuel Buses to Zero Emission Electric Buses for the NFTA in The Buffalo-Niagara Falls MSA". Applied Economics Theses. 44., available online. last accessed October 2024, https://digitalcommons.buffalostate.edu/economics_theses/44

Mohamed, M. A., & Mohamed, F. A. (2020). Design and Simulate an Off-Grid PV System with a Battery Bank for EV Charging. *Universal Journal of Electrical and Electronic Engineering*, 7(5), 273–288. DOI: 10.13189/ujeee.2020.070502

Musti, K. S. S. (2023a). Multicriteria Decision Analysis for Sustainable Green Financing in Energy Sector. In *Green Finance Instruments, FinTech, and Investment Strategies Sustainable Finance* (pp. 3–25). Springer International., DOI: 10.1007/978-3-031-29031-2_1

Musti, K. S. S. (2023b). Energy Sector Growth and Sustainability: Role of Government. In Leadership and Governance for Sustainability (pp. 186-203). IGI Global. DOI: 10.4018/978-1-6684-9711-1.ch010

Ningbo (2024), Ningbo Deye Inverter Technology Co., Ltd. | Solar Components | China. Available online, Retrieved October, 2024, from https://www.enfsolar.com/deye-inverter

Panta, U., Gairola, P., & Nezamuddin, N. (2024). Modelling benefit-to-cost ratio for initial phase electrification using battery electric bus. *Transport Policy*, 145, 137–149. DOI: 10.1016/j.tranpol.2023.10.016

Potkány, M., Hlatká, M., Debnár, M., & Hanzl, J. (2018). Comparison of the Lifecycle Cost Structure of Electric and Diesel Buses. *Nase More (Dubrovnik)*, 65(4), 270–275. DOI: 10.17818/NM/2018/4SI.20

Preetha Yesheswini, B., Jai Iswarya, S., Amani, B., Prakash, P., & Sindhu, M. R. (2020). Solar PV Charging Station for Electric Vehicles. *2020 International Conference for Emerging Technology (INCET)*, 1–7. DOI: 10.1109/INCET49848.2020.9154187

Shariff, S. M., Alam, M. S., Ahmad, F., Rafat, Y., Asghar, M. S. J., & Khan, S. (2020). System Design and Realization of a Solar-Powered Electric Vehicle Charging Station. *IEEE Systems Journal*, 14(2), 2748–2758. DOI: 10.1109/JSYST.2019.2931880

Shirunda O.S., Kanyimba K., Musti K.S.S (2023), Policy perspective for developing electric vehicle ecosystem, in Ecosystem Dynamics and Strategies for Startups Scalability, pp. 175 – 195, DOI: DOI: 10.4018/979-8-3693-0527-0.ch009

Śmieszek, M., & Mateichyk, V. (2021). Determining the fuel consumption of a public city bus in urban traffic. *IOP Conference Series. Materials Science and Engineering*, 1199(1), 012080. DOI: 10.1088/1757-899X/1199/1/012080

Surya, S., & Patil, V. (2019). Cuk Converter as an Efficient Driver for LED. 2019 4th International Conference on Electrical, Electronics, Communication, Computer Technologies and Optimization Techniques (ICEECCOT), 49–52. DOI: 10.1109/ICEECCOT46775.2019.9114648

Tian, M., Wei, Y., Huang, X., & Ding, Z. (2022). Integrated Charging Station and Mixed Fleet Planning for Electric Taxi and Electric Bus. 2022 IEEE/IAS Industrial and Commercial Power System Asia (I&CPS Asia), 687–692. DOI: 10.1109/ICPSAsia55496.2022.9949794

Verbrugge, B., Hasan, M. M., Rasool, H., Geury, T., El Baghdadi, M., & Hegazy, O. (2021). Smart Integration of Electric Buses in Cities: A Technological Review. *Sustainability (Basel)*, 13(21), 12189. DOI: 10.3390/su132112189

Volvo 9700 | Volvo Buses. (2024). Available online, Retrieved October 2024, from https://www.volvobuses.com/en/coaches/coaches/volvo-9700.html

Wenz, K.-P., Serrano-Guerrero, X., Barragán-Escandón, A., González, L. G., & Clairand, J.-M. (2021). Route prioritization of urban public transportation from conventional to electric buses: A new methodology and a study of case in an intermediate city of Ecuador. *Renewable & Sustainable Energy Reviews*, 148, 111215. DOI: 10.1016/j.rser.2021.111215

Zhou, Y., Liu, X. C., Wei, R., & Golub, A. (2021). Bi-Objective Optimization for Battery Electric Bus Deployment Considering Cost and Environmental Equity. *IEEE Transactions on Intelligent Transportation Systems*, 22(4), 2487–2497. DOI: 10.1109/TITS.2020.3043687

KEYWORDS AND DEFINITIONS

Battery Electric Bus (BEB): A battery electric bus is similar to that of a diesel-powered bus but essentially is driven by an electric motor and obtains energy from on-board batteries. They need to charged periodically, as and when batteries are discharged.

Carbon Emissions: Gases, such as carbon dioxide, that are discharged or vented into the environment by automobiles. Internal combustion engine buses emit pollutants like carbon dioxide, nitrogen oxides, and particulate matter, but electric buses do not produce any emissions during operation.

Charging infrastructure: Charging infrastructure encompasses the necessary equipment and technologies for delivering charging services, such as electrical supply circuits, charging stations, control and management systems, and data transmission devices for overseeing, controlling, and facilitating payment. Electric buses necessitate the establishment or enhancement of charging infrastructure, whereas internal combustion engine buses depend on pre-existing refuelling infrastructure.\

Energy consumption: refers to the overall energy demand for a specific process, and it is quantified in kilowatt-hours (kWh). This includes the utilization of electricity, gas, fuel, oil, and biomass. Energy consumption per bus for a given distance.

Environmental impacts: refer to alterations in the natural or constructed surroundings caused by specific activities, which might have negative consequences on the atmosphere, soil, water, aquatic life, wildlife, or the ecosystem's occupants. Electric buses exhibit a reduced environmental footprint when taking into account the complete life cycle of the vehicle and energy sources.

Fuel efficiency: The term "efficiency" refers to the extent to which a vehicle can convert the energy from fuel into kinetic energy for propulsion. Electric buses are more energy-efficient than internal combustion engine buses, resulting in lower fuel efficiency for the latter.

Operating cost: The costs related to the continuous supervision and organization of your firm. Electric buses typically have reduced operational expenses as a result of the comparatively lower expense of electricity in comparison to diesel or gasoline.

Range: Variation range refers to the extent of difference between the upper and lower bounds on a specific scale. Internal combustion engine buses possess greater ranges compared to the majority of electric buses as a result of constraints in battery technology.

Chapter 2
Communication Requirements and Performance Metrics for Electric Vehicle Charging: A Comprehensive Review

Pranshu Saxena
Bennett University, India

Parita Jain
KIET Group of Institutions, India

Puneet Kumar Aggarwal
https://orcid.org/0000-0001-5616-8564
ABES Engineering College, India

Mandeep Singh
https://orcid.org/0000-0003-3278-4877
Bennett University, India

Sachin Goel
Raj Kumar Goel Institute of Technology, Ghaziabad, India

Monica Batra
ABES Engineering College, Ghaziabad, India

ABSTRACT

Electric vehicles, often known as EVs, are often touted as more environmentally friendly and less noisy alternatives to gasoline-powered automobiles. EVs are gaining popularity due to several factors, including declining prices and increased awareness of climate change and environmental issues. This chapter examines the recent developments in electric vehicles (EVs) concerning the trends in battery technology, performance matrix, charging techniques, new challenges in research, and unexplored possibilities. To be more specific, there is an analysis of the current market for electric vehicles across the globe and the potential for the market's growth in the following years. This article presents an in-depth review of the various

DOI: 10.4018/979-8-3693-6844-2.ch002

battery technologies now available, bearing in mind that the battery is one of the most essential components of electric vehicles. In addition, we investigate the several standards presently accessible for charging electric automobiles and the concepts of power regulation and battery energy management.

1. INTRODUCTION

Electric vehicles have recently come to the forefront as a potential solution to the environmental issues caused by traditional modes of transportation. These issues include carbon emissions and the amount of gasoline that is consumed. Developing a reliable infrastructure for charging these vehicles is critical to facilitating the widespread adoption of electric vehicles (EVs). To efficiently manage the charging process, there needs to be reliable and timely communication between the charging systems and the EVs. In the research that has been done on EVs, many different radio access technologies (RATs) have been thought of as ways to make communication between highly mobile EVs and their charging subsystems easier and to capture and transfer data about the state of charge (SoC), user locations, and charging decisions. This informatison can be used to set up charging schedules and choose the best routes for electric cars (Sanguesa et al., 2021, Y. yan et al., 2012). This study aims to conduct a literature evaluation on vehicular communications to control and organise EV charging. The first part of the research focuses on analysing the requirements for communication and the beneficial communication technologies applicable to vehicular communication (Yan et al., 2012).

It looks at the necessity of quick and dependable connectivity to enable easy and effective charging of EVs. High mobility and various network circumstances are two issues and challenges unique to vehicle communication that are addressed. After reviewing existing literature, the research offers insights into the data exchange techniques, network architectures, and communication protocols needed for efficient EV charging management. The research explores the significance of physical layer security techniques in safeguarding communication channels and preserving the integrity of EV charging systems and communication requirements. It addresses privacy and cyber-physical security issues by examining different methods and strategies to protect communication and stop unauthorised access. It is carefully examined how various RATs contribute to coordinating and managing EV charging. The study looks at the latency, throughput, dependability, and security properties of several RATs. It investigates how these technologies may improve communication between EVs and the infrastructure for charging, allowing for effective data interchange and coordination (Y. He et al., 2012, D. Patil et al., 2018). The research also thoroughly examines the standards, potential technologies, and communication requirements for the Internet of Electric Vehicles (IoEV). In order to facilitate smooth communication and efficient management of the EV ecosystem, the IoEV seeks to combine EVs, charging infrastructure, and other network elements into a cohesive system. The survey sheds light on ongoing developments and potential future paths for electric vehicle demand management by examining the existing research efforts in this field. This assessment of the literature concludes by emphasising the significance of effective communication systems for organising and managing EV charging (S. habib et al., 2018, N. Javaid et al., 2018). This survey is a significant resource for parties developing and deploying EV charging infrastructure by summarising the communication requirements, helpful technology, and ongoing research initiatives. The results of this study increase demand management for electric vehicles and aid in our understanding of effective communication tactics.

1.1 Need of Study

Recent trends with the price of petrol, diesel, CNG, and autogas are increasing rapidly. While the price of gasoline typically ranges between 9 and 10 rupees per km, the typical operating cost of an electric vehicle is between 1.2 and 1.4 rupees per mile. If one takes this into account over a more extended period, say six to eight years, the sum that will ultimately spend on battery charging will not come close to exceeding one hundred thousand rupees because of how cheap power is. Figure 1 shows a comparative analysis of different fuelling from Jun-2017 to Mar-2023 (Saguesa et al., 2021, KN. Mude 2014, E. Ghanam et al., 2021, Y. J. jang, 2016).

1.2 The Motivation of the Study

There are several reasons which motivated to write this study, is are illustrated below;

- The economic motivation for the participation of electric vehicles in the electricity market is highly beneficial as per km cost is drastically reduced compared to petrol, Diesel, CNG, and LPG.
- The environment benefits greatly from using EVs, which produce no pollutants from their tailpipes.
- The driving experience provided by EVs is superior because they can generate swift torque. This enables EVs to shoot off of starting lines and deliver acceleration and deceleration that are both smooth and responsive. Additionally, the center of gravity is lower in EVs, contributing to enhanced manoeuvrability, responsiveness, and ride comfort.

- EVs can reduce the use of oil. The study suggested that the utilisation of oil results in various issues, and despite this, the majority of energy for transportation is still derived from oil. Electric vehicles can shake up the current system and provide a cleaner and more effective solution for everyone's transportation needs. Overall, the usage of electric vehicles in the United States has the potential to reduce daily oil consumption by 1.5 million barrels by the year 2035.

Figure 1. Analysis of trends in the price of Petrol, Diesel, CNG, and LPG between Jun-2017 to March-2023. (Source: Petrol Planning and Analysis Cell)

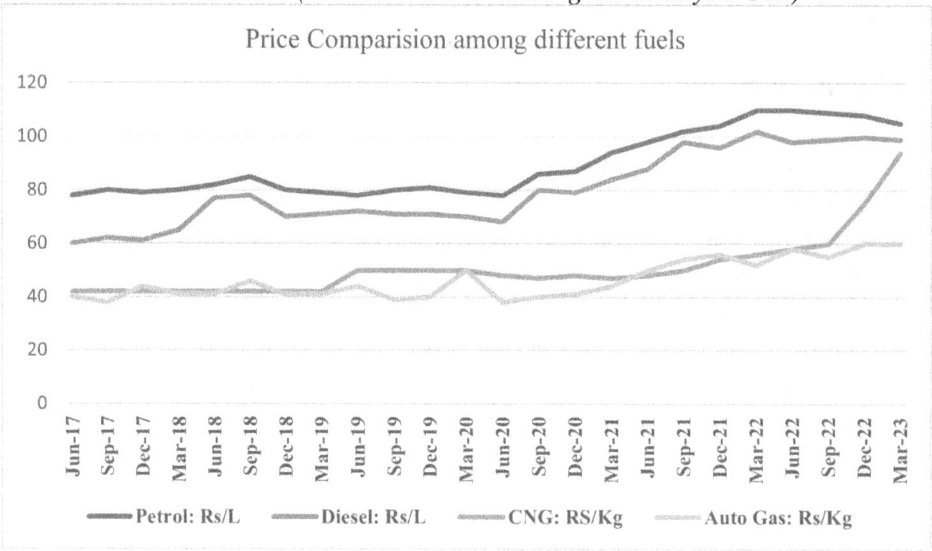

This chapter is broken down further in the following manner. In section 2, an analysis of the market is conducted for different kinds of EVs and different power technologies is used and followed by the technical requirement of electric vehicles for their widespread adoption in the market in section 3. A review of the intention of buyers regarding the acquisition of EVs is done in section 4. Section 5 summarises different communication requirements and performance matrices used in electric vehicles. Different types of battery requirement in different EV category is explained in section 6. Section 7 will explain the different types of batteries in the market and their compatibility with EVs. Section 8 summarises the power needed while charging the batteries and analyses the cost and other measures. A critical review of different types of connectors and their respective speed is placed in section 9, followed by various opportunities and challenges EVs faces in section 10. The paper is summarised, and futures advancement is placed in sections 11 & section 12, respectively.

2. AN ANALYSIS OF THE MARKET FOR EV AND ELECTRIC POWER TECHNOLOGIES

An example of an energy source with an intermittent output derives its energy from renewable sources such as wind and photovoltaics. Research has switched its focus toward energy storage technology (EST) as a means of stabilising sources of energy that can only produce energy intermittently (Y. J Jang et al., 2018, EA El-Ghanam et al., 2020, S Jeong et al., 2015, D. Kosmanos et al., 2018, L maglaras et al., 2015). Compressed air energy storage (CAES), flywheel energy storage (FES), lead-acid (Pb-A), and nickel-cadmium (Ni–Cd), to mention a few, are among the most popular types of energy storage technologies (ESTs) found in renewable power plants (E. ElGhanam et al., 2021). The vast majority of renewable power resources use pumped hydro storage as their primary energy storage technology. The proportion of Li-ion batteries is equal to four. According to (O smiai et al., 2015), the technological elements used in evaluating these ESTs for renewable energy power plants include power density, power range, round trip efficiency, operational cost, and energy density. These are the aspects of technology that are considered in the evaluation of these ESTs. However, the necessary performance in various technological aspects changes substantially depending on the applications of the ESTs (for example, power control, bridging power, and energy management). This is because different applications call for different combinations of elements. This is because the applications of the ESTs are each unique. The fact that the evaluation of ESTs is dependent on the application should not come as much of a surprise (L maglaras et al., 2015, O Smiai et al., 2015, W. Su. Et al., 2012). When assessing ESTs, It is common practice to consider a wide range of elements, including those related to technology, the economy, the environment, and society (W. Su. Et al., 2012). This is done to combine the EST's technology capabilities and market goals more effectively.

Energy storage technology research for electric vehicles has not been as thorough as research for power plants. As a result, the assessment of ESTs for energy production is much more thorough. In addition, the subject matter of these studies is virtually exclusively technological improvements, and the researchers rarely make any connections between the qualities and capabilities of technological advancements and customers' preferences. Establishing a connection between a product's technical performance and the preferences of its target market is essential to the product's success from the consumer acceptance point of view (Z. Liu et al., 2019, SK Singh et al., 2024).

Lithium-ion (Li-Ion) batteries are used to power the majority of hybrid and pure electric vehicles on the road today due to their high energy density and ease of recharging. Major manufacturers have invested in Li-Ion rechargeable batteries and fuel cell technologies as substitutes (O. Smiai et al., 2015). Plug-in hybrids are an additional choice that can be made (W. Su, et al., 2012). These vehicles combine the power of a traditional internal combustion engine with that of batteries. However, combustion engines are completely outlawed from use in the transportation sector. There is an alternative known as the plug-in hybrid (IS Bayram et al., 2012). According to research (OECD), the power systems of future electric vehicles will likely consist primarily of fuel cells, rechargeable batteries, or a combination of these technologies.

In order to meet the requirements of the ever-expanding market for electric vehicles, the technologies of fuel cells and rechargeable Li-Ion batteries have been continuously improved. This has allowed manufacturers to meet these demands. This has been done in order to meet the demands of the market. According to (IS Bayram et al., 2012), each potential technology has unique technological and economic characteristics, such as cost, energy density, refuelling/recharging time, installation flexibility, and so on (IS Bayram et al., 2011). The degree to which these two primary competing and complementary technologies differ in their techno-economic features can influence the rate at which the market will adopt the technology (E Hossain et al., 2012). The techno-economic features of these two technologies will be examined in the following parts in greater detail than they were covered in the sections that came before them.

3. TECHNOLOGICAL REQUIREMENTS OF EVS FOR THEIR WIDESPREAD SUSTAINABILITY IN THE MARKET

The mass market adoption process of electric vehicles can be researched with the help of established theories on the spread of innovation. According to the innovation diffusion hypothesis developed by (Saxena P, and Singh SK, 2013), the factors influencing the adoption rate include relative advantage, trialability, complexity, observability, and compatibility. In addition to the factors that impact, the various kinds of customers contribute to the product distribution process uniquely. It is possible to divide early adopters of innovation into several distinct categories based on the degree of zeal with which they take on board novel concepts. A key milestone is reached when there is successful dissemination to the early majority, also known as the critical mass. According to (Saxena P et al., 2013), the two primary groups that make up the market for electric vehicles in the United States are commuters and those who run errands. Even though these two groups might have distinct transportation requirements, their assumed operating costs are almost identical (Saxena P, and Singh SK, 2012). This includes the cost of gasoline, taxes, maintenance charges, and purchase price. According to a survey by (Saxena P, and Goyal A., 2023), the key factors contributing to the proliferation of electric vehicles are operation cost, purchase price, recharging time, power, driving range, emission, incentives for adoption, and refuelling network density. Incentives for adoption include tax credits and rebates. (S. Aman et al., 2013) performed research into the elements that restrict the general use of electric vehicles and found that consumers' top concerns were linked to the vehicles' battery range, charging infrastructure, cost, safety, and reliability. (S. Aman et al., 2013) also revealed that several problems impede the widespread use of electric vehicles. Price, product range, performance, environmental impact, longevity, and convenience were ranked as the most important decision-making factors for potential buyers in a survey of social preferences among German consumers (T Bohn et al., 2012). Although cost was a primary consideration in that study, the order of importance placed on different aspects of social choice varied depending on the probable customer base.

The findings of a more recent study (K. Gowri et al., 2011) conducted in China from the point of view of many stakeholders suggested the level of technology advancement, rules and regulations. Additionally, a number of essential aspects go into customer acceptance of electric vehicles, including consumer expectation and price, as well as market structure and competitiveness. Automakers are faced with whether to equip electric vehicles with a lithium-ion or fuel cell powertrain, on top of considerations relating to the government's policies and the public sector's infrastructure. According to research (SAE vehicle 2014), the decision between these two competing technologies is influenced by several criteria, including compatibility, technological superiority, and the reputation and trustworthiness of the brand. As an example, the researchers used BMW. On the other hand, research that evaluates the obstacles to the widespread adoption of electric vehicles identifies numerous hurdles, including technological, infrastructural, economical, behavioural, and external (L jian et al., 2013). According to a survey conducted in Poland on individual electric automobiles using a somewhat sophisticated alternative performance model, the most critical factor in user approval is technological performance requirements.

Multi-criteria research methodologies are typically utilised in energy storage-related research. This is because various energy storage choices share numerous technological characteristics. According to a study (Z Ma et al., 2013), an energy storage system's lifetime, storage capacity, risk, and safety are some of the most significant technological issues. According to (L Gan et al., 2011), the criteria most frequently employed in research models include financial aspects such as operating expenses, capital expenditures, and energy costs. When conducting research that is related to the socioeconomic scenery. In addition to health and safety concerns, it is crucial to consider social factors such as cultural norms, levels of acceptance, and financial incentives offered by the government (Z Ma et al., 2010).

This chapter employs a multiple evaluation approach to take into account all of the concerns that have been raised and to take into account the various aspects that have an effect. The model incorporates not only technological elements but also the social needs of various market categories for the objectives of segmentation and positioning in the growing market for EVs.

4. INTENTIONS OF BUYERS REGARDING THE ACQUISITION OF EVS

It has been discovered that purchasing new energy cars in China is favourably associated with several demographic parameters, including living area, social ties, and occupation (E sortomme et al., 2011). These demographic parameters can potentially influence consumers' motivation to purchase electric automobiles. Both national policy and infrastructure can potentially influence the amount of motivation that customers have. According to a study conducted in Ireland (A kwasinski et al., 2012), the demographic factors that have the most significant impact on customer motivation are the level of education, level of income, daily driving range, and age of consumers.

The performance parameters of electric vehicles are crucial aspects that play an essential part in determining the intention to make a purchase, in addition to the incentives provided by national laws such as petrol or annual auto taxes (K valentine et al., 2011). Consumers decide which model of electric car to purchase based on several factors, including the purchase price as well as the resale price, the fuel efficiency (P. Saxena and A Goyal, 2019), the driving range (P. Saxena and A Goyal, 2022), and the mechanical or environmental performance (P. Saxena et al., 2023). According to the research findings, customers value these aspects highly.

The evaluation of the many power technologies that are currently accessible will, as was mentioned in the introduction, assist established automobile manufacturers in selecting the most promising emerging technologies in which to invest. The findings of this research will make such tactical and strategic choices easier to make. From the perspective of academic research, the bulk of the studies found in the literature either analyse the progression pattern of a technology's development or forecast the market acceptability of new technology. Both of these types of studies may be found in the literature. Very little research considers the market's desires when analysing the qualities of competing technologies. There are two schools of thought regarding how the market will accept a new technology: the Technology Push school and the

Market Pull school. On the other hand, this study aims to build a bridge between power technologies and developing markets for electric vehicles.

5. COMMUNICATION REQUIREMENTS AND PERFORMANCE METRICS

Many different methods can be used to charge an electric vehicle (EV), and these methods change depending on the circumstances of the surrounding area as well as the needs of the driver. As a direct result, the infrastructure for charging electric vehicles can take on various forms, and each is optimal for a specific set of circumstances unique to that particular form. Depending on the country in which you are located, electric vehicle chargers, also frequently referred to as electric vehicle supply equipment (EVSE), have several different requirements and specifications than those found in other countries. These differences are the consequence of several factors coming together, the most significant of which are the features of the electrical grid and the many different models of electric vehicles that are now on the market. In this chapter, the technical fundamentals of electric vehicle charging infrastructure are broken down, and it is underlined how important it is to apply a contextual approach when designing and putting into action EV charging networks at the local level. Contextual methods are highly significant for creating and implementing electric vehicle charging networks at the local level (P. Saxena et al., 2023).

Electric vehicle supply equipment, more commonly referred to as EVSE (electric vehicle supply equipment), is an essential part of the infrastructure for charging electric automobiles. At the neighbourhood level, the electric vehicle supply equipment, also abbreviated as EVSE for its acronym in full, creates a connection to the power grid. After that, it combines this connection with a control system and a hardwired link to make charging electric automobiles as safe and dependable as possible. User authentication, authorisation for charging, information recording and exchange for network management, data privacy and security, and authorisation for charging are only some services control systems for EVSEs offer. It is strongly suggested that Electric Vehicle Supply Equipment (EVSEs) that are outfitted with at least the most fundamental control and management functions be utilised for any endeavour that calls for charging.

Plug-in charging, often known as cable charging, is the charging technology that is now in use at the highest frequency. This method of charging is also called conductive charging in some circles. The requirements for an EVSE for conductive charging are contingent on a wide range of factors, some of which include the type of vehicle, the capacity of the battery, the charging technique, and the power ratings of the EVSE.

5. BATTERY REQUIREMENTS FOR SEVERAL ELECTRIC VEHICLE CATEGORIES

The technical specifications of the batteries installed in electric cars (EVs) are the basis for establishing the requirements for charging electric vehicles (EVs). The battery must receive power at the correct voltage and current levels for the charging process to be successful. The many varieties of electric cars, as shown in Table 1, have particular needs that must be met by their batteries' capacity and voltage levels. The power needed to run E-2Ws and e-3Ws comes from low-voltage batteries. Batteries that could only produce a modest voltage output powered the initial generation of electric vehicles powered by electric automobiles. Even though these will continue to be utilised in specific applications like taxis, they may be phased out at some time in the future. This is something that is a possibility. This is something that might take place at some point in the future. The following types of electric vehicles, also called e-cars, are members of the second generation of electric vehicles. High-voltage batteries power these e-cars, members of the second generation of electric vehicles. In the future, there will be a need for both low-voltage and high-voltage electric LCVs. This is because the amount of weight each type of vehicle can carry will define which type is the most appropriate option.

Table 1. Battery specification among different kinds of Vehicle Category

Types of Vehicles	Battery Capacity in kWh	Voltage capacity of Battery (V)
EV-2 Wheeler	1.2-3.3 kWh	48-72 V
EV-3 Wheeler	3.6-8 kWh	48-60 V
EV-Cars (1st generation)	21 kWh	72 V
EV-Cars (2nd generation)	30-80 kWh	350-500 V
Buses, Heavy Commercial vehicle	204 kWh	700-800 V

Table 1 explains the battery specification among different kinds of vehicle categories and battery capacity in kWh.

6. TYPES OF BATTERY USED IN DIFFERENT VEHICLE CATEGORY

In figure 2, battery requirement is prerequisites along with the evolutionary growth of battery capacity in kWh. Later in the section, types of different kinds of batteries and their specifications are presented (Z Ma et at, 2010).

Figure 2. Evolutionary growth of battery capacity from 1980 to 2023.

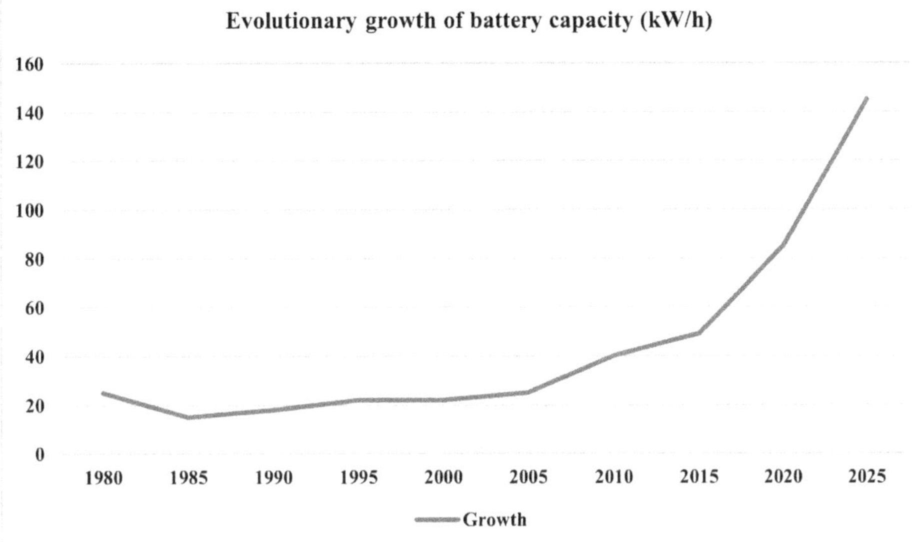

6.1 *Lead Acid*

When lead-acid batteries were the predominant type used in electric vehicles when they were first introduced to consumers. The hold that the lead-acid batteries maintained was, up to this point, the longest and most secure of any other type. These batteries have an electrode made of lead that has a harmful porosity and is placed on a plate. The plate is kept immersed in a diluted form of sulfuric acid. This is the location where charging and discharging takes place during the process of converting electrical energy into chemical energy and vice versa. They have a low energy density and give an operational temperature range that is relatively greater than other types. Recycling these batteries is made much simpler by recent developments. It has been demonstrated that the battery, as a whole, is better for the environment because almost all of its components can be recycled and reused.

These batteries have a relatively poor energy density and are much heavier than their counterparts. However, the lead-acid batteries in question have meagre production costs and should not count on being able to use them for an extended period. Because of the shallow discharge depth, this directly affects the lifetime of the product as a whole. Lead-acid batteries have a short operating life, which means they need to be replaced more frequently between four to fifteen years, depending on the nature of the usage and the type of battery. As a result, lead-acid batteries are more expensive in the long term. In addition, lead-acid batteries never drain more than 30–40 per cent, which, in the long run, contributes to the battery being damaged.

6.2 Lithium-ion Batteries

Following the success of lead-acid batteries, the market was successfully overtaken by lithium-ion batteries. The majority of today's electric two-wheelers still use lithium-ion batteries despite the introduction of newer battery technologies. These are relatively denser in their natural state and are based on more recent technological developments.

A lithium-ion battery, as its name suggests, is made up of lithium ions that have a positive charge and are transferred through a liquid electrolyte from the opposing end, also known as the anode, to the positive end, also known as the cathode, and vice versa, in order to generate the necessary charge. This process is known as ion transfer. Lithium-ion batteries are less heavy but more expensive than the lead-acid batteries used in previous generations of electronics. In addition, they provide the choice between a rapid charging pace and a gradual discharge rate for the battery.

The lithium-ion batteries provide a better ratio of power to weight, increased performance at higher temperatures, and a lower self-rate than their conventional counterparts. These lithium-ion batteries have a greater capacity than traditional batteries for maintaining a full charge for an extended period, making them superior to conventional batteries. Because of how these batteries are constructed, they have a substantially longer cycle life and can also lose up to 90 per cent of their total capacity. Because many components and parts of these batteries may be recycled, using them is an environmentally responsible choice.

Despite their diminutive size, lithium-ion batteries deliver superior performance over their nickel-cadmium and lithium-metal counterparts. They have a higher capacity for rapid discharge, which allows them to propel the vehicle in question at a higher rate. The increased pushing power, continuous supply of increased electricity, and rapid discharge benefit HEV.

6.3 Nickel Metal Hydride

Although these kinds of batteries have been available for purchase for a sizeable amount of time, most of their applications have been in hybrid vehicles with electric and internal combustion engine (ICE) motors. This is the reason why we are not entirely familiar with its name. In this configuration, charging and discharging occur within the electrodes, composed of a unique alloy that absorbs hydrogen and uses a unique alkaline electrolyte. Most applications for these batteries involve acting as a suitable replacement for other types of non-rechargeable alkaline batteries with a comparable shape. In addition, the risk of leakage and explosion associated with these nickel-metal hydride batteries is significantly lower than that associated with other types of batteries. On the other side, these come at a more significant financial investment.

These batteries have been constructed so that one terminal contains nickel alloy while the other contains nickel oxyhydroxide. The electrolyte in these batteries is composed of potassium hydroxide. This particular battery has a relatively sluggish charging and discharging pace and a poorer power-to-weight ratio than comparable batteries. Because of this, the charging process takes a significant amount of time.

When exposed to extremely high temperatures, Ni-MH batteries have a greater propensity to degrade considerably more quickly. As a result of this characteristic, they are located on one of the lower rungs of the quality and usability ladder. However, their temperature range of operation is exceptionally astounding, and it is also highly safe and dependable. On average, a Ni-MH battery's life cycle lasts around 3000 cycles. They have a voltage output of 1.2 volts and are very kind to the environment. Ni-Cud technology has been rendered obsolete due to this battery's adoption of a newer, more advanced technology.

These batteries are utilised in electric vehicles and laptops, car batteries, electric razors, various medical instruments, and other similar products.

6.4 Solid State

When talking about electric vehicle batteries, solid-state batteries represent the most recent technological progress in this category. Manufacturing the central unit of a solid-state battery has been accomplished by utilising the unique technology developed by Gogoro and ProLingum. To condense a lengthy explanation and avoid excessive technical jargon, it is sufficient to point out that, in place of a liquid electrolyte, it uses a solid electrolyte, which is how it got its name. Even in this case, the ions continue moving back and forth between the cathode and the anode, increasing energy density. These batteries are designed to have a lifespan that is intended to be significantly longer, and they are significantly safer and considerably more portable than their competitors. The technology is not yet fully developed and is still being improved; hence, it is still considered in the developmental stage.

6.5 The nickel-cadmium Battery

If rechargeable batteries were used throughout the 90s, then people are already familiar with the nickel-cadmium technology developed during that time. The "Ni-Cd" accumulators offered many benefits, including a considerable storage density and a lifespan of approximately 500 to 1,000 charging cycles.

However, they did have a memory effect, a physical phenomenon in which the battery's performance decreases if it is subjected to partial "charge-drain" cycles. These batteries did have this problem. As a result of the toxicity of cadmium, the use of Ni-Cd batteries, which were common in the construction of electric vehicles in the 1990s, is now outlawed (Z Ma et al., 2010)..

7. POWER CONSUMPTION DURING CHARGING PROCEDURE

Direct current (DC) is the electricity used to charge an electric car's battery pack. A converter is required to supply DC to the battery because energy distribution systems deliver power as an alternate current (AC) (Z Ma et al., 2010).

Conductive charging can be accomplished with either alternating or direct current. The onboard charger of an electric vehicle equipped with alternate current electric vehicle supply equipment (AC EVSE) is the component that is responsible for receiving the AC electricity and subsequently converting it to DC. By converting the power outside and supplying DC electricity directly to the battery, direct current electric vehicle supply equipment, also known as DC EVSE, eliminates the need for the vehicle to have an onboard charger. Both alternating and direct current charging can be subdivided into four different charging modes, with Modes 1-3 referring to AC charging and Mode 4 referring to DC charging (L Gan et al., 2011).

Using a cable and plug, an electric vehicle (EV) can be successfully connected to a standard socket outlet in either Mode 1 or Mode 2, depending on the user's preference. Because Mode 1, also known as "dumb charging," does not permit any communication between the electric car and the EVSE, it is not recommended for use. Instead, Mode 2 is recommended. The portable cable utilised in Mode 2 already has a protection and control capability built into it, and this wire is frequently used for charging gadgets in the home (L Gan et al., 2011). Modes 3 and 4, which each give a separate charger device to transmit electricity to the electric car, are utilised for charging electric vehicles in commercial or public settings because their management systems have been enhanced. The power rating and compatibility with the various EV segments are both stated in Table 2.

Table 2. Power Rating, along with compatible EV segments

	Different levels of power	**Different types of current**	**Compatible Electrical vehicle segments**
Normal-power charging	**power < = 7kW**	**Alternative & Direct Current**	**EV-2Wheelers, EV-3Wheelers, EV-cars, other LCVs (up to 1 ton)**
	7kW < power < = 22kW	**Alternative & Direct Current**	
High-power charging	22kW < power < = 50kW	Direct Current	EV-cars, LCVs, and MCVs (1-6 tons)
	50kW < power < 200kW	Direct Current	

8. EV CHARGING: TYPES OF CONNECTORS AND SPEED

Charging cables for electric vehicles typically feature two connectors: one that fits into the socket on the vehicle and the other that plugs directly into the charge point (L Gan et al., 2011),.

The type of connector required can be determined, in part, by the type of car driven and in part by the power rating (speed) of the charging point.

Electric vehicles can be equipped with either a Type 1 or Type 2 connector and for DC rapid charging, either CHAdeMO or CCS. Slow charging can be done with a Type 1 plug, while fast charging can be done with a Type 2 socket.

Most slow and fast charge points have a Type 2 plug installed. Occasionally, a cable will be fastened to them rather than attached to them directly. Every DC quick charging station has a cable predominantly fitted with a CHAdeMO and a CCS connector.

Most people who drive electric vehicles purchase a portable charging cable for their cars that are compatible with either the Type 1 or Type 2 sockets on their vehicles. This allows them to charge their cars utilising public networks. In Figure 3, a demo shows communication between an electrical socket and an electrical vehicle.

Figure 3. Communication Diagram between Electrical socket and EV

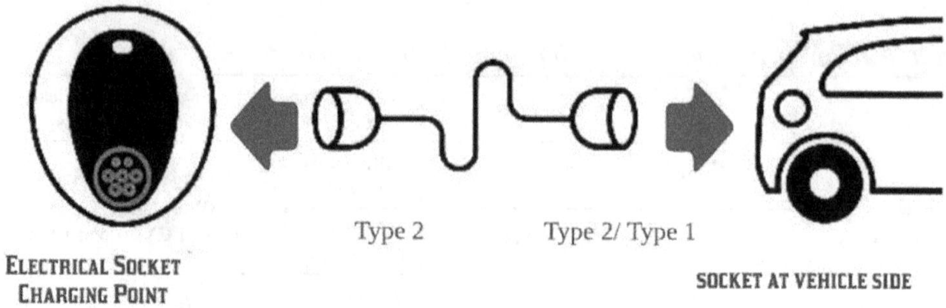

Typically, two charging modes are available to charge the electric vehicle. One is with alternate current and another in direct current.

8.1 Charging through Alternative Current (AC charging)

Two distinct varieties of AC connectors are installed on the vehicle's side, and both can be used for topping off a charge at home, place of work, or destination. Table 3, Table 4, and Table 5 explain different types of AC and DC connectors along with the power required to charge the EV. Moreover, different tables also explain kms runs per hour charging and significant features of connectors.

Table 3. Comparative analysis among types of AC connector, power required, kms runs after an hour of charging for AC Charging.

Type of AC Connector	Power Required	Approx Kms at per hour charging	Other Features
Type-1 (5-pins Connector)	3.7 kW	20.1 kms	• five-pins connector • Connectors are used universally in the US and other parts of the world. • No mechanism for locking provided • Used only a single phase of electricity.
	7 kW	40.2 kms	
Type-2 (7-pins Connector)	3.7 kW	20.1 kms	• 7-pins • Standard EU connector • Inbuilt locking mechanism • Can carry three-phase power
	7 kW	40.2 kms	
	22 kW (Three Phase Supply)	120 kms	

8.2 Charging through Direct Current (DC charging)

There are three distinct types of DC car-side connectors; most of these connectors are used for rapid charging while the vehicle is in motion. Most DC rapid charging stations will consist of cables equipped with a CHAdeMO and a CCS connector at the end of each. This means that will need to choose which connector is compatible with the socket in the vehicle. In order to prevent damage to the battery, quick chargers are not designed to charge at their maximum power rating constantly.

Table 4. Comparative analysis among types of AC connector, power required, km runs after an hour of charging for DC charging

<table>
<tr><th>Type of AC Connector</th><th>Power Required</th><th>Approx Kms at per hour charging</th><th>Other Features</th></tr>
<tr><td rowspan="2">CHAdeMO</td><td>50 kW</td><td>120 kms</td><td rowspan="2">• five-pins connector
• Connectors are used universally in the US and other parts of the world.
• No mechanism for locking provided
• Used only single-phase electricity.</td></tr>
<tr><td>100 kW</td><td>241 kms</td></tr>
<tr><td rowspan="3">Combined Charging System (CSS)</td><td>50 kW</td><td>120 kms</td><td rowspan="3">• 7-pins
• Standard EU connector
• Inbuilt locking mechanism
• Can carry three-phase power</td></tr>
<tr><td>150 kW</td><td>362 kms</td></tr>
<tr><td>350 kW</td><td>845 kms</td></tr>
<tr><td rowspan="2">Type 2</td><td>150 kW</td><td>241 kms</td><td rowspan="2">• Only Tesla Superchargers provide DC via a Type 2 connector.</td></tr>
<tr><td>250 kW</td><td>603 kms</td></tr>
</table>

8.3 Charging Point Side Electric Car Connector Types

There is essentially just one kind of ChargePoint socket, typically used for topping off a charge at home, work, and travelling. However, some people may charge their devices in an emergency by plugging them into a standard three-prong plug located in a power socket (A kwasinski et al., 2012).

The socket for the type 2 charge points is universal and can be thought of similarly to the wall socket for charging phones (that is, the socket is the same for each, but the cable is specific to the model of the car or phone). In other words, the socket is universal, but the cable must be purchased separately. To put it another way, the socket is compatible with a wide variety of devices, but the cable must be purchased separately (A kwasinski et al., 2012).

Table 5. Comparative analysis among types of AC connector, power required, km run after an hour of charging.

Type of AC Connector	Power Required	Approx Kms at per hour charging	Other Features
Type 2	7 kW (single-phase electricity)	40.2 kms	• The connector is universal and is compatible with all standard ChargePoint sockets. • The driver is responsible for bringing the appropriate cable. • It functions similarly to a wall plug while charging a smartphone.
	22 kW (three-phase electricity)	120 kms	
3-pin plug	2.3 kW	12.8 kms	• Slow charging • Not in the best practices • only be used when there is an emergency

9. ELECTRICAL VEHICLE OPPORTUNITY AND CHALLENGES

The transformation from cars powered by internal combustion engines (ICEs) to electric vehicles (EVs) presents a one-of-a-kind opportunity for the utility industry. Electric vehicles (EVs) present a new variable load that has the potential to be incorporated into the electric system in a manner that will maximise the benefits for users of EVs as well as those who operate the grid. Suppose the charging of electric vehicles is synchronised with the production of electricity without carbon emissions. In that case, there is the potential for a reduction in carbon emissions not only in the transportation sector but also in the electric sector. However, on the flip side of this possibility is a warning that if the integration of EVs is not carefully managed, it could add to the load during inconvenient times, resulting in increased prices and carbon emissions (A kwasinski et al., 2012).

This document provides a guide for regulators to follow as they integrate electric vehicles (EVs). First, we provide an overview of the essential issues that authorities must consider to determine the most effective way to integrate electric vehicles. After that, we will talk about how performance-based regulation (PBR), which is the process of developing high-level goals, performance criteria, and clear and measurable metrics to assess whether goals and criteria are satisfied, may be used to make it easier for electric vehicles to be integrated into society. A significant quantity of greenhouse gas (GHG) emissions in the United States may be attributed to the transportation industry. These emissions accounted for around 29% of the total in 2017. Most of these emissions are caused by light-duty vehicles, passenger cars and trucks. Heavy-duty vehicles are the primary contributors to this problem. This is responsible for 59% of the total emissions. Between 1990 and 2017, the absolute number of greenhouse gas emissions from the transportation industry climbed more than from any other sector, while the number of GHG emissions from the electric sector decreased. This contrast highlights the possibilities for vehicle electrification to produce favourable results by moving transportation energy demand from polluting and high-carbon fuels to increasingly lower-carbon and less polluting forms of electricity. This would result in a shift from dirty fuels to electricity, which would reduce pollution and carbon emissions.

These advantages are multiplied by the fact that electric vehicles are superior to internal combustion engines in the efficiency of transforming energy from electric current into miles driven. Suppose we have an internal combustion engine (ICE) and an electric vehicle (EV) and wish to drive 100 miles. In that case, we will need to acquire more energy in the form of petrol to operate the ICE than we would need to purchase electricity to drive the same number of miles in the EV. In other words, we will need to buy more energy to run the ICE than we will need to purchase electricity to run the EV. Changing to electric vehicles will enable us to reduce our overall energy consumption for our transportation needs.

The fundamental efficiency of electric vehicles (EVs) in converting energy to miles travelled (and producing less pollution as a result) will continue to promote both the adoption of EVs and the decrease of pollutants. When efficiency is measured in terms of emissions, a full-sized electric vehicle (EV) has 53% lower emissions than a comparable petrol automobile throughout the vehicle's lifetime. After an average of 4,900 miles driven, this efficiency compensates for the more significant emissions produced during the manufacturing process of EVs. Even if an electric vehicle's battery is charged solely by generation powered by coal, the EV will nevertheless create fewer emissions per mile than a gasoline-powered vehicle of equivalent size and performance. Red reductions are significantly more significant when charging with a more environmentally friendly generation mix. Similar reductions in greenhouse gas emissions can be achieved by substituting diesel for electric power in fleet vehicles such as buses and other vehicles.

Changing from petrol to electricity as a fuel source can reduce the amount of money spent on energy by the power system and the transportation sector. In addition, the adaptable nature of the load associated with the electrification of transportation provides an opportunity to incorporate EVs into the system in a manner that results in the lowest possible cost for customers, ratepayers, utilities, and the general public. In order to take advantage of these cost savings, Regulatory Assistance Project (RAP) has devised a three-pronged strategy, which is described in greater depth elsewhere:

- Innovative pricing, which will encourage and appropriately reward smart charging;

- Innovative technology, which will permit data sharing and automatic reactions to permit smart charging, and
- Intelligent infrastructure will involve the strategic deployment of charging and grid infrastructure.

Utilising performance-based regulation is one way for regulators to implement a strategy to manage and make it easier for electric vehicles (EVs) to integrate into the electric system. PBR makes it possible for regulators to create goals, performance criteria, and indicators, which they may use to ensure that the integration of electric vehicles is handled effectively.

10. CONCLUSION

This study examined a wide variety of topics, such as the many distinct categories of electric vehicles (EVs), the technology that drives them, the benefits that EVs provide over conventional automobiles that are propelled by internal combustion engines (ICE), the growth of EV sales throughout the past few years, the many different methods by which they can be charged, and the developments that are on the horizon for the industry. In addition to this, a synopsis of the most significant challenges presented in research, as well as the opportunities made available by these challenges. Batteries are an indispensable component of electric vehicles because the range an electric vehicle can drive between charges is related to the battery's storage capacity.

We analysed a few distinct kinds of batteries by contrasting and comparing their attributes in light of the criteria that were presented, as well as their compatibility with various kinds of electric vehicles. This chapter sheds light on the various types of charging connectors currently on the market, specifically regarding the rates at which they can charge electric vehicles. The chapter also contains an analysis of the current state of the electric vehicle market and a discussion of the technological needs associated with electric vehicles.

11. FUTURE SCOPE

After going over the latest findings from research on the status of EVs, it has been determined that new techniques have the potential to help overcome the challenges facing the development of EVs. Aside from that, it is physically impossible to discuss everything of significance in a single study. The research requires some future recommendations for illuminating its usefulness, which are outlined in the following paragraphs for convenience.

- The technology for storing energy in batteries needs to be developed for EVs to become more widespread, and there is also a need to make the conventional charging connections more user easy.
- Even when connected to the grid, charging electric vehicle batteries produces undesirable side effects. It may take some time for these impacts to be diminished, but once they are, there will be a far better opportunity to integrate electric vehicles with renewable energy sources.
- Sometimes recycling the components that go into electric vehicle batteries might be difficult. Therefore, it is necessary to discover a new method for the storage of energy.
- In EV smart cities, with the development of renewable energy sources, information and communication should be further along in their progression. In order to come up with the best strategy, we need to read additional literature or take more online surveys, and the concept can come from developed countries that have already created electric vehicles.
- Bring the temperature of the electric vehicle battery down; an air-cooled medium technology, such as using water or PCM (phase change material), could be used for this purpose. A lot of incidents of electrical vehicle caught fire are reported across the globe.
- The acceptance of EVs on a global scale still need more time. Implementation of electric vehicles can be improved by modelling after the practices of nations that have already accepted EVs.

REFERENCES

Aman, S., Simmhan, Y., & Prasanna, V. K. (2013). Energy management systems: State of the art and emerging trends. *IEEE Communications Magazine*, 51(1), 114–119. DOI: 10.1109/MCOM.2013.6400447

Bayram, I. S., Michailidis, G., Devetsikiotis, M., & Bhattacharya, S. (2011). Local energy storage sizing in plug-in hybrid electric vehicle charging stations under blocking probability constraints. In *Proceedings of the IEEE International Conference on Smart Grid Communications* (pp. 78–83). DOI: 10.1109/SmartGridComm.2011.6102396

Bayram, I. S., Michailidis, G., Devetsikiotis, M., Granelli, F., & Bhattacharya, S. (2012). Smart vehicles in the smart grid: Challenges, trends, and application to the design of charging stations. In Chakrabortty, A., & Ilic, M. D. (Eds.), *Control and Optimisation Methods for Electric Smart Grids* (Vol. 3, pp. 133–145). Springer. DOI: 10.1007/978-1-4614-1605-0_6

Bohn, T., & Chaudhry, H. (2012). Overview of SAE standards for plug-in electric vehicles. In *Proceedings of IEEE PES Innovative Smart Grid Technologies* (pp. 1–7). DOI: 10.1109/ISGT.2012.6175597

Electric Power Research Institute. (2019). *Interoperability of public electric vehicle charging infrastructure.* Retrieved June 20, 2021, from https://www.eei.org

ElGhanam, E., Hassan, M., & Osman, A. (2021). Design of a high-power, LCC-compensated, dynamic, wireless electric vehicle charging system with improved misalignment tolerance. *Energies*, 14(4), 885. DOI: 10.3390/en14040885

ElGhanam, E. A., Hassan, M. S., & Osman, A. H. (2020). Deployment optimization of dynamic wireless electric vehicle charging systems: A review. In *Proceedings of the 2020 IEEE International IoT, Electronics, and Mechatronics Conference (IEMTRONICS)* (pp. 1–7). DOI: 10.1109/IEMTRONICS51293.2020.9216415

Gan, L., Topcu, U., & Low, S. (2011). Optimal decentralized protocol for electric vehicle charging. In *Proceedings of the IEEE Decision and Control and European Control Conference* (pp. 5798–5804). DOI: 10.1109/CDC.2011.6161220

Garg, H., Singh, M., Sharma, V., & Agarwal, M. (2022). Decentralized application (DAPP) to enable e-voting system using blockchain technology. In *2022 Second International Conference on Computer Science, Engineering and Applications (ICCSEA)* (pp. 1–6). https://doi.org/DOI: 10.1109/ICCSEA54677.2022.9936413

Gowri, K., Pratt, R. G., Tuffner, F. K., & Kintner-Meyer, M. C. W. (2011). *Vehicle-to-grid communication standards development, testing, and validation: Status report*. Pacific Northwest National Laboratory. DOI: 10.2172/1029430

Habib, S., Khan, M. M., Abbas, F., Sang, L., Shahid, M. U., & Tang, H. (n.d.). A comprehensive study of implemented international standards, technical challenges, impacts, and prospects for electric vehicles.

He, Y., Venkatesh, B., & Guan, L. (2012). Optimal scheduling for charging and discharging of electric vehicles. *IEEE Transactions on Smart Grid*, 3(3), 1095–1105. DOI: 10.1109/TSG.2011.2173507

Hossain, E., Han, Z., & Poor, H. V. (2012). *Smart grid communications and networking*. Cambridge University Press. DOI: 10.1017/CBO9781139013468

Jang, Y. J. (2018). Survey of the operation and system study on wireless charging electric vehicle systems. *Transportation Research Part C, Emerging Technologies*, 95, 844–866. DOI: 10.1016/j.trc.2018.04.006

Jang, Y. J., Suh, E. S., & Kim, J. W. (2016). System architecture and mathematical models of electric transit bus system utilizing wireless power transfer technology. *IEEE Systems Journal*, 10(2), 495–506. DOI: 10.1109/JSYST.2014.2369485

Javaid, N., Hafeez, G., Iqbal, S., Alrajeh, N., Alabed, M. S., & Guizani, M. (2018). Energy-efficient integration of renewable energy sources in the smart grid for demand-side management. *IEEE Access : Practical Innovations, Open Solutions*, 6, 77077–77096. DOI: 10.1109/ACCESS.2018.2866461

Jeong, S., Jang, Y. J., & Kum, D. (2015). Economic analysis of the dynamic charging electric vehicle. *IEEE Transactions on Power Electronics*, 30(7), 6368–6377. DOI: 10.1109/TPEL.2015.2424712

Jian, L., Xue, H., Xu, G., Zhu, X., Zhao, D., & Shao, Z. Y. (2013). Regulated charging of plug-in hybrid electric vehicles for minimizing load variance in household smart microgrid. *IEEE Transactions on Industrial Electronics*, 60(8), 3218–3226. DOI: 10.1109/TIE.2012.2198037

Kosmanos, D., Maglaras, L. A., Mavrovouniotis, M., Moschoyiannis, S., Argyriou, A., & Janicke, H. (2015). Information and communication technology research opportunities in dynamic charging for electric vehicles. In *Proceedings of the Euromicro Conference on Digital System Design* (pp. 297–300).

Kosmanos, D., Maglaras, L. A., Mavrovouniotis, M., Moschoyiannis, S., Argyriou, A., Maglaras, A., & Janicke, H. (2018). Route optimization of electric vehicles based on dynamic wireless charging. *IEEE Access : Practical Innovations, Open Solutions*, 6, 42551–42565. DOI: 10.1109/ACCESS.2018.2847765

Kwasinski, A., & Kwasinski, A. (2012). Signal processing in the electrification of vehicular transportation: Techniques for electric and plug-in hybrid electric vehicles on the smart grid. *IEEE Signal Processing Magazine*, 29(5), 14–23. DOI: 10.1109/MSP.2012.2186645

Liu, Z., Lee, H., Ali, G., Pesch, D., & Xiao, P. (2019). A survey on resource allocation in vehicular networks. *arXiv, 1909.13587.*

Ma, Z., Callaway, D. S., & Hiskens, I. A. (2010). Decentralized charging control for large populations of plug-in electric vehicles. In *Proceedings of the 49th IEEE Conference on Decision and Control (CDC)* (pp. 206–212). DOI: 10.1109/CDC.2010.5717547

Ma, Z., Callaway, D. S., & Hiskens, I. A. (2013). Decentralized charging control of large populations of plug-in electric vehicles. *IEEE Transactions on Control Systems Technology*, 21(1), 67–78. DOI: 10.1109/TCST.2011.2174059

Maglaras, L., Jiang, J., Maglaras, A., Topalis, F., & Moschoyiannis, S. (2015). Dynamic wireless charging of electric vehicles on the move with mobile energy disseminators. *International Journal of Advanced Computer Science and Applications*, 6(5), 239–251.

Mude, K. N., Dashora, H. K., Bertoluzzo, M., & Buja, G. (2014). From wired to in-moving charging of electric vehicles. In *Proceedings of the WSEAS International Conference on Development, Energy, Environment, Economics (DEEE)* (pp. 33–42).

Patil, D., McDonough, M. K., Miller, J. M., Fahimi, B., & Balsara, P. T. (2018). Wireless power transfer for vehicular applications: Overview and challenges. *IEEE Transactions on Transportation Electrification*, 4(1), 3–37. DOI: 10.1109/TTE.2017.2780627

Pranshu, S., & Goyal, A. (2019). Study of computerized segmentation & classification techniques: An application to histopathological imagery. *Informatica (Vilnius)*, 43(4), 561–572.

Pranshu, S., & Goyal, A. (2022). Two-stage binary classification of follicular histology using support vector machine. *Zhonghua Yi Xue Yi Chuan Xue Za Zhi*, 31(3), 258–268.

Pranshu, S., & Goyal, A. (2022). Computer-assisted grading of follicular lymphoma: A classification based on SVM, machine learning, and transfer learning approaches. *Imaging Science Journal*, 70(1), 30–45. DOI: 10.1080/13682199.2022.2162663

Sanguesa, J. A., Torres-Sanz, V., Garrido, P., Martinez, F. J., & Marquez-Barja, J. M. (2021). A review on electric vehicles: Technologies and challenges. *Smart Cities*, 4(1), 372–404. DOI: 10.3390/smartcities4010022

Saxena, P., & Goyal, A. (2023). Accurate demarcation of a biased nucleus from H&E-stained follicular lymphoma tissue samples. *Imaging Science Journal*, 71(8), 715–727. DOI: 10.1080/13682199.2023.2192550

Saxena, P., Singh, S. K., & Agrawal, P. (2013). Texture classification of biased cytoplasmic tissue samples from histopathological imagery by Gabor application. *Journal of Network Innovation and Computation*, 1(1), 248–259.

Saxena, P., Singh, S. K., & Sinha, A. (2024). Computer-assisted interpretation, in-depth exploration, and single-cell type annotation of RNA sequence data using k-means clustering algorithm. *Computer Methods in Biomechanics and Biomedical Engineering*, 1–11. Advance online publication. DOI: 10.1080/10255842.2023.2300685 PMID: 38235728

Saxena, P., Sinha, A., & Singh, S. K. (2023). The outbreak of coronavirus (2019-nCoV) in India: A statistical case study. *Journal of International Science and Technology*, 11(4), 570.

Singh, S. K., Rashid, M., Alshamrani, S. S., Alnfiai, M. M., Saxena, P., & Khamparia, A. (2024). Efficient transfer learning approach for acute lymphoblastic leukemia diagnosis: Classification of lymphocytes and lymphoblastic cells. *TS. Traitement du Signal*, 41(4), 1749–1761. DOI: 10.18280/ts.410409

Sortomme, E., Hindi, M. M., MacPherson, S. D. J., & Venkata, S. (2011). Coordinated charging of plug-in hybrid electric vehicles to minimize distribution system losses. *IEEE Transactions on Smart Grid*, 2(1), 198–205. DOI: 10.1109/TSG.2010.2090913

Su, W., Eichi, H., Zeng, W., & Chow, M. (2012). A survey on the electrification of transportation in a smart grid environment. *IEEE Transactions on Industrial Informatics*, 8(1), 1–10. DOI: 10.1109/TII.2011.2172454

Valentine, K., Temple, W. G., & Zhang, K. M. (2011). Intelligent electric vehicle charging: Rethinking the valley-fill. *Journal of Power Sources*, 196(24), 10717–10726. DOI: 10.1016/j.jpowsour.2011.08.076

Yan, Y., Qian, Y., Sharif, H., & Tipper, D. (2012). A survey on smart grid communication infrastructures: Motivations, requirements, and challenges. *IEEE Communications Surveys and Tutorials*, 15(1), 1–16.

Section 2
Battery Systems and Energy Storage Solutions

Chapter 3
Advanced Battery Cooling Systems for Enhanced Efficiency and Longevity

Naman Sharma
KIET Group of Institutions, India

Swati Sharma
KIET Group of Institutions, India

ABSTRACT

Electric vehicles rely on lithium-ion batteries, which require precise temperature control to ensure durability, performance, and safety. Battery Thermal Management Systems (BTMS) play a crucial role in maintaining optimal battery temperature. Various cooling methods are used, including phase-change materials (PCM), liquid cooling, and air cooling. Liquid cooling enhances vehicle performance but adds weight, while air cooling is lighter but less effective. A hybrid cooling plate has been developed to reduce weight, stabilize temperature, and improve EV efficiency (Hwang et al., 2024). Researchers are exploring novel cooling techniques to address the thermal limitations of high-power density batteries (Zhang et al., 2022). This chapter reviews current and future cooling methods, emphasizing the importance of effective thermal management for next-generation EVs and ensuring battery performance across temperature ranges.

DOI: 10.4018/979-8-3693-6844-2.ch003

1.0 INTRODUCTION

A vehicle which moves or gets into running state with the help of one or more electric motors is known as an electric vehicle. A battery, fuel cells, or a collector system are the components which can all be used to power the car independently. It has been recognised and made sure that all the developed and developing countries have long-term plans and strategies to use EVs instead of those internal combustion vehicles that produce smoke and pollution and to use renewable energy to generate electricity (Taghizad-Tavana et al., 2023). Other than electricity the vehicle also runs from other sources. Electric vehicles include a wide range of transportation types, such as electric cars and trains, electric spacecraft, electric airplanes, as well as electric boats and underwater vehicles. With the electrification brought forth by the Second Industrial Revolution, early electric automobiles initially appeared. Since electricity offers a level of comfortness, quietness, and ease of operating a gasoline-powered cars of the time could not match, using it was the preferred method for motor vehicles (Taghizad-Tavana et al., 2023). The storage limitation offered by contemporary battery technologies caused range anxiety, which hindered the broad adaptability of electric vehicles throughout the 20th century. Internal combustion engines, including those running on gasoline and diesel, dominated automobile and truck power for almost a century. Worldwide, more than ninety percent of the transport sector is powered by fuels derived from oils (van Vliet et al., 2011). Some vehicle types, however, continue to depend on electricity for propulsion. These include compact, slow in speed, short in range battery-powered personal vehicles like mobility scooters and mass transit vehicles like electric trains, trams, monorails, and trolley buses.

Towards the ending of the 20th century, there was an upsurge in interest and adaption in electric transportation due to the adverse environmental impacts of traditional transportation as well as concerns about peak oil. EVs are different from traditional vehicles as the electricity they use can come from a diversity of sources, such as nuclear power, fossil fuels and some renewable energy sources like solar and wind power, or any mix of those. Provided by the fuel and power generation technologies, electric vehicles have different carbon footprints and other pollutants. The car may use a flywheel, supercapacitors, or batteries for storing the electricity. Internal combustion engine vehicles majorly only use one or a few energy sources, most commonly non-renewable and non-continual fossil fuels. Regenerative braking, which restores power to the on-board battery while recovering kinetic energy lost as heat during friction braking, is an advantage of these vehicles over the traditional ones.

Understanding Environmental Impacts

An electric vehicle's environmental impact is significantly influenced by the source of the power used to charge it. EVs with renewable energy propulsion have a substantially lower carbon footprint than cars that run on gasoline. However, the advantages to the environment are lessened if the electricity is generated by facilities that burn fossil fuels. The environmental benefits of electric vehicles extend beyond their reduced tailpipe emissions (Nordelöf et al., 2014).

- **Enhanced Efficiency:** Electric motors convert a larger percentage of energy into usable power, resulting in less energy being lost, as compared to internal combustion engines.
- **Braking with regeneration:** A crucial component of electric vehicles, recovers the kinetic energy lost during regular braking and redo it into electrical power, which enhances the overall efficiency.

The Future of Electric Vehicles: A Brighter Electric Horizon

The transportation industry has a bright future thanks to electric vehicles. Widespread EV adoption will depend on developments in battery technology to extend threshold and the growth of renewable energy sources for the production of power. More and more people agree that these vehicles will take over from traditional internal combustion engine (ICE) vehicles soon. (Ntombela et al., 2023). EVs have the power to completely transform transportation and pave the way for a more sustainable and eco-friendly future as these technologies advance. For ages, inventors have been fascinated by the idea of electric cars, or EVs. However, a crucial component was missing in making these machines function: a means of storing the energy which needs to drive them. Rechargeable batteries is a technology which was providing the solution in the early days of electric vehicles.

Lead-acid batteries were the first batteries that were rechargeable too for EVs that were successfully sold in the market. Despite their drawbacks, lead-acid batteries offered a practical means of storing the electrical energy needed to power the first electric cars. Lead-acid batteries were in use for a long time even before the popularity of electric vehicles (EVs). This body of existing knowledge made their reasonably simple integration into electric vehicles conceivable. Because lead-acid batteries are globally available, early manufacturers considered them a cost-effective solution. When compared to other battery technologies of that time, lead-acid battery production was somewhat less expensive.

These considerations led to the dominance of lead-acid batteries in the early electric vehicle market. Although they had restricted weight and range, they paved the way for further advancements in battery technology, which eventually shaped the future of electric vehicles (Liu et al., 2022).

Problem Statement and Objective

The effective control of battery temperature in electric cars is still a difficulty, despite developments in battery technology, particularly when operating circumstances are variable. The temperature which can not be controlled can seriously affect the safety, longevity, and performance of batteries in EV's. Thus, this chapter's line up is to investigate advanced Battery Thermal Management Systems (BTMS) that upgrade lifetime and efficiency, with an emphasis on cutting-edge methods and approaches that guarantee ideal temperature control for future electric vehicles.

1.1 Battery Thermal Management System

The electric vehicles (EVs) are gaining allurement in the current market as they are more efficient and advantageous over gasoline-powered cars. Maintaining the longevity, safety, and optimal performance of EV battery packs is a crucial task. The Battery Thermal Management System (BTMS) plays an essential role in this effort. Various battery thermal management techniques are discussed, highlighting their advantages and disadvantages. A comprehensive review of experimental and theoretical studies on different battery cooling techniques is also presented (Tete et al., 2021). Look at Figure 1.

Figure 1. Battery Thermal Management System

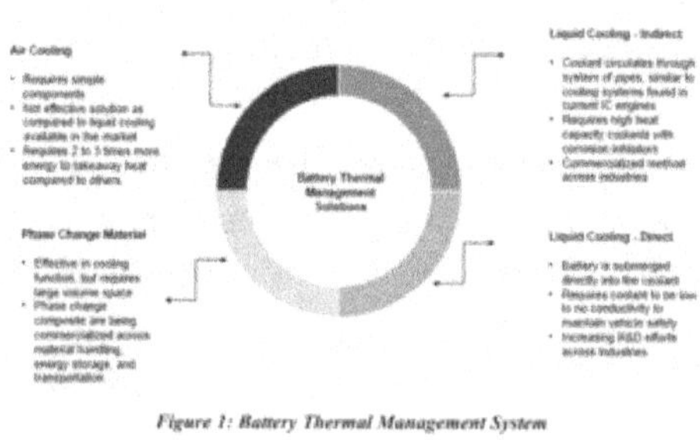

The battery being used must be kept within a certain essential temperature range, usually between 15°C and 35°C, by the BTMS. When you operate outside of this range, bad things can happen:

- **Performance:** Temperatures above a certain range have harmful effects on battery life. The high temperatures are limiting the range and acceleration of these electric cars or electric vehicles, which also impair efficiency and power output. On the other hand, low temperatures also prevent the battery from operating at their maximum capacity, which can be shown in their performance also.
- **Safety:** One of the main safety concerns is overheating. It speeds up the failure of these batteries, which increases the risk of thermal runaway—a hazardous situation in which the battery warms up quickly and emits combustible gasses. Thermal runaway is the potential cause of fire in severe situations. It takes a well-designed BTMS to stop these kinds of situations from happening.
- **Battery Life:** Preserving the battery pack at ideal temperatures greatly increases its longevity. Working in extremely high or low temperatures speeds up the process of degradation and eventually lowers the battery's capacity to store a charge. An effective BTMS contributes to delaying this deterioration, extending the EV's battery life.

The design, development, and testing of a BTMS for electric vehicles are the foremost objectives of this project. The primary goal is that we have to design a system that efficiently controls battery temperature while taking into account many aspects such as cost, weight, efficiency, and overall system complexity.

2.0 LITHIUM-ION BATTERIES

Lithium-ion batteries are growing in importance as a technology for developing low-carbon transportation systems since their first commercial release in 1991. Lithium plays a crucial role in the shift toward low-carbon energy. While there have been efforts to outline what sustainability means in relation to lithium, the guidelines remain largely unclear. (Petavratzi et al., 2022). However, it is becoming politically contentious to discuss information on how sustainably batteries are made. In actuality, business procedures related to the extraction of raw materials give rise to questions about the responsibility and openness of extractive resource management. Diverse parties that represent business growth, public interest organizations, and political giants who are worried about the results of the technology's manufacture and use are involved in the intricate supply chain of lithium technology. Establishing globally accepted norms for the governance of technology was one way to ensure balanced extraction processes.

To address these concerns, the framework for the Assessment of Sustainability in Supply Chains can be used to evaluate the standards' compliance. Thus, social, environmental, and governance commitments are the main topics of the qualitative assessment. Social and environmental indicators, compliance, and management systems and standards are the indicators used in the quantitative assessment (Günther et al., 2015). According to one source, by 2035, more than 25% of the Li and over 67% of the Co required for electrical vehicles is going to come from recycled sources.In contrast, electric cars can be seen as significantly reducing the requirement for raw materials when one considers the enormous amounts of hydrocarbons that traditional automobiles absorb over the course of their lifetime. This lack of clarity leads to vastly different estimates for future lithium demand, with projections for 2050—aligned with scenarios targeting a 50% reduction in global emissions—ranging from 184,100 to 989,300 tons per year. (Speirs et al., 2014).

Despite these advancements, it is crucial to understand that the production of electric vehicles still has notable environmental impacts. In 2022, an EV's production produced, on average, approximately 48% more CO2 emissions than that of a similar vehicle with an internal combustion engine. However, this contrast is mitigated by the significantly higher emissions an internal combustion engine vehicle produces over the course of its lifespan when collated with the emissions produced during the process of producing the electricity needed to power the EV.Greenpeace released a movie in 2023 disputing the notion that these vehicles are a "silver bullet for climate," contending that there is a significant environmental impact during the development phase (Becker and Shukla, 2022).For example, Hyundai's transition to electric vehicles, while a positive step for environmental sustainability, is partially offset by the company's increasing sales of SUVs, even electrified ones. Despite the potential benefits of electric SUVs, their environmental impact is significant due to the high energy and raw material consumption during manufacturing (Ortiz et al., 2024). Greenpeace has proposed a substitute mobility model focusing on cycling, public transportation, and ride-sharing to mitigate these concerns. Solving the thermal issues associated with electric vehicles requires advanced battery thermal management systems (BTMS). Unlike conventional internal combustion engine vehicles that rely on air conditioning for cooling, electric vehicles often require more complex liquid cooling systems, these systems circulate a coolant, such as a mixture of water and glycol, through channels or plates surrounding the battery cells to effectively absorb heat and transfer it to a radiator or heat exchanger.

The ease of liquid cooling systems include:

- It controls the temperature ranges quite efficiently and the way is more accurate than the air cooling technique.
- It allows for faster charging and hence escalates the driving range.
- Extending the life and health of batteries.

However, there are cons related to it:

- Increased vehicle weight and complexity.
- Requires more fans and pumps, which raises parasitic energy usage.
- Possibility of leakage, which, if ignored, could harm battery cells.

Looking ahead, even more sophisticated cooling methods, such as phase-change materials that absorb heat while operating and immersion cooling, which submerges batteries in coolant, are being investigated by researchers. These advanced cooling systems are expected to significantly improve performance and efficiency as EV technology develops.

3.0 COOLING STRATEGIES

By allowing control over a battery's temperature properties under both regular and harsh operating situations, a battery thermal management system ensures both performance and safety. The operating system of these batteries can be made better by preventing the freezing of electrolytes, lithium plating and thermal runaways by using the effective system of battery thermal management. Traditional cooling strategies in the battery thermal management system used in commercial settings include air cooling and indirect liquid cooling. Another of the more promising cooling technologies is Liquid-cooled BTMS, where due attention is needed to the sealing cover during the design stage to prevent leakages. (Thakur et al., 2020).

3.1. Air Cooling

Since it is less expensive, has a simpler design, is easier to install and maintain, weighs less overall, and does not leak as opposed to other cooling methods, therefore has become a popular choice nowadays. Look at Figure 2. This air-cooled battery thermal management system is mostly appropriate in situations requiring little heat dispel, though, due to the poor thermophysical qualities of the air. It is unable to regulate temperature rise and keep battery pack temperature constant when working in harsh environments.

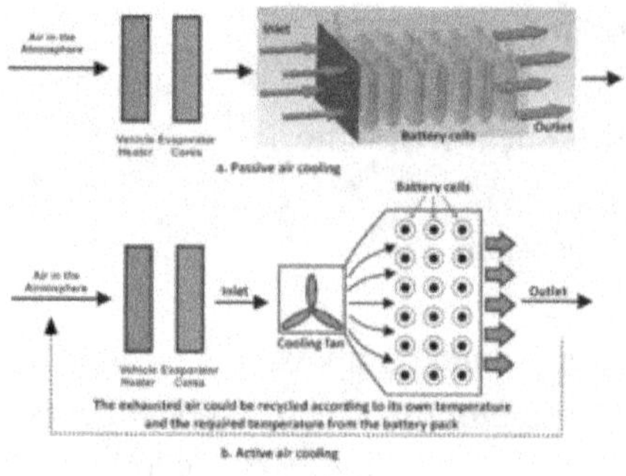

Figure 2

The better thermophysical characteristics of liquid over air have made indirect liquid cooling a common substitute for air cooling in commercial battery thermal management applications(Thakur et al., 2020). When the air cooling is opposed, the indirect liquid cooling provides greater influence over temperature rise and guarantees a significant boost in temperature conformity for the battery.

3.2. Indirect Liquid Cooling

Heat exchange between the coolant and the battery is indirect, in a liquid-based battery thermal management system, through tubing, cold plate, or jacket. The coolant does not have direct contact with the surface of the battery cell. Water, nanofluids, liquid metals, and glycol/water mixtures are some of the coolants used. As shown in Figure below

Figure 3. Indirect Cooling System

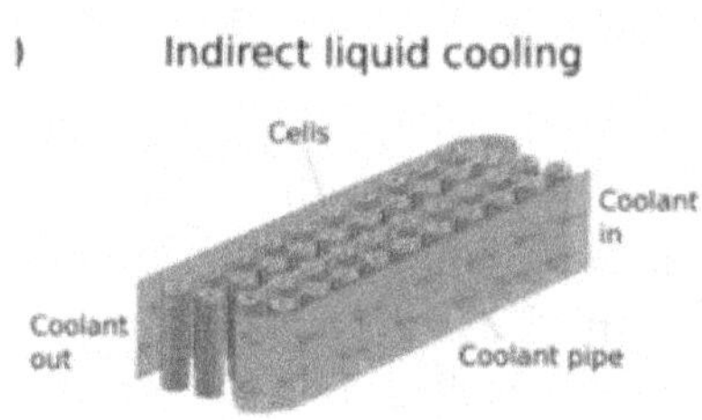

However, since this employs cold plate as well as jacket cooling for heat exchange, their high thermal resistance limits their ability to transfer heat efficiently. Moreover, the cooling system has a more intricate structure, is more expensive, and weighs more. Specifically, coolant leaks are a cause for concern. To increase overall cooling performance, it is crucial to optimize the cooling system in case of indirect liquid's design or structure. Although an indirect liquid cooling provides a better heat transfer coefficient in comparison to the the air cooling and has undergone through many structural modifications, its efficacy for the coming generation of electric vehicle battery packs with greater power densities has been hampered by three major restrictions. This issue necessitates the pursuit of cutting-edge cooling solutions that can function in tandem with next-generation battery thermal management methodologies.

4.0 ADVANCED COOLING STRATEGIES

1. Direct Liquid Cooling

A suggested cooling system offers the battery exceptional efficiency in terms of thermal management, with the paramount battery's surface temperature decreasing because the coolant flow rate surges. Because the NOVEC 7000 coolant boils in the process of cyclic charging and discharging therefore the battery surface temperature is regulated in the range from 35°C to 38°C. The maximum temperature and temperature uniformity of the battery should be restricted to 21–46°C and 41 °F, respectively, to achieve systematic and secure operations of the battery in the electric transportations (Garud et al., 2023).

To propose an effective battery management system, jilte compared a liquid-filled cooling system for the battery with a fluid flowing through the battery thermal management system (btms). When the battery's charge level decreases at a moderate pace and the surrounding temperature is low, the liquid-filled cooling method is suitable. In contrast, a fluid-circulated battery cooling system allows the battery to function at higher discharge rates while ensuring that the maximum temperature remains within a safe range. Due to the absence of components like heat exchangers and fluid circulation pumps, the fluid-based battery cooling system is a more of a economical choice compared to the fluid-circulated system. Nevertheless, the fluid-circulated system offers better temperature consistency for the battery.

Zhou et al. suggested phase change liquid cooling based on heat pipes to enhance battery thermal management system performance as well as prevent thermal runaway. For the given hybrid cooling model, the utmost heat limit of the battery and temperature contrast are regulated at 117.5 $\pm$ 32.9°F and 35.78 °F, respectively, at high discharge rates and high-power cycles. The hybrid cooling system successfully inhibits heat propagation under thermal runaway circumstances, maintaining a steady temperature above 14 °C for 14 seconds and an utmost cell temperature below 184 ± 1°C(Garud et al., 2023). The average temperature of the cell is maintained at 23 °c through immersion cooling using ampcool ac-100 dielectric coolant, which is cooler than the temperature achieved by forced air convection. Bhattacharjee and his team created an improved cooling technique for lithium-ion battery stacks, which helps regulate their temperature. When air cooling is used, the battery stack's maximum temperature is 49.76 °C, when immersion cooling is used at a 2c discharge rate, the temperature reduces to 27.43 °C, a reduction of 44.87%. In order to estimate the cooling efficiency of traditional liquid cooling against direct two-phase refrigerant cooling for EV batteries, even in the most extreme circumstances, direct two-phase refrigerant cooling kept the highest cell temperature below 45 °C, while liquid cooling went beyond the limit in typical environmental circumstances. Under severe climatic circumstances, direct two-phase refrigerant cooling produced a 15.0% reduction in internal resistance and a 16.1% increase in battery capacity during battery aging as collated to liquid cooling. Due to its small shape and tiny channels, direct two-phase refrigerant cooling lowered the cooling module's weight by up to 56% when compared to liquid cooling. Pulugundla and colleagues made it possible for 21,700 cylindrical cells to be cooled by submersion, allowing for direct heat transfer to a dielectric coolant that is flowing. The battery cell's temperature profile is positively impacted by the submerged-cell cooling method's high heat transfer efficiency and decreased total thermal resistance. Also the outer surface is in direct contact with the dielectric coolant flowing which hence reduces the thermal gradient in the vertical direction of the cells.

Since using single-phase mode and forced convective heat transfer of the dielectric fluid, the maximum temperature of the battery module is maintained at 35.10 °C at a discharge rate of 5 °C (Hong et al., 2020), the nucleate boiling increases the heat transfer of this dielectric fluid and improves the temperature consistency of the direct liquid-cooled battery module. The research on sophisticated cooling methods has been driven by the desire to maximize battery life and extend the driving range of electric vehicles (EVs). Direct liquid cooling is an improved technique that offers better temperature control than the current indirect methods. By directly connecting the battery cells to the coolant channels, the direct liquid cooling technology eliminates these barriers and further permits convective heat transfer, a far more effective strategy. This allows the coolant to instantly absorb heat from the cell walls as it passes through the channels during convection. since a result, the battery pack's temperature distribution becomes more consistent since the coolant and cells experience a smaller temperature differential.The main battery technology included in electric vehicles (EVs), lithium-ion batteries, operate optimally and have the longest lifespans in a particular temperature range. Battery life and capacity are decreased by overheating (Mahmood et al., 2023). Several scientific advantages are noted by using direct cooling to maintain a more stable and ideal temperature profile:

- **Reduced thermal stress:** Heat transmission helps to lessen the temperature gradients that exist inside the battery pack. Battery stability is increased and the likelihood of damaging lithium plating development is reduced by reducing the heat stress applied to individual cells.
- **Enhanced electrochemical performance:** Temperature plays a significant role on the electrochemical processes that takes place within the cells of the battery. The utilization of direct cooling guarantees that cells function within the ideal temperature range, resulting in enhanced battery performance, power delivery, and efficiency.

  Despite the advantages, implementing direct liquid cooling presents significant scientific and engineering challenges:
- **Leakage prevention:** A totally leak-proof system is required since the coolant and battery cells come into direct contact with each other. Any leak can cause coolant contamination, electrical shorts, or fires. Batteries can also be harmed in this . In order to guarantee system integrity, innovative sealing materials and careful design is a must.
- **Coolant properties:** Conventional coolants have an electrical conductivity when utilized in indirect systems. New coolants with excellent thermal conductivity for effective heat transfer are needed for direct cooling. To ensure electrical safety inside the battery pack, these coolants must also have a very low electrical conductivity. The development of these sophisticated coolants is still under research.
- **Packaging constraints:** The battery pack's integration of a network of cooling tubes requires significant design considerations. One major and everlasting problem is to maximize the battery space maintaing the weight and achieving effective heat transmission. In order to achieve this equilibrium, computational modeling and optimization approaches are also must.

Direct liquid cooling holds immense potential for the future of EVs(Tete et al., 2021). Current difficulties are the focus of ongoing scientific and technological efforts:

- **Advanced materials:** Ensuring the dependability and security of direct cooling systems will necessitate the development of novel, high-performing sealing materials and leak detection systems.
- **Nanofluid coolants:** Exploration is now conducted on nanofluids, or suspensions of nanoparticles in coolants. The relatively low electrical conductivity

of these nanofluids, in addition to possibly better thermal conductivity, makes them appealing choices for direct cooling applications (Abdelkareem et al., 2022).

- **Manufacturing advancements:** More widespread implementation of direct cooling across various EV models would require efficient and economical production procedures as it gains traction.

For example, Tesla's Model S and Model X use cutting edge liquid-fluid cooling systems for preserving the best possible battery performance, particularly when driving at high speeds or charging quickly. By effectively dispersing heat and averting thermal runaway, these systems move coolant fluid through tubes around the battery cells. Similar to this, the BMW iX uses an advanced liquid cooling system to control the heat produced by its sizable battery pack, guaranteeing effective and safe running. In the end direct liquid cooling provides an efficient scientific method for managing thermal heating of the EV . This kind of technology has the ability to majorly affect the performance, limits, and overall battery longevity of the electric vehicle by promoting effective heat transport and maintaining ideal battery temperatures. To overcome the current challenges and make the way for the day when direct liquid cooling is standard in electric vehicles, the continuation of research and development initiatives is must.

2. Phase change material cooling system

Lithium-ion battery longevity safety and performance in electric vehicles (EVs) appear to depend on maintaining the right battery temperature. Overheating not only reduces capacity and speeds up deterioration but it can also cause thermal runaway a potentially fatal situation that could result in fire. Phase change material (PCM)-based advanced cooling systems give us a useful means of controlling the temperature. As the temperature increases PCM absorbs and stores energy first as sensible heat and then as latent energy once it reaches the phase transition temperature. The temperature drops below the phase transition temperature releasing the stored energy and causing the PCM to return to its initial state (Wazeer et al. in 2022). The thermal energy during a solid-to-liquid phase transition is absorbed and released by PCMs. PCMs absorb heat during battery operation and subsequently melt it to prevent temperature increases. Heat is absorbed as latent heat without causing a noticeable rise in temperature once PCM reaches its melting point. Conversely the PCM solidifies and releases the latent heat that has been stored when the battery cools. The use of PCMs in Tesla car battery packs is one well-known instance. When the battery cells are being charged or used extensively these materials absorb heat from them to prevent overheating and ensure optimal performance. PCMs are commonly integrated into the design of battery packs and act as a thermal barrier between the battery cells and the external environment. By keeping the operating temperature constant, this tactic improves battery endurance and security.

Advantages of PCM-based Cooling:

- **Improved Temperature Uniformity:** Temperature gradients that might accelerate worsening can be lessened and heat can be absorbed systemically by carefully arranging PCMs around a battery pack(Wazeer et al., 2022).
- **Passive Cooling Capability:** During moderate-rate charging and discharging, PCMs can offer passive cooling, which lessens the need for energy-intensive active cooling systems.
- **Weight Reduction Potential:** PCMs can provide weight reductions over traditional liquid cooling systems since large parts like pumps and radiators are eliminated(Wazeer et al., 2022).

Limitations and Future Directions:

- **Limited Operating Range:** The PCMs' melting point limits their efficacy. The PCM's effectiveness in high-power applications is limited beyond this point because it loses its capacity to absorb heat as latent heat.
- **Low Thermal Conductivity:** Heat transfer efficiency may be hampered by PCMs' intrinsically low thermal conductivity. Researchers are looking at ways to increase thermal conductivity, such as adding chemicals with high conductivity.
- **Cost and Complexity:** While studies are still being conducted, adding more cooling methods to PCMs can make systems more complex and expensive. Finding a balance between cost-effectiveness and performance is still difficult.

A viable method for improving battery temperature management in EVs is PCM-based cooling. The goal of research is to overcome constraints such as heat conductivity and operating range. There is hope for the expansion of PCM use in commercial EV battery heat management systems. In battery systems the two primary methods for managing thermal energy are direct liquid cooling and phase-change materials (PCMs). Direct liquid cooling offers excellent cooling capacity and precise temperature control through the constant flow of a coolant fluid but it also introduces complexity and leakage risk. On the other hand PCMs passionately absorb and release the heat during phase transitions which serves as a buffer against changing of temperature. The portability and lack of requirement of the external power are desirable attributes for some specific applications. Also, their cooling capacity is usually at limit when compared to traditional liquid cooling and in some major extreme situations they may require extra cooling systems. The best thermal management solutions can actually be achieved by hybrid systems that combine the two approaches as this balances the advantages of each distinct approach while minimizing its drawbacks. One may decide to use PCMs or direct liquid cooling depending on a number of factors including battery chemistry operating conditions, system constraints and safety concerns.

5.0 BATTERY THERMAL MANAGEMENT SYSTEM

An essential part of the battery system is the battery thermal management system (BTMS) which keeps the battery temperature constant within operating parameters. Numerous cells are connected in series, parallel and other configurations to form a battery system. Due to their exceptional performance, lithium-ion batteries are the most sold in the business. However, temperature is crucial to LIB operation, and uneven temperatures seriously jeopardize their functionality.Thermal management of lithium-ion batteries (LIBs) involves two types: cooling and heating. Battery packs running under extreme temperature conditions or during high-rate charging or discharging will need to be cooled, while battery packs working under low temperatures must be heated to sustain a reasonably high energy output (Fayaz et al., 2022) For this reason, temperature homogeneity is essential to BTMS. In order for the effective cooling system to be in control, the highest temperature (Tmax) should typically be less than 41 °C and the least temperature (Tmin) should be significant than 15°C.

Previous studies confirmed the non-uniformity of temperature by showing that the Tmax is concentrated near the electrodes and at the core of the battery. Ohmic heat, enthalpy heating, internal heat production and electrochemical reaction entropy change, and other reasons are some of the causes of this non-uniform temperature(-Fayaz et al., 2022). Furthermore, a different study found that certain cells have varying life cycles at different temperatures, which results in capacity loss. Furthermore, the researchers found that the capacity loss decreases from 1.5% to 2% with a 5 °C increase in cell temperature. Consequently, an efficient BTMS can solve the heat generating issue. In general, there are two main approaches to regulate the heat produced by Li-ion batteries: external management and internal management.By using various anode and cathode material types with varying wideness, the internal management lowers the battery temperature. The second method makes utilisation of external management systems that improve battery heat dissipation through the use of liquid cooling, heat pipes, air circuits, phase-change materials, and hybrid loops. The external management technology in BTMS is vastly used as it is more efficient.

Overview of Cooling Methods of BTMS

Active as well as passive cooling systems are the two types of cooling techniques used in BTMS. Refer to the given Figure

Figure 4. Active and Passive Cooling System

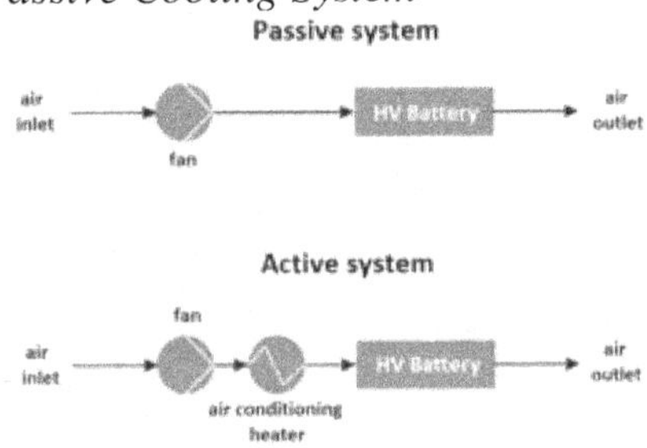

It was necessary for the active cooling system to have auxiliary parts including compressors, pumps, fans, pipes, and motors.Those studies for liquid-cooled battery packs mainly adopted a component-level approach with a specific battery pack design (Chung and Kim, 2019).These components' main drawback is that, in energy storage applications, they need an imprudent quantity of energy requirements, moving parts, and an appropriate mass balancing technique. The inability of this cooling system to efficiently equipoise the heat load at extremely high temperatures is another issue.

Battery thermal management uses an active cooling system that includes thermoelectric coolers, blower fans with varying directions, and pumps that drive liquid into the battery. This air cooling solution is a bulk conventional way in releasing the heat from a battery among battery cooling methods. Its unique benefits—cheap cost, straightforward design, light in weight, low specific heat capacity, and extended service mean life—make it a popular technique. By removing heat buildup inside the battery, the air cooling BTMS can keep the battery's temperature within the operational range of 25 to 40 °C. However, because of its low heat transfer coefficient, the natural air convection approach is slender in utmost weather situations. Hence, in order to keep the temperature of the battery within the safe range, BTMS uses a forced air system. Furthermore, since it would be ineffective, it's not advised to utilize an active air cooling system when the temperature drops below 20°C.Similar to this, the system would need additional fan power to maintain a steady temperature above 35°C, which would result in excessive energy use.

Liquid-based BTMS is another type of active cooling system. Liquid cooling is more efficient than air-based cooling systems because it requires less pumping power due to its overhead heat transfer coefficient, large cooling capacity, along with low maintenance cost. Additionally, observers have suggested liquid-cooled systems that employ water, glycol, nanofluid, and other substances because air-based BTMSs have low cooling efficiency at high C-rates and high output temperatures (Chung and Kim, 2019).

Figure 5. active cooling system is liquid-based BTMS

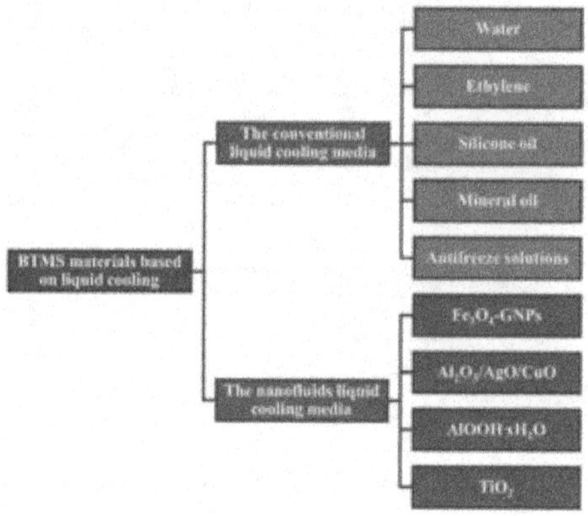

Passive cooling systems fall into the second group since they run without the need for an external power source. It makes use of surroundings to take heat out of the batteries. The most popular approach for creating thermal management systems is passive technique because it doesn't require an external power source. These include their extended lifespan, affordability, safety, compact size, and lack of sophistication. The most popular passive BTMSs typically use mixtures of these two methods as well as PCMs and Heat Pipes. Phase transitions in the PCM method store latent heat energy at almost isothermal temperatures. Due to its high energy density and thermal stability, latent heat thermal energy storage (LHTES) is superior while comparing with the sensible heat thermal energy storage (SHTES).When PCM melts, it can release energy into the surrounding air, and when it freezes, it releases an equal proportion of energy. When needed, this thermal energy would be released at any moment and through many media. PCM stores sensible heat when heat is applied externally, but when it reaches its melting point, a process known as endothermic reaction breaks the chemical bond, causing the material to start storing latent heat through a solid-to-fluid phase transition that continues at a steady temperature until the phase transition is finished. This undertaking is called PCM charging. Then, in the opposite process, PCM is released when its temperature drops to below freezing, and the chemical bond is renewed by the exothermic reaction.

Limitations of BTMS

Although cooling techniques are frequently used in the battery thermal management, there are certainly drawbacks too. The literary text indicates that the BTMS natural air cooling system is reasonably priced, easy maintenance and dependable; however, because of its low coefficient of heat transfer, it is not very effective at cooling. According to a study, employing natural air flow BTMS to produce an even distribution of airflow across the battery is challenging. Therefore, in natural-air convection systems, the larger temperature difference between cells and insufficient heating or cooling efficiency cannot be allowed.It is noted that the desired operating temperature for EVs is between 15 °C and 35 °C to achieve optimum output from the battery (Siddique et al., 2018). As a result, this method should only be utilized with small loads and batteries. Additionally, forced air cooling systems for batteries work better to handle heavy loads, but they come with a severe negative in the shape of space requirements and high power consumption fans. Additionally, analysis shows that when velocity rises, so does the forced air flow's cyclic cost. In order to tame this shallow transfer heat capacity, a cooling system based on liquid is used. It is more expensive, has a more intricate design, and is harder to integrate into the system when collated to air-cooled and PCM-cooled BTMS(Siddique et al., 2018). Additionally, there is a chance that cooling techniques will leak, which could cause the electrolyte and coolant to combine. Battery explosions could result from this, especially if the coolant is water or glycol.In order to address this risk, more precautions and steps must be taken. Furthermore, adding more components raises the system's cost. Moreover, the higher thermal resistance in between the cell and coolant is because of the presence of cooling tubes. which could pose a problem for fast-charging applications. The topic of fluid was covered by Ashok B. et al. Strong viscosity fluids can occasionally result in increased pumping losses; nevertheless, there are some fluids with low viscosity and strong thermal conductivity, such as mercury, although they are expensive. The total system complexity is significantly higher than for air cooling systems because of additional parts (such as radiators and chillers) and intricate plumbing.This leads to increased weight, coolant leakage dangers, and potential failure spots. Rapid charging circumstances may also be hampered by the thermal resistance across the cell and coolant, which is sometimes increased by the presence of cooling tubes or plates. Additionally, insufficient electrical insulation makes it difficult to execute liquid cooling in practice. This is particularly true for water and glycol cooling systems. PCM-based passive cooling systems could be used to get around the active cooling system's higher power consumption and space constraints. PCM offers a number of benefits, including a high latent heat capacity, excellent thermal performance in battery applications, and encouraging results. Unfortunately, because of its low thermal conductivity, it is unable to tolerate the constant thermal load. Apart from thermal conductivity, another issue with BTMS is PCM leakage, which reduces the energy storage capacity of PCM by decreasing its volume (Wu et al., 2019). In the case of inorganic PCM, this shift in volume phase transition is noticeably more pronounced. Additionally, because PCM is flammable, its use in BTMS is limited because the battery pack may suffer from the spread of thermal failure. In addition, PCM has problems with supercooling and volume change related to inorganic PCM. Researchers have also overlooked the cost analysis limitation, specifically the impact of pricey eutectic and specific inorganic salts on PCM's techno-economic analysis. Finally, while the Heat Pipe cooling system for BTMS uses less energy when collated with the air cooling, the challenge is temperature management. Furthermore, a separate unit for the condenser is required for the design of the heat pipe TMS for cylindrical cells or battery packs, which adds to the system's weight and volume. For a single thermal heat pipe (THP) the contact area and length are limited. The use of a high-density battery with a THP array will make the BTMS heavier and more costly necessitating a large battery pack. Utilizing the THP technique for BTMS in EVs has been hindered by certain limitations. THP for instance has a tiny volume is ineffective and has little surface contact with the battery.

7.0 CONCLUSION:

In order to optimize the longevity security and efficiency of electric vehicles (EVs) sophisticated battery cooling systems are necessary. Through effective temperature control these innovations help to improve energy efficiency extend battery life and reduce the risk of thermal runaway. Examples of conventional cooling methods with restrictions on temperature uniformity and heat transfer effectiveness are indirect liquid cooling and air cooling. Phase change material (PCM) cooling and direct liquid cooling are two examples of advanced cooling techniques that offer more promising alternatives. Direct liquid cooling which comes into direct contact with battery cells through coolant channels provides better temperature control than PCM cooling which uses the latent heat of phase transitions to absorb and release thermal energy. Nevertheless there are obstacles to overcome when implementing advanced cooling systems such as leak prevention coolant properties and packaging constraints. In order to overcome these obstacles research has concentrated on developing new materials improving manufacturing processes and streamlining system architecture. Future research looks at how to combine battery cooling technologies with overall vehicle thermal control to achieve the best possible balance between comfort and energy efficiency. The creation of novel coolants and materials with enhanced thermal and electrical conductivity properties is another intriguing field.

Advanced simulation techniques can be used to improve cooling system design while reducing development time and costs. Additionally efforts should be made to develop cost-effective strategies to expand the availability of better cooling systems for a wider range of EV models. Ultimately the successful implementation of cutting-edge battery cooling systems will be crucial to realizing the full potential of electric vehicles (EVs) as a sustainable and efficient form of transportation.

Ongoing Challenges in Battery Cooling Technology

There are still a number of issues with battery cooling solutions despite major developments. Particularly with high-energy-density batteries, thermal runaway—a catastrophic occurrence where unchecked heat generation causes a sharp rise in temperature and maybe a fire—remains a major worry. Another issue is keeping the battery packs temperature uniform since fluctuations could lead to premature depreciation or issues with performance. To avoid excessive energy draw and reduced system efficiency overall a balance between energy consumption and cooling efficiency is crucial. The cost of installing advanced cooling systems may also have an effect on the economics of electric cars. The seamless integration of cooling systems into battery packs while maintaining safety and structural integrity is a difficult design task. Additionally ensuring the long-term reliability and performance of cooling systems under a variety of operating conditions is essential to their widespread adoption. These problems need to be fixed if battery-powered technologies are to be developed and marketed further.

8.0 FUTURE SCOPE:

- **Integration with Vehicle Thermal Management Systems:** Integration of the battery cooling systems with the general vehicle thermal management system may be possible for optimal comfort and energy efficiency.
- **Advanced Materials and Coolants:** The development of innovative coolants and materials with enhanced thermal stability electrical conductivity and environmental friendliness can lead to progress.
- **Modeling and Simulation:** By employing advanced simulation techniques to optimize cooling system performance and design development time and expenses can be decreased.
- **Economical Solutions:** By investigating methods to reduce the price of creating and deploying advanced cooling systems so that a wider range of EV models can make use of them.

- **Real-World Testing and Validation:** To guarantee sophisticated cooling systems' dependability and performance in a range of settings, thorough testing and validation should be carried out under varied operating situations.

8.0 REFERENCES

Becker, J., & Shukla, V. P. (2022, November). Electric Vehicle Battery Thermal Management Systems Current Trends and Future Directions. *SSRAML*, 5(2), 47–66.

Chung, Y., & Kim, M. S. (2019, September). Thermal analysis and pack level design of battery thermal management system with liquid cooling for electric vehicles," *. *Energy Conversion and Management*, 196, 105–116. DOI: 10.1016/j.enconman.2019.05.083

Fayaz, H., Afzal, A., Samee, A. D. M., Soudagar, M. E. M., Akram, N., Mujtaba, M. A., Jilte, R. D., Islam, M. T., Ağbulut, Ü., & Saleel, C. A. (2022). Optimization of Thermal and Structural Design in Lithium-Ion Batteries to Obtain Energy Efficient Battery Thermal Management System (BTMS): A Critical Review. *Archives of Computational Methods in Engineering*, 29(1), 129–194. DOI: 10.1007/s11831-021-09571-0 PMID: 33935484

Garud, K. S., Tai, L. D., Hwang, S.-G., Nguyen, N.-H., & Lee, M.-Y. (2023). A Review of Advanced Cooling Strategies for Battery Thermal Management Systems in Electric Vehicles. *Symmetry*, 15(7), 1322. DOI: 10.3390/sym15071322

Günther, H.-O., Kannegiesser, M., & Autenrieb, N. (2015, March). The role of electric vehicles for supply chain sustainability in the automotive industry," *. *Journal of Cleaner Production*, 90, 220–233. DOI: 10.1016/j.jclepro.2014.11.058

Hawkins, T. R., Gausen, O. M., & Strømman, A. H. (2012). Environmental impacts of hybrid and electric vehicles—A review. *The International Journal of Life Cycle Assessment*, 17(8), 997–1014. DOI: 10.1007/s11367-012-0440-9

Hong, S. H., Jang, D. S., Park, S., Yun, S., & Kim, Y. (2020, June). Thermal performance of direct two-phase refrigerant cooling for lithium-ion batteries in electric vehicles," *. *Applied Thermal Engineering*, 173, 115213. DOI: 10.1016/j.applthermaleng.2020.115213

Hwang, F. S., Confrey, T., Reidy, C., Picovici, D., Callaghan, D., Culliton, D., & Nolan, C. (2024). Review of battery thermal management systems in electric vehicles. *Renewable & Sustainable Energy Reviews*, 192, 114171. Advance online publication. DOI: 10.1016/j.rser.2023.114171

Liu, W., Placke, T., & Chau, K. T. (2022). Overview of batteries and battery management for electric vehicles. *Energy Reports*, 8, 4058–4084.

Abdelkareem, M. A., Maghrabie, H. M., Abo-Khalil, A. G., Adhari, O. H. K., Sayed, E. T., Radwan, A., & Olabi, A. G. (2022). Battery thermal management systems based on nanofluids for electric vehicles. *Journal of Energy Storage*, 50, 104385.

Mahmud, M., Rahman, K. S., Rokonuzzaman, M., Habib, A. K. M. A., Islam, M. R., Motakabber, S. M. A., Channumsin, S., & Chowdhury, S. (2023, December). Lithium-ion battery thermal management for electric vehicles using phase change material: A review," *. *Results in Engineering*, 20, 101424. DOI: 10.1016/j.rineng.2023.101424

Nordelöf, A., Messagie, M., Tillman, A. M., Ljunggren Söderman, M., & Van Mierlo, J. (2014). Environmental impacts of hybrid, plug-in hybrid, and battery electric vehicles—What can we learn from life cycle assessment? *The International Journal of Life Cycle Assessment*, 19(11), 1866–1890. DOI: 10.1007/s11367-014-0788-0

Ntombela, M., Musasa, K., & Moloi, K. (2023). A comprehensive review for battery electric vehicles (BEV) drive circuits technology, operations, and challenges. *World Electric Vehicle Journal*, 14(7), 195. DOI: 10.3390/wevj14070195

Ortiz, Y., Arevalo, W., Peña, D., & Jurado, F. (2024). Recent advances in thermal management strategies for lithium-ion batteries: A comprehensive review. *Batteries*, 10(3), 83. DOI: 10.3390/batteries10030083

Petavratzi, E., Sanchez-Lopez, D., Hughes, A., Stacey, J., Ford, J., & Butcher, A. (2022). The impacts of environmental, social and governance (ESG) issues in achieving sustainable lithium supply in the Lithium Triangle. *Mineral Economics*, 35(3-4), 673–699. DOI: 10.1007/s13563-022-00332-4

Siddique, A. R. M., Mahmud, S., & Van Heyst, B. (2018, October). A comprehensive review on a passive (phase change materials) and an active (thermoelectric cooler) battery thermal management system and their limitations," *. *Journal of Power Sources*, 401, 224–237. DOI: 10.1016/j.jpowsour.2018.08.094

Speirs, J., Contestabile, M., Houari, Y., & Gross, R. (2014). The future of lithium availability for electric vehicle batteries. *Renewable & Sustainable Energy Reviews*, 35, 183–193.

Taghizad-Tavana, K., Alizadeh, A., Ghanbari-Ghalehjoughi, M., & Nojavan, S. (2023). A comprehensive review of electric vehicles in energy systems: Integration with renewable energy sources, charging levels, different types, and standards. *Energies*, 16(2), 630. DOI: 10.3390/en16020630

Tete, P. R., Gupta, M. M., & Joshi, S. S. (2021, March). Developments in battery thermal management systems for electric vehicles: A technical review," *. *Journal of Energy Storage*, 35, 102255. DOI: 10.1016/j.est.2021.102255

Tete, P. R., Gupta, M. M., & Joshi, S. S. (2021, March). Developments in battery thermal management systems for electric vehicles: A technical review," *. *Journal of Energy Storage*, 35, 102255. DOI: 10.1016/j.est.2021.102255

Thakur, A. K., Prabakaran, R., Elkadeem, M. R., Sharshir, S. W., Arıcı, M., Wang, C., Zhao, W., Hwang, J.-Y., & Saidur, R. (2020, December). A state of art review and future viewpoint on advance cooling techniques for Lithium–ion battery system of electric vehicles," *. *Journal of Energy Storage*, 32, 101771. DOI: 10.1016/j.est.2020.101771

van Vliet, O., Brouwer, A. S., Kuramochi, T., van den Broek, M., & Faaij, A. (2011, February). Energy use, cost and CO2 emissions of electric cars. *Journal of Power Sources*, 196(4), 2298–2310. DOI: 10.1016/j.jpowsour.2010.09.119

Wazeer, A., Das, A., Abeykoon, C., Sinha, A., & Karmakar, A. (2022, September). Phase change materials for battery thermal management of electric and hybrid vehicles: A review," *. *Energy Nexus*, 7, 100131. DOI: 10.1016/j.nexus.2022.100131

Wu, W., Wang, S., Wu, W., Chen, K., Hong, S., & Lai, Y. (2019, February). A critical review of battery thermal performance and liquid based battery thermal management," *. *Energy Conversion and Management*, 182, 262–281. DOI: 10.1016/j.enconman.2018.12.051

Zhang, X., Li, Z., Luo, L., Fan, Y., & Du, Z. (2022, January). A review on thermal management of lithium-ion batteries for electric vehicles. *Energy*, 238, 121652. Advance online publication. DOI: 10.1016/j.energy.2021.121652

Chapter 4
Life Cycle Assessment of Emerging Battery Technologies for Electric Vehicles

Himadri Sekhar Das
https://orcid.org/0000-0002-3509-3388
Haldia Institute of Technology, India

ABSTRACT

The life cycle assessment (LCA) of emerging battery technologies for electric vehicles (EVs) evaluates the environmental impacts from raw material extraction to end-of-life disposal. This study examines lithium-ion, solid-state, and next-generation battery chemistries, highlighting their potential to reduce carbon footprints and resource consumption. By assessing production, usage, and recycling stages, the LCA provides insights into sustainability and areas for improvement. The findings underscore the importance of advancing battery technologies to achieve greener transportation solutions, emphasizing the need for eco-friendly materials, efficient manufacturing processes, and effective recycling methods to minimize environmental impacts and enhance the sustainability of EVs.

1. INTRODUCTION

Life Cycle Assessment (LCA) is a critical tool in evaluating the environmental impacts of these emerging battery technologies across their entire life cycle—from raw material extraction to end-of-life disposal or recycling. The global transportation sector is undergoing a significant transformation, driven by the need to not only

DOI: 10.4018/979-8-3693-6844-2.ch004

reduces the greenhouse gas emissions also decrease dependence on fossil fuels. At the prominence of this change is the extensive adoption of electric vehicles (EVs), which offer a detergent and bearable alternative to traditional combustion engine vehicles (Alanazi and others., 2023). Central to the performance and viability of EVs is the development of advanced battery technologies, which provide the necessary energy storage and power.

Battery technologies for EVs have evolved rapidly over the past few decades, with Lithium-Ion Batteries (LIBs) currently dominating the market. However, the search for better performance, higher energy density, improved safety, and lower environmental impact has led to the exploration and development of several emerging battery technologies (Liu et al, 2022). Each of these technologies offers unique advantages and faces distinct challenges, making it essential to comprehensively assess their environmental impacts to guide sustainable development.

Significance of Life Cycle Assessment for Battery Technologies

The importance of conducting a comprehensive LCA for emerging battery technologies cannot be overstated (Itani et al., 2023). Figure.1 shows the life cycle assessment of battery Technology. As the EV market continues to expand, understanding the full environmental implications of battery production, use, and disposal is crucial for several reasons. LCA helps identify the most environmentally sustainable battery technologies, guiding research and development towards solutions that minimize adverse impacts. By evaluating the resource requirements and potential depletion associated with different battery technologies, LCA supports more efficient and responsible use of raw materials. Providing consumers with information on the environmental impacts of battery technologies can drive demand for greener products and encourage manufacturers to adopt more sustainable practices.

Figure 1. Life Cycle Assessment of battery technology.

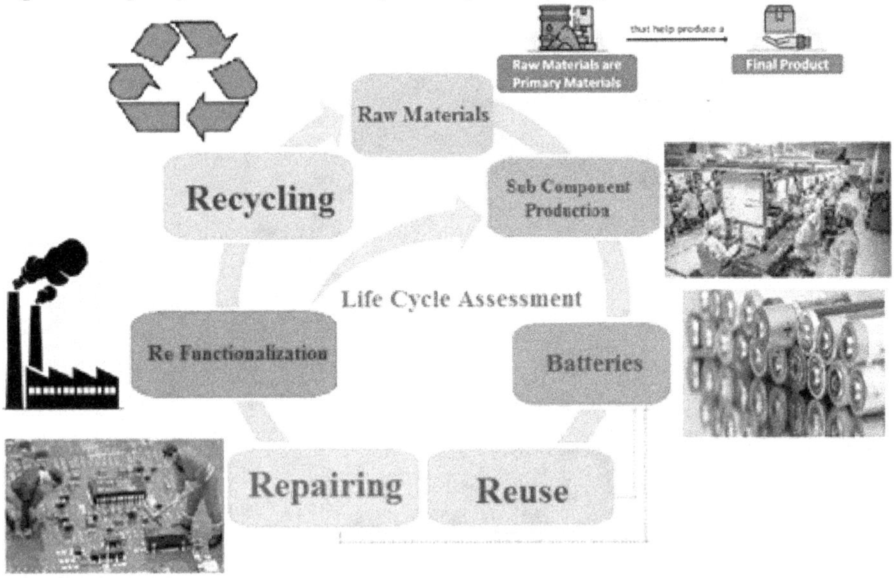

This chapter provides to the understanding of the referring to practices or policies that do not negatively affect the environment impacts of arising battery technologies for electric vehicles (EVs) through a particularized Life Cycle Assessment (LCA). The chapter offers valuable insights for policymakers and researchers by identifying sustainable alternatives to lithium-ion batteries, guiding future innovations and policy decisions.

Overview of Battery Technologies for EVs

The rapid evolution of battery technologies is revolutionizing the automotive industry, particularly in the realm of electric vehicles (EVs). As societies to reduce greenhouse gas emissions, mitigate climate change impacts, and move away from fossil fuel dependency, the development of advanced batteries plays a vital role in enabling cleaner and more sustainable transportation solutions (Panda et al., 2023). This introduction provides an overview of the significance, evolution and key considerations of battery technologies for EVs. Figure.2 shows the overview of battery technologies for EVs.

Figure 2. Overview of battery Technologies for EVs.

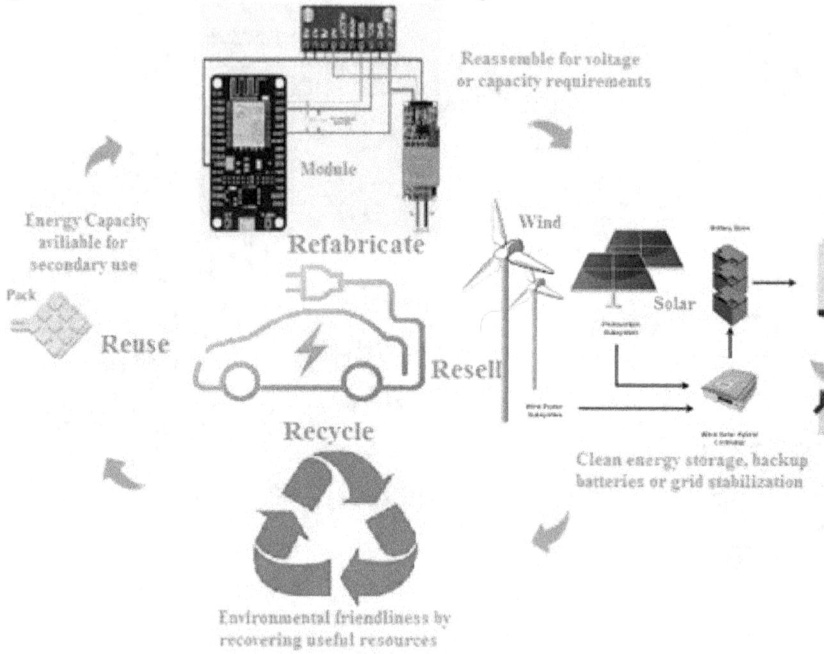

Importance of Battery Technologies in EVs

Electric vehicles represent a life-changing shift towards sustainable mobility. Unlike usual internal combustion engine vehicles, EVs utilize electric motors powered by onboard battery packs. These batteries serve as the primary energy storage device, providing the necessary power for propulsion and auxiliary systems (Olabi et al., 2022). Electric busses represent a life-changing shift towards tenable mobility. Unlike usual within combustion tool instruments, EVs utilize energetic motors stimulate by onboard battery packs. These batteries symbolize the primary strength depository units, providing the essential capacity for propulsion and auxiliary wholes (Olabi and others., 2022). The performance, range, and persistence of EVs densely rely on the capacities of their artillery technologies.

Evolution of Battery Technologies

Early Developments: The journey towards viable EV batteries began with lead-acid batteries, which offered limited range and performance. These were followed by Nickel-Metal Hydride (NiMH) batteries, that provided revised energy mass and cycle history, making them appropriate for early mixture electric cars.

Dominance of Lithium-Ion Batteries (LIBs): The advent of Lithium-Ion Batteries (LIBs) marked a significant advancement in EV technology. LIBs offer higher energy density, longer lifespan, and faster charging capabilities compared to previous technologies.

Emerging Technologies: Beyond LIBs, ongoing research and development are focused on next-generation battery technologies. These include Solid-State Batteries (SSBs), Lithium-Sulfur (Li-S) Batteries, Sodium-Ion Batteries (NIBs), and various other innovative approaches.

Future Directions and Challenges

Looking ahead, the future of battery technologies for EVs holds promise and challenges. Innovations are expected to focus on improving energy density, reducing costs, enhancing safety, and advancing sustainable practices.

The ambition behind this episode stems from the need to comprehensively evaluate the life cycle environmental impacts of arising battery technology, to a degree lithium-sulfur, complete-state, sodium-ion, and metallic zinc-air batteries, which are being grown as substitutes to unoriginal lithium-ion batteries.As EV adoption accelerates, it is crucial to understand the broader implications of these technologies on resource usage, carbon footprint, recyclability, and long-term sustainability.

1.1. Lithium-Ion Batteries (LIBs)

Lithium-Ion Batteries (LIBs) stand as the predominant battery technology in the realm of electric vehicles (EVs), driven by their exceptional characteristics and widespread adoption. LIBs offer a extreme strength density, message they can store a meaningful amount of strength relating to their size and weight (Azizighalehsari et al., 2024).Compared to traditional lead-acid batteries and even some other types of rechargeable batteries like Nickel-Metal Hydride (NiMH), LIBs are more effective in agreements of strength adaptation and discharge. This efficiency translates into better overall vehicle performance and responsiveness.LIBs are created to have a comparably long lifespan, worthy lasting numerous charge-discharge eras before important degradation occurs. This longevity contributes to the economic viability of EVs by reducing the frequency of battery replacements.

Materials

LIBs are composed of several key materials, each playing a critical role in the battery's performance and characteristics:

Lithium: lithium aids the change of ions between the cathode and anode all along charging and discharging phases.

Cobalt, Nickel, and Manganese: These metals are commonly used in the cathode materials of LIBs to optimize specific performance metrics. Enhances battery stability and energy density, although concerns about ethical mining practices and supply chain sustainability have prompted efforts to reduce cobalt content or find alternative materials. Improves energy density and capacity, making batteries more powerful and efficient. Enhances battery safety and stability, contributing to the overall reliability of LIBs. These materials are chosen and combined in varying ratios to achieve desired performance characteristics, balancing factors like energy density, cost, safety, and environmental impact.

1.2. Solid-State Batteries (SSBs)

Solid-state batteries utilize advanced materials in their construction, focusing primarily on the development of solid electrolytes and compatible electrode materials:

Solid Electrolytes: The key component that distinguishes SSBs from conventional LIBs is the solid electrolyte. This material plays a critical part in facilitating the activity of ions between the battery's electrodes (cathode and anode) all the while charging and discharging cycle.

Examples of solid electrolytes include:

Ceramics: Such as lithium ceramics, which offer high ionic conductivity and stability under various operating conditions.

Polymers: Polymer-based electrolytes are flexible and easier to process, potentially reducing manufacturing costs and enabling novel battery designs.

Electrode Materials: SSBs typically use similar electrode materials as LIBs, such as lithium-based compounds for the cathode and various forms of lithium or other materials for the anode. These materials are chosen to optimize specific performance metrics like energy density, cycle life, and cost-effectiveness.

Solid-State Batteries (SSBs) show a significant progress in battery technology for electric vehicles, contribution higher strength density, upgraded safety, and conceivably more interminable lifespan distinguished to traditional Lithium-Ion Batteries (LIBs). The development and commercialization of SSBs hold promise for accelerating the adoption of EVs by addressing key performance and safety concerns associated with current battery technologies (Frith et al., 2023).

1.3. Lithium-Sulfur Batteries (Li-S)

Lithium-Sulfur Batteries (Li-S) represent a alternative to conventional Lithium-Ion Batteries (LIBs), offering significantly higher theoretical energy densities (Gicha et al., 2024). This potential is primarily due to the use of sulfur, which is more abundant and less expensive compared to the transition metals (like cobalt, nickel, and manganese) used in LIBs. Li-S batteries have a hypothetical strength density that is to say various times above that of LIBs. This wealth they can potentially store more energy per unit weight, which is important for reaching the driving range of electric vehicles (EVs) and enhancing their overall act. Sulfur is plentiful in character and widely applicable, making it a economical alternative to the scarce and expensive metals used in LIBs. This take care of conceivably weaken the production costs of Li-S batteries and raise their financial feasibility.

1.4. Sodium-Ion Batteries (NIBs)

Sodium-Ion Batteries (NIBs) are promising alternative to Lithium-Ion Batteries (LIBs), primarily due to the abundance and cost-effectiveness of sodium compared to lithium. Sodium is widely available in natural deposits worldwide, making it more abundant and less expensive than lithium. This abundance reduces concerns related to resource availability and price volatility, potentially lowering the overall cost of battery production. The use of sodium in NIBs offers a cost-effective alternative to lithium-based technologies, which is crucial for scaling up battery manufacturing and supporting the extensive adoption of electric vehicles (EVs) and strength depository systems (Sayahpour et al.,2022).

Efforts in research and development for Sodium-Ion Batteries (NIBs) are focused on addressing key performance metrics to enhance their competitiveness with LIBs:

One of the primary research goals is to increase the energy density of NIBs to levels comparable to or exceeding those of LIBs. This involves optimizing electrode materials, electrolyte formulations, and battery cell design to enhance the storage capacity of sodium ions. Another critical focus area is improving the cycle life of NIBs. This includes developing electrode materials that can withstand repeated charge-discharge cycles without significant degradation, as well as optimizing the electrolyte composition to minimize side reactions and enhance battery stability. Research efforts also aim to improve the safety and reliability of NIBs, addressing issues related to electrode stability, electrolyte compatibility, and overall battery performance under different operating conditions.

1.5. Other Emerging Technologies

Flow Batteries: Flow batteries show a promising class of energy storage technologies popular for their scalability and long phase life, making acceptable for energy storage applications:

Utilization of Liquid Electrolytes: Flow batteries employ liquid electrolytes stored externally in tanks, allowing for decoupling of energy storage capacity and power output. This design facilitates easy scalability by adjusting the size of electrolyte tanks, making flow batteries adaptable to various energy storage requirements (Olabi et al., 2023).

Long Cycle Life: Flow batteries are from their long cycle life, that is generally attributed to the synthetic support of the liquid electrolytes. This longevity form makes suitable for uses needing frequent charge-discharge cycles, to a degree energy from renewable source integration and grid stabilization.

Lower Energy Densities: Despite their advantages in scalability and cycle life, flow batteries typically exhibit lower energy densities compared to conventional Lithium-Ion Batteries (LIBs). This limitation makes them less suitable for applications where compact size and high energy density are critical factors (Aaqib et al.,2024).

Metal-Air Batteries: Metal-air batteries handle oxygen from the air as the cathode material, contribution the potential for very extreme strength densities:

High Energy Density: Metal-air batteries have the theoretical potential to achieve significantly higher energy densities than conventional LIBs. This is because the oxygen used in the cathode reaction is abundant and lightweight, enabling greater energy storage capacity per unit weight.

Technical Challenges:

Despite their high energy density potential, metal-air batteries face significant technical challenges that have hindered their widespread commercialization:

Air Electrode Design: Developing efficient air electrodes that facilitate rapid oxygen diffusion and electrochemical reactions is crucial but remains a complex engineering task.

Cycle Life and Stability: Ensuring long cycle life and stability while preventing degradation of electrode materials, particularly during oxygen reduction and evolution reactions, poses ongoing challenges.

Environmental Factors: Managing the environmental impact of metal-air battery technologies, including potential issues related to electrolyte composition and metal resource availability, requires careful consideration.

Table.1 provides a high-level comparison and highlights the trade-offs between different emerging battery technologies in terms of their environmental footprint and technological challenges.

Table 1. shows the comparison between different emerging battery Technologies.

Battery Technology	Key Materials	Energy Density (Wh/kg)	Environmental Impact	Recycling Potential	Current Challenges

Table 2

Lithium-Ion (Li-Ion)	Lithium, Cobalt, Nickel, Graphite	150-250	High carbon footprint due to energy-intensive production and material extraction	Moderate: Recycling of cobalt and nickel is feasible, but lithium recovery is inefficient	High reliance on cobalt and lithium; limited recycling infrastructure; hazardous waste management

Table 3

Lithium-Sulfur (Li-S)	Lithium, Sulfur	300-500	Lower material extraction impact due to sulfur abundance but higher environmental risk from lithium	Low: Sulfur is abundant, but recycling technology is still underdeveloped	Limited cycle life; lithium resource scarcity; complex end-of-life recycling

Table 4

Solid-State Batteries	Solid electrolytes, Lithium, Nickel	300-500	Lower environmental impact due to absence of liquid electrolytes, reducing safety risks	Low: Recycling processes not yet mature due to novel materials used	Scalability challenges; high production costs; lack of mature recycling pathways

Table 5

Sodium-Ion (Na-Ion)	Sodium, Carbon, Iron, Manganese	100-150	Lower impact than Li-ion due to abundant and non-toxic materials like sodium	Moderate: Recycling is simpler compared to Li-ion, but still in early development	Lower energy density; immature supply chain; requires further research for performance improvement

Table 6

Nickel-Metal Hydride (NiMH)	Nickel, Rare Earth Elements	60-120	Lower impact than Li-ion but high in rare earth metal usage	High: Nickel and rare earth metals are recoverable and highly recyclable	Limited energy density compared to Li-ion; higher weight-to-energy ratio

Table 7

Redox Flow Batteries	Vanadium, Iron	20-40	Low impact due to scalability with renewable energy but low energy density	High: Vanadium can be recycled, and flow battery systems are designed for reuse	Very low energy density; large footprint; currently better suited for stationary applications

Table 8

Zinc-Air Batteries	Zinc, Air	200-300	Low impact due to abundance of zinc and air-based cathode	Low: Recycling of zinc is feasible, but current technology requires further development	Limited commercial viability; lower efficiency and power output; air electrode degradation

Applications and Future Developments

Flow Batteries: Ideal for fixed energy depository applications, in the way that energy from renewable source integration, grid-scale storage and auxiliary power schemes where scalability and unending reliability are detracting.

Metal-Air Batteries: Potential applications include electric vehicles (EVs) seeking to maximize driving range, portable electronics requiring lightweight and high energy density solutions, and grid-scale storage systems where energy density and cost-effectiveness are paramount.

2. METHODOLOGY OF LIFE CYCLE ASSESSMENT

Life Cycle Assessment (LCA) is a orderly methods used to judge the environmental impacts of a product, process, or help across allure complete life cycle. It includes stages from raw material ancestry, through result and use, completely-of-growth transfer or recycling.LCA quantifies resource use, energy consumption, emissions, and waste generation to assess environmental burdens such as global warming potential, resource depletion, and toxicity. By providing a holistic view, LCA informs decision-making to promote sustainable practices and optimize environmental performance throughout the lifecycle of products and technologies.Figure.3 shows the methodology of life cycle Assessment.

Figure 3. Methodology of life cycle Assessment.

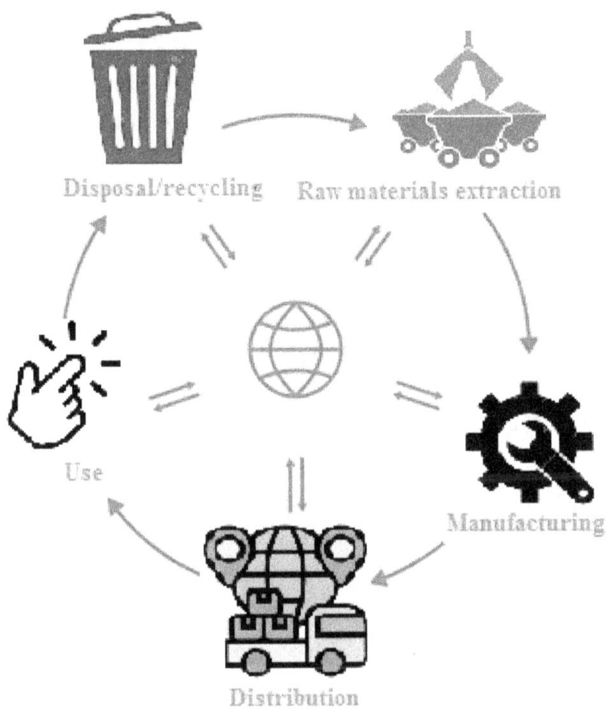

2.1 Goal and Scope Definition

The objective of transporting a Life Cycle Assessment (LCA) for battery technologies used in electric vehicles (EVs) is to completely judge and equate their environmental impacts across the whole lifecycle. This estimate aims to determine insights into the sustainability depiction of various battery types, guiding administrative towards more environmentally intimate strength storage resolutions (Das and others., 2024).

Objective: By quantifying these impacts, stakeholders can understand the full environmental footprint of each technology and identify opportunities for improvement. Functional Unit: The functional unit for this LCA typically includes: 1 kWh of battery capacity: This unit allows for a direct comparison of different battery technologies based on their energy storage capabilities. Operation of an EV over a certain distance (e.g., 100,000 km): This unit considers the operational lifespan and performance of batteries in real-world driving conditions, reflecting their usage over a significant period.

2.2 Inventory Analysis

Data Collection: Detailed data collection involves collect information on materials, energy consumption, emissions, and waste generation at each stage of the battery life cycle:

Raw Material Extraction: Includes the extraction of metals such as lithium, cobalt, nickel, and manganese, evaluating impacts on land use, water resources, and biodiversity (Zhou et al., 2024).

Material Processing: Covers processes like refining and purification of raw materials, with considerations for energy consumption and emissions from chemical reactions.

Battery Manufacturing: Involves assembly processes, energy use in manufacturing facilities, transportation of components, and packaging materials.

Use Phase: Considers energy consumption during charging and discharging cycles, efficiency losses, and operational emissions associated with electricity generation.

End-of-Life Treatment: Addresses recycling processes, waste management, and environmental impacts of disposal methods for spent batteries.

2.3 Impact Assessment

Impact Categories: Key impact categories typically assessed in LCA studies of battery technologies for EVs include:

Global Warming Potential (GWP): Measures the emissions of greenhouse gases (e.g., CO_2, methane) contributing to climate change.

Resource Depletion: Evaluates the depletion of non-renewable resources, such as minerals and fossil fuels, used in battery production.

Acidification and Eutrophication: Assesses the release of acidic compounds and nutrients into soil and water bodies, affecting ecosystems.

Human Toxicity: Analyzes the potential impacts of hazardous substances on human health, considering exposure pathways and toxicity endpoints (Meegoda et al., 2024).

Tools and Software: LCA software tools like SimaPro, GaBi, or OpenLCA are utilized to conduct impact assessments. These tools facilitate the calculation of environmental indicators based on input data and predefined impact assessment methods (e.g., IPCC, CML) to quantify impacts across different categories.

2.4 Interpretation

Results Analysis: Interpretation of LCA results involves:

Identifying Key Impact Drivers: Analyzing data to identify stages of the battery life cycle with the highest environmental impacts (e.g., raw material extraction, manufacturing).

Comparative Analysis: Comparing different battery technologies to assess their relative environmental performance based on impact categories.

Hotspot Identification: Pinpointing specific processes or materials contributing significantly to environmental burdens, guiding prioritization of mitigation strategies.

Sensitivity and Uncertainty Analysis: Conducting sensitivity analysis to:

Assess Robustness: Evaluate the sensitivity of results to variations in input parameters and assumptions, ensuring reliability and accuracy of findings.

Address Uncertainties: Quantify uncertainties in data or methodologies used in the LCA, providing insights into the reliability and confidence level of results.

3. LIFE CYCLE ASSESSMENT RESULTS

LCA results facilitate informed decision-making by identifying environmental hotspots, guiding improvements in sustainability practices, and supporting the development of policies and technologies that reduce environmental footprints and promote sustainable development.

3.1 Environmental Impacts of Lithium-Ion Batteries

Lithium-Ion Batteries (LIBs) have revolutionized energy storage technology, enabling the adoption of electric vehicles (EVs) and portable electronics. However, their production, use, and disposal present several environmental challenges that need to be addressed:

Resource Depletion

Scarce Metals: LIBs require materials such as lithium, cobalt, nickel, and manganese for their electrodes and electrolytes. These metals are often mined from limited global reserves, leading to concerns about resource depletion and geopolitical implications.

Environmental Footprint: Mining activities associated with these metals can lead to habitat destruction, water contamination, and soil degradation, impacting local ecosystems and biodiversity (Mehmet et al., 2024).

Energy Consumption

Material Processing: The origin, refining, and handle of natural resources (e.g., lithium origin from marinade or ore, sky mining and cleansing) demand significant amounts of strength. This provides to greenhouse gas emissions and loud noises, infuriating climate change impacts.

Battery Manufacturing: The manufacturing process of LIBs involves complex chemical processes, electrode fabrication, and assembly, all of which consume substantial energy and resources. Energy-intensive manufacturing contributes to the overall environmental footprint of LIBs.

End-of-Life Considerations

Recycling Challenges: LIBs contain valuable materials like lithium, cobalt, nickel, and copper that can be recovered and reused. However, current recycling technologies are still developing and face challenges in cost-effectiveness and material recovery rates.

Environmental Risks: Improper disposal or recycling of LIBs can lead to environmental contamination from hazardous materials, like lithium and cobalt, seeping into soil and water sources. This poses risks to human health and ecosystem integrity (Zanoletti et al., 2024).

Sustainable Practices and Solutions

Addressing the environmental impacts of LIBs requires concerted efforts across the entire life cycle:

Material Sourcing: Promoting responsible sourcing practices and increasing transparency in the supply chain can mitigate environmental and social risks associated with mining and extraction.

Energy Efficiency: Advancing energy-efficient technologies and processes in material extraction, battery manufacturing, and recycling can reduce the overall energy footprint of LIBs.

Recycling Innovations: Investing in test of progressive recycling sciences, to a degree hydrometallurgical and pyrometallurgical processes, can enhance material improvement rates and economic viability.

Policy and Regulation: Implementing stringent environmental regulations and incentives to promote sustainable practices, circular economy principles, and extended producer responsibility can drive industry-wide adoption of environmentally sound practices.

3.2 Solid-State Batteries

Solid-State Batteries (SSBs) are a promising next-generation technology that offers several potential environmental benefits compared to conventional Lithium-Ion Batteries (LIBs):

Material Efficiency

Reduced Material Use: SSBs have the potential to reduce material consumption compared to LIBs. Liquid electrolyte replaced with a solid electrolyte, SSBs eliminate the need for separator materials and reduce the overall volume of active materials required for electrodes. This efficiency can lower the incidental impact associated with natural resources distillation, alter, and waste production

Improved Safety: The use of solid electrolytes in SSBs enhances battery safety by mitigating risks associated with electrolyte leakage, thermal runaway, and fire hazards (Xiqian et al., 2022). This reduces the need for safety features and protective measures during battery design and manufacturing, potentially streamlining production processes and minimizing environmental impacts related to safety testing and compliance.

Manufacturing Processes

Environmental Impact: Emerging manufacturing processes for SSBs, such as advanced deposition techniques and solid-state electrolyte fabrication methods, have the potential to defeat material impacts distinguished to usual LIB manufacturing (Prasad et al, 2023). These processes may require fewer chemicals, solvents, and energy-intensive steps, leading to lower greenhouse gas emissions, energy consumption, and waste generation.

Limited Data: However, current data on the environmental footprint of SSB manufacturing is limited. As the technology matures and scales up, comprehensive life cycle assessments (LCAs) will be necessary to evaluate and optimize environmental performance across all stages of production.

Sustainable Development

Research and Innovation: Continued research and innovation in SSB technology are crucial for advancing material efficiency, enhancing manufacturing processes, and improving overall environmental sustainability. This includes developing novel electrode materials, optimizing solid electrolyte compositions, and integrating sustainable practices into battery design and production.

Policy and Regulation: Regulatory frameworks that promote sustainable manufacturing practices, encourage investments in clean energy technologies, and incentivize circular economy principles will be essential for driving industry-wide adoption of SSBs and ensuring their environmental benefits are maximized.

3.3 Lithium-Sulfur Batteries

Lithium-Sulfur Batteries (Li-S) represent an innovative approach in battery technology, offering unique advantages and facing specific challenges:

Raw Materials

Abundant and Less Toxic: Li-S batteries employ sulfur as a basic component in the cathode material. Sulfur is plentifully feasible in type, making it a economical and sustainable alternative to the transition metals used in normal Lithium-Ion Batteries (LIBs).Moreover, sulfur is less toxic compared to some of the metals used in LIBs, reducing not only potential environmental but also the health risks associated with battery manufacturing and disposal.

Challenges

Shorter Cycle Life: One of the basic challenges of Li-S batteries is their shorter cycle life compared to LIBs. During frequent charge-discharge phases, the sulfur cathode tends to humiliate on account of the dissolution of polysulfides and volume growth/shortening effects. This degradation leads to reduced battery capacity and performance over time, necessitating more frequent replacements.

Impact of Replacement Frequency: The shorter cycle life of Li-S batteries may result in higher frequency of battery replacements, which can have several associated environmental impacts:

- **Resource Intensity**: Increased demand for raw materials and energy-intensive manufacturing processes to produce replacement batteries.
- **Waste Generation**: Disposal of spent batteries and associated waste management challenges, including potential environmental contamination from battery components.
- **Economic Costs**: Higher costs associated with frequent battery replacements, impacting consumer affordability and overall economic sustainability of Li-S technology.

Mitigation Strategies

Material Innovation: Ongoing research focuses on developing advanced sulfur cathode materials and electrolyte formulations to enhance cycle stability and performance of Li-S batteries. Strategies include incorporating nanostructured materials, functional additives, and protective coatings to mitigate polysulfide dissolution and improve electrode stability.

Life Cycle Management: Implementing comprehensive life cycle assessments (LCAs) to evaluate environmental impacts across all stages of Li-S battery production, use, and disposal. This includes optimizing recycling technologies to recover valuable materials and reduce environmental footprint associated with battery end-of-life.

Regulatory Support: Policy frameworks that promote sustainable battery manufacturing practices, encourage research and development investments, and incentivize recycling and circular economy principles will be essential for advancing the environmental sustainability of Li-S batteries (Yiju et al., 2021).

3.4 Sodium-Ion Batteries

Sodium-Ion Batteries (NIBs) are emerging as a promising alternative to Lithium-Ion Batteries (LIBs), offering distinct environmental advantages and facing specific challenges:

Resource Availability

Abundance of Sodium: Sodium, the primary material used in NIBs, is abundant in Earth's crust and widely available globally. This abundance reduces concerns related to resource depletion compared to lithium, which is mined from limited global reserves (Zhao et al., 2023). The availability of sodium supports more sustainable and stable supply chains for battery manufacturing, potentially mitigating geopolitical risks associated with critical materials.

Performance Challenges

Energy Density: Enhancing the energy density of NIBs remains a key research focus. While sodium-based electrodes have shown promising electrochemical properties, achieving energy densities comparable to LIBs is critical for expanding the application scope of NIBs, particularly in electric vehicles (EVs) and energy storage systems.

Cycle Life: Improving the cycle life of NIBs is essential for maximizing their environmental performance. Research efforts are directed towards developing advanced electrode materials, optimizing electrolyte formulations, and refining battery cell designs to enhance durability and stability over numerous charge-discharge cycles.

Environmental Considerations

Lifecycle Impacts: Comprehensive life cycle assessments (LCAs) are essential for evaluating the material impacts of NIBs across their complete lifecycle. This includes determining strength use during natural resources origin, manufacturing processes, assault movement, and end-of-growth disposal or reusing.

End-of-Life Management: Developing adept and economical recycling technologies for NIBs is important for underrating waste generation and improving valuable matters. Effective reusing processes can contribute to lowering incidental impacts guide resource distillation and advancing a circular economy approach to battery manufacturing and reuse.

Sodium-Ion Batteries (NIBs) offer significant environmental advantages due to the abundance of sodium and lower resource depletion impacts compared to lithium-based technologies. Addressing depiction challenges through mechanics change and adopting holistic approaches to lifecycle administration will be essential for achieving the entire environmental potential of NIBs in numbering sustainable energy storage solutions.

3.5 Comparative Analysis

Impact Trade-offs

Different battery technologies present varying trade-offs across key factors such as energy density, material availability, and environmental impacts:

Energy Density: Technologies like LIBs and Li-S typically offer higher energy densities compared to Sodium-Ion Batteries (NIBs) and Flow Batteries. Higher energy density is advantageous for applications requiring compact size and longer operational range, like electric vehicles (EVs) and portable electronics.

Material Availability: LIBs rely on lithium, cobalt, nickel, and manganese, which are sourced from limited global reserves and face concerns related to resource depletion and environmental impact (Natsuki et al., 2022). In contrast, technologies like NIBs benefit from the abundance of sodium, reducing dependency on scarce materials and mitigating supply chain risks.

Environmental Impacts: The material footprint of battery technologies changes established factors to a degree raw material origin, manufacturing processes, strength use, and end-of-life management. Technologies with lower material intensity and energy consumption during manufacturing, such as NIBs and potentially SSBs, may offer advantages in reducing overall environmental impacts compared to conventional LIBs.

Long-Term Outlook

The future environmental performance of battery technologies will be influenced by ongoing innovations and improvements in several areas:

Manufacturing Processes: Advances in production techniques, to a degree cleaner result systems and energy-effective processes, can minimize the referring to practices or policies that do not negatively affect the environment footprint of artillery manufacturing. This involves lowering greenhouse smoke emissions, waste production, and resource devouring during terminal lie, cell congregation, and module unification.

Material Sourcing: Responsible sourcing practices and efforts to diversify material supply chains can mitigate risks associated with resource scarcity and promote sustainable extraction practices. Recycling initiatives for critical battery materials, such as lithium, cobalt, and nickel, are crucial for closing the loop and reducing dependency on primary raw materials.

Recycling Technologies: Developing and scaling up efficient recycling technologies will be pivotal in recovering valuable materials from spent batteries, reducing landfill waste, and minimizing environmental contamination. Innovations in battery recycling processes, including hydrometallurgical and pyrometallurgical methods, aim to achieve higher material recovery rates and improve economic viability.

As battery technologies continue to evolve, their comparative environmental impacts will be shaped by technological advancements, regulatory frameworks, and market dynamics. Addressing trade-offs between energy density, material availability, and environmental sustainability requires a holistic approach that integrates technological innovation, sustainable practices, and policy support.

4. END-OF-LIFE CONSIDERATIONS

4.1 Recycling Technologies

Current Practices: Recycling methods for batteries vary depending on the type of battery chemistry and the materials used in their construction:

I. **Lithium-Ion Batteries (LIBs)**:
   - **Mechanical Separation**: Current recycling practices often involve shredding batteries to separate components like electrodes, casing materials, and electrolytes mechanically.
   - **Hydrometallurgical Processes**: Techniques such as leaching are used to dissolve metals like lithium, cobalt, nickel, and manganese from shredded battery components (Sean et al., 2023). These metals can then be recovered and reused in new battery production.
   - **Pyrometallurgical Processes**: Some recycling facilities use high-temperature processes to melt and separate metals from other battery components, allowing for recovery and purification of valuable metals.

II. **Lead-Acid Batteries**:
   - **Smelting**: Lead-acid batteries are commonly recycled through smelting processes, where lead is separated from sulfuric acid and other materials. The lead can then be purified and used to manufacture new batteries or other products.

III. **Nickel-Cadmium (Ni-Cd) Batteries**:
   - **Chemical Treatment**: Ni-Cd batteries are often recycled using chemical methods to extract and separate cadmium and nickel. These recovered metals can be reused in various industrial applications.

IV. **Lithium-Ion Phosphate (LiFePO4) Batteries**:
   - Similar to LIBs, LiFePO4 batteries can undergo recycling processes involving mechanical separation, followed by chemical treatments to

recover lithium, iron, and phosphate components. Figure.4. shows the recycling Technology of batteries.

Figure 4. Recycling Technology of batteries.

Future Prospects: Advances in recycling technologies are focused on improving material recovery rates, reducing environmental impacts, and enhancing the economic viability of battery recycling:

High-Efficiency Processes: Research is underway to develop more efficient and cost-effective recycling techniques, such as:

- o **Direct Recycling**: Methods that aim to reuse battery components without fully breaking them down, reducing energy consumption and waste generation.
- o **Selective Dissolution**: Techniques to selectively dissolve and recover specific metals from battery components, optimizing resource recovery and minimizing environmental impact.
- o **Electrochemical Methods**: Innovations in electrochemical processes for recovering metals directly from battery electrodes, offering potential advantages in energy efficiency and environmental sustainability.

Automation and Scale-Up: Scaling up recycling technologies through automation and industrial integration can enhance processing efficiency and reduce labor costs, making recycling more economically viable for large-scale battery waste streams.

Circular Economy Initiatives: Integrated approaches that promote closed-loop systems and circular economy principles are gaining traction. This includes designing batteries for easier disassembly, improving material traceability, and incentivizing battery manufacturers to use recycled materials in new battery production.

Environmental Benefits

Effective assault reusing not only conserves valuable resources but further reduces the environmental footprint guide natural resources extraction and production processes. By recovering and reusing metals like lithium, cobalt, nickel, and others, recycling technologies contribute to:

Resource Conservation: Minimizing reliance on primary raw materials and reducing environmental impacts from mining activities.

Waste Reduction: Mitigating landfill waste and preventing hazardous materials from entering the environment through proper disposal and recycling practices.

Energy Savings: Lowering energy consumption compared to primary production methods, thereby reducing greenhouse gas emissions and contributing to climate mitigation efforts.

4.2 Second-Life Applications

Second-life applications involve repurposing batteries, particularly from electric vehicles (EVs), for alternative uses after their initial automotive service life. This approach offers environmental and economic benefits by extending the usefulness of batteries beyond their primary application:

Reusing Batteries

I. Stationary Energy Storage:
   - o **Grid Integration**: Used EV batteries maybe repurposed for stationary strength depository systems, in the way that grid-scale batteries or auxiliary power provisions for marketing and residential constructions.
   - o **Peak Demand Management**: Stationary storage applications help manage peak electricity demand by storing surplus energy during off-peak periods and supplying it during peak hours, thus enhancing grid stability and reliability (Meegoda et al., 2024).
   - o **Renewable Energy Integration**: Pairing second-growth batteries with energy from undepletable source beginnings like solar and wind farms

aids smoother unification into the grid, reducing confidence on non renewable energy-based capacity generation.

II. **Industrial Applications**:
    - o **Industrial Energy Management**: Batteries can support industrial operations by providing reliable power backup or smoothing energy fluctuations in manufacturing processes.
    - o **Off-Grid Solutions**: Deploying repurposed batteries in off-grid settings, such as remote communities or mining operations, enhances energy independence and reliability.

Environmental Benefits

I. **Delaying Recycling**: Extending the functional existence of batteries in second-growth applications delays the need for reusing or disposition, lowering the next demand for natural resources and underrating environmental impacts guide battery result and reusing processes.

II. **Resource Conservation**: By reusing batteries, valuable matters such as lithium, cobalt, and nickel in exchange for larger remain common longer, checking the referring to practices or policies that do not negatively affect the environment footprint of basic material ancestry and processing.

III. **Emission Reductions**: Second-life applications influence greenhouse gas issuance reductions by promoting strength efficiency and simplifying the unification of renewable energy beginnings into the energy mix.

IV. **Waste Reduction**: Utilizing batteries in second-growth uses reduces waste production and landfill disposal, advancing a tolerable approach to directing end-of-history batteries.

Challenges and Considerations

Performance and Safety: Ensuring the reliability and safety of batteries in second-life applications requires thorough assessment of battery health, capacity degradation, and potential risks associated with extended use.

Technological Adaptation: Adapting battery management systems and operational strategies to suit new applications and usage scenarios is essential for maximizing performance and longevity.

Economic Viability: Assessing the economic feasibility of second-life applications involves evaluating costs associated with battery refurbishment, integration into new systems, and operational maintenance against potential benefits such as energy savings and environmental gains.

5. POLICY AND REGULATORY FRAMEWORK

5.1 Global Regulations

Standards and Directives: Regulations governing battery manufacturing, use, and recycling vary globally but typically focus on environmental protection, resource conservation, and product safety. Key standards and directives include:

I. **Battery Directive (EU)**:
   - o **Purpose**: Regulates the manufacture, use, and disposal of batteries and accumulators within the European Union (EU).
   - o **Requirements**: Sets limits on the use of hazardous substances in batteries, advances reusing and recovery of matters, and orders branding for services information and security.

II. **Waste Electrical and Electronic Equipment (WEEE) Directive (EU)**:
   - o **Purpose**: Requires producers to take responsibility for managing the collection, recycling and disposal of electrical and electronic equipment, including batteries.
   - o **Requirements**: Establishes collection targets, promotes eco-design principles to facilitate recycling, and encourages separate collection of batteries for efficient recycling.

III. **UN Regulation on the Transport of Dangerous Goods (UNTDG)**:
   - o **Purpose**: Provides international guidelines for the safe transport of batteries and battery-powered equipment.
   - o **Requirements**: Specifies packaging, labeling, and handling requirements to mitigate risks associated with the transportation of hazardous materials, including lithium batteries.

IV. **RoHS Directive (EU)**:
   - o **Purpose**: Restricts the use of certain hazardous substances (e.g., lead, mercury, cadmium) in electrical and also in electronic equipment, including batteries (Robert et al, 2022).
   - o **Requirements**: Ensures that batteries placed on the market in the EU comply with restrictions on hazardous substances to protect human health and the environment.

V. **Environmental Protection Agency (EPA) Regulations (USA)**:
   - o **Purpose**: Sets standards for battery manufacturing processes, emissions control, and hazardous waste management under the Resource Conservation and Recovery Act (RCRA) and Clean Air Act (CAA).
   - o **Requirements**: Mandates agreement accompanying air and water quality principles, waste management procedures, and contamination stop

measures to underrate environmental impacts from battery result and operation.

Impact on Life Cycle Assessment (LCA): Regulations play a crucial role in shaping the environmental performance of battery technologies through several mechanisms:

I. **Environmental Standards**: Standards and directives, such as those under the EU Battery Directive and RoHS Directive, impose restrictions on hazardous substances and promote eco-design principles. Compliance with these standards ensures that batteries are manufactured using cleaner processes, reducing environmental impacts from raw material extraction, manufacturing, and disposal.
II. **Recycling Requirements**: Regulations mandating battery collection and recycling targets, as seen in the WEEE Directive, promote the efficient recovery of valuable materials from end-of-life batteries. Improved recycling rates contribute to resource conservation, minimize landfill waste and also decrease the environmental footprint of battery technologies.
III. **Safety and Transport Regulations**: International guidelines under the UNTDG ensure the safe handling and transport of batteries, particularly lithium-ion batteries, which pose risks of fire and explosion if mishandled. Compliance with transport regulations enhances safety and reduces incidents that could lead to environmental and human health hazards.
IV. **Economic Incentives and Penalties**: Some regulations provide economic incentives, such as subsidies for eco-friendly battery technologies or penalties for non-compliance with recycling targets. These incentives influence market behavior and encourage investments in sustainable battery manufacturing and recycling practices.

Future Trends

Harmonization and Global Alignment: Efforts to harmonize battery regulations across regions and promote global standards are gaining momentum. Enhanced international cooperation can streamline compliance requirements, facilitate cross-border trade in battery technologies, and promote sustainable development goals.

Technological Innovation: Regulatory frameworks that encourage innovation in battery design, recycling technologies, and sustainable practices will drive continuous improvement in environmental performance and resource efficiency.

Circular Economy Principles: Embracing circular economy principles, such as device talk over again, remanufacturing, and recycling, will guide future supervisory growths towards a more sustainable and flexible battery manufacturing.

5.2 Incentives and Support

Government Initiatives: Governments about the globe implement various programs and inducements to advance the adoption of sustainable battery technologies and enhance reusing practices:

I. **Research Funding and Grants**:
   - o **Purpose**: Governments provide funding opportunities and grants to support research and development (R&D) in advanced battery technologies, including materials innovation, energy storage systems, and recycling technologies.
   - o **Examples**: Programs like the U.S. Department of Energy's Advanced Battery Materials Research (ABMR) Program and the European Union's Horizon Europe framework allocate funding to consortia, universities, and research institutions to accelerate technological advancements.

II. **Tax Incentives and Subsidies**:
   - o **Purpose**: Tax credits, rebates, and subsidies incentivize investments in sustainable battery manufacturing facilities, renewable energy integration, and electric vehicle (EV) adoption.
   - o **Examples**: Incentives such as the U.S. Federal Investment Tax Credit (ITC) for energy storage projects and subsidies for EV purchases in countries like Norway and Germany stimulate market demand and reduce upfront costs for consumers and businesses.

III. **Regulatory Support**:
   - o **Purpose**: Governments establish regulatory frameworks that encourage eco-design principles, product stewardship, and lifecycle management of batteries. These regulations promote environmental sustainability, resource conservation, and safe handling practices.
   - o **Examples**: The EU Battery Directive mandates collection and recycling targets for batteries, while initiatives like California's Advanced Clean Trucks regulation incentivize zero-emission vehicle technologies, including battery-electric trucks.

IV. **Public Procurement Policies**:
   - o **Purpose**: Governments influence their purchasing power to obtain sustainable output and sciences.

     Examples: Public procurement policies in the EU prioritize products and services with eco-labels and certifications, encouraging suppliers to adopt sustainable practices in battery production and recycling.

Industry Collaboration

Industry partnerships and collaborations play a crucial role in advancing sustainable battery technologies through joint research, technology transfer, and market deployment:

I. **Research and Development (R&D) Consortia**:
   - o **Purpose**: Industry consortia bring together companies, research institutions, and academia to pool resources, share expertise, and accelerate innovation in battery materials, manufacturing processes, and recycling technologies.
   - o **Examples**: Consortia such as the Battery Innovation Center (BIC) in the United States and the Faraday Institution in the UK facilitate collaborative R&D projects to address technical challenges and commercialize new battery technologies.

II. **Supply Chain Integration**:
   - o **Purpose**: Collaborative supply chain initiatives optimize resource efficiency, promote transparency, and enhance sustainability across the battery value chain, throughout the process from raw material extraction to end-of-life recycling.
   - o **Examples**: Partnerships between battery manufacturers, materials suppliers, and recycling companies foster closed-loop systems and circular economy practices, guaranteeing that valuable materials are renewed and reused in new battery production.

III. **Policy Advocacy and Standards Development**:
   - o **Purpose**: Industry associations and trade organizations advocate for supportive policies, standards, and regulatory frameworks that facilitate market growth, investment certainty, and technological innovation in sustainable battery technologies.
   - o **Examples**: Organizations like the Global Battery Alliance (GBA) collaborate with stakeholders to develop sustainability principles, promote responsible sourcing of battery materials, and advocate for harmonized regulations to address global challenges.

Future Directions

Technological Innovation: Continued cooperation between administration instrumentalities, industry collaborators, and research organizations will drive advancements in artillery act, safety, and material sustainability.

Market Adoption: Scaling up production of sustainable battery technologies and expanding recycling infrastructure will depend on ongoing investments, supportive policies, and market demand for clean energy solutions.

Global Cooperation: Enhanced international cooperation and knowledge sharing will accelerate the transition towards a low-carbon economy, ensuring that sustainable battery technologies contribute to climate resilience and resource efficiency goals.

CONCLUSION:

The life cycle assessment (LCA) of emerging battery technologies for electric vehicles (EVs) provides a comprehensive framework to evaluate and compare their environmental impacts across various stages, from raw material extraction to disposal. This assessment is crucial for informing sustainable development strategies and guiding decision-making processes in the transition towards cleaner mobility solutions. In conclusion, the LCA of emerging battery technologies for EVs provides valuable insights into their environmental performance and sustainability implications. By addressing key challenges through technological innovation, regulatory support, and collaborative efforts, stakeholders can promote the adoption of sustainable energy storage solutions and contribute to a greener and more resilient transportation sector.

REFERENCES:

Aaqib., ... (2024) MXenes-enhanced vanadium redox flow batteries: A promising energy solution,Journal of Energy Storage,96,112711,ISSN 2352-152X,DOI: 10.1016/j.est.2024.112711

Alanazi, . (2023). Electric Vehicles: Benefits, Challenges, and Potential Solutions for Widespread Adaptation. *Applied Sciences (Basel, Switzerland)*, 13(10), 6016. DOI: 10.3390/app13106016

Almudena., ... (2022) Recent advances in lithium-sulfur batteries using biomass-derived carbons as sulfur host, Renewable and Sustainable Energy Reviews, 154,111783,ISSN 1364-0321,DOI: 10.1016/j.rser.2021.111783

Azizighalehsari, , Venugopal, P., Pratap Singh, D., Batista Soeiro, T., & Rietveld, G. (2024). Empowering Electric Vehicles Batteries: A Comprehensive Look at the Application and Challenges of Second-Life Batteries. *Batteries*, 10(5), 161. DOI: 10.3390/batteries10050161

Das, , Bhat, M. Y., & Sajith, S. (2024). Life cycle assessment of electric vehicles: A systematic review of literature. *Environmental Science and Pollution Research International*, 31(1), 73–89. DOI: 10.1007/s11356-023-30999-3 PMID: 38038907

Frith, , Lacey, M. J., & Ulissi, U. (2023). A non-academic perspective on the future of lithium-based batteries. *Nature Communications*, 14(1), 420. DOI: 10.1038/s41467-023-35933-2 PMID: 36702830

Gicha, , Tufa, L. T., Nwaji, N., Hu, X., & Lee, J. (2024). Advances in All-Solid-State Lithium–Sulfur Batteries for Commercialization. *Nano-Micro Letters*, 16(1), 172. DOI: 10.1007/s40820-024-01385-6 PMID: 38619762

Itani, , & De Bernardinis, A. (2023). A. Review on New-Generation Batteries Technologies: Trends and Future Directions. *Energies*, 16(22), 7530. DOI: 10.3390/en16227530

Liu et al, (2022) Tobias Placke, K.T. Chau,Overview of batteries and battery management for electric vehicles,Energy Reports,8,pp. 4058-4084,ISSN 2352-4847,DOI: 10.1016/j.egyr.2022.03.016

Machín, , Morant, C., & Márquez, F. (2024). Advancements and Challenges in Solid-State Battery Technology: An In-Depth Review of Solid Electrolytes and Anode Innovations. *Batteries*, 10(1), 29. DOI: 10.3390/batteries10010029

Meegoda, , Charbel, G., & Watts, D. (2024). Sustainable Management of Rechargeable Batteries Used in Electric Vehicles. *Batteries*, 10(5), 167. DOI: 10.3390/batteries10050167

Meegoda, , Charbel, G., & Watts, D. (2024). Second Life of Used Lithium-Ion Batteries from Electric Vehicles in the USA. *Environments (Basel, Switzerland)*, 11(5), 97. DOI: 10.3390/environments11050097

Mehmet., ... (2024) A review on the lithium-ion battery problems used in electric vehicles,Next Sustainability,3,100036,ISSN 2949-8236,DOI: 10.1016/j.nxsust.2024.100036

Mori, . (2023). Cathode materials for lithium-sulfur battery: A review. *Journal of Solid State Electrochemistry*, 27(4), 813–839. DOI: 10.1007/s10008-023-05387-z

Natsuki Nakamura et al. (2022) Future potential for lithium-sulfur batteries,Journal of Power Sources, 558, 232566,ISSN 0378-7753,DOI: 10.1016/j.jpowsour.2022.232566

Olabi, , Allam, M. A., Abdelkareem, M. A., Deepa, T. D., Alami, A. H., Abbas, Q., Alkhalidi, A., & Sayed, E. T. (2023). Redox Flow Batteries: Recent Development in Main Components, Emerging Technologies, Diagnostic Techniques, Large-Scale Applications, and Challenges and Barriers. *Batteries*, 9(8), 409. DOI: 10.3390/batteries9080409

Olabi et al., (2022) Battery electric vehicles: Progress, power electronic converters, strength (S), weakness (W), opportunity (O), and threats (T),International Journal of Thermofluids,16,100212,ISSN 2666-2027,DOI: 10.1016/j.ijft.2022.100212

Panda, (2023) Comprehensive Review of Electric Vehicle Technology and Its Impacts: Detailed Investigation of Charging Infrastructure, Power Management, and Control Techniques. *Appl. Sci.13*, 8919. .DOI: 10.3390/app13158919

Prasad et al., (2023) Environmental life cycle assessment of emerging solid-state batteries: A review, Chemical Engineering Journal Advances,13,100439,ISSN 2666-8211,DOI: 10.1016/j.ceja.2022.100439

Robert, . (2022). The Regulatory Environment for Lithium-Ion Battery Recycling. *ACS Energy Letters*, 7(2), 736–740. DOI: 10.1021/acsenergylett.1c02724

Sayahpour, , Hirsh, H., Parab, S., Nguyen, L. H. B., Zhang, M., & Meng, Y. S. (2022). Perspective: Design of cathode materials for sustainable sodium-ion batteries. *MRS Energy & Sustainability : a Review Journal*, 9(2), 183–197. DOI: 10.1557/s43581-022-00029-9

Sean et al., (2023) Designing lithium-ion batteries for recycle: The role of adhesives,Next Energy,1,2,2023,100023,ISSN 2949-821X,DOI: 10.1016/j.nxener.2023.100023

Xiqian et al., (2022) Battery Safety: From Lithium-Ion to Solid-State Batteries,Engineering, 21,pp.9-14,ISSN 2095-8099, https://doi.org/DOI: 10.1016/j.eng.2022.06.022

Yiju., ... (2021) Material design and structure optimization for rechargeable lithium-sulfur batteries, Matter,4,4,2021,pp.1142-1188,ISSN 2590-2385,DOI: 10.1016/j.matt.2021.01.012

Zanoletti, , Carena, E., Ferrara, C., & Bontempi, E. (2024). A Review of Lithium-Ion Battery Recycling: Technologies, Sustainability, and Open Issues. *Batteries*, 10(1), 38. DOI: 10.3390/batteries10010038

Zhao et al, (2023) Engineering of Sodium-Ion Batteries: Opportunities and Challenges, Engineering, 24,pp. 172-183,ISSN 2095-8099,DOI: 10.1016/j.eng.2021.08.032

Zhou et al., (2024) Life cycle assessment of recycling lithium-ion battery related mineral processing by-products: A review,Minerals Engineering,208,108600,ISSN 0892-6875,DOI: 10.1016/j.mineng.2024.108600

Chapter 5
Advancements in Renewable Energy Harvesting for EV Charging Infrastructure

Mandeep Singh
https://orcid.org/0000-0003-3278-4877
Bennett University, India

Rahul Sharma
Raj Kumar Goel Institute of Technology, Ghaziabad, India

Ramveer Singh
https://orcid.org/0009-0003-8773-3863
Raj Kumar Goel Institute of Technology, Ghaziabad, India

Aruna Malik
Dr. B.R. Ambedkar National Institute of Technology, Jalandhar, India

Puneet Kumar Aggarwal
https://orcid.org/0000-0001-5616-8564
ABES Engineering College, India

ABSTRACT

Renewable energy sources (RES) show great promise in providing sustainable power for electric vehicle (EV) charging infrastructure, offering cleaner alternatives to traditional fossil fuels. Solar-powered and wind-powered EV charging stations tap into the abundant energy from the sun and wind to replenish EV batteries, significantly reducing carbon emissions and promoting environment-friendly transportation.

DOI: 10.4018/979-8-3693-6844-2.ch005

However, integrating RES into EV charging infrastructure presents certain technical challenges. The main issue lies in the variability of renewable energy generation, as it heavily relies on weather conditions. To address this, advanced energy storage solutions and cutting-edge Vehicle-to-Grid (V2G) technology have been explored to ensure a stable and reliable energy supply for EV charging. Additionally, effectively integrating RES with existing power grids requires sophisticated power management and grid balancing strategies to optimize energy usage and prevent overloading. Renewable energy sources' roles in EV charging station infrastructure will discuss in this chapter.

1. INTRODUCTION

There are several reasons why the demand for electric vehicles is growing, from environmental friendliness with no pollution to a reliance on fossil fuel energy, less noise, and efficiency. Due to this, the demand for electric energy is very high globally. From 2010, when EVs that were on the roads barely went into hundreds to about 17 million by 2021. With the use of EVs, climatic impacts can be significantly reduced along with the reduction of pollutants. Implementation of electric vehicles is not going ahead due to low range and long hours required for recharging (Von Jouanne et. al., 2005). EVs are not expensive to operate and consume less energy than any other vehicle.

We can reduce the local power grid load by employing wind and solar energy in the charging infrastructure of EVs. Renewable energy is pure and abundant, and to balance demand while decreasing emissions by using it for the recharge of EVs in a more sustainable and efficient way to charge, W. Kempton et. al., 2016. Solar and wind-based charging infrastructure development can be considered as a starting point for research for renewable energy-based Electric vehicle charging infrastructure, T. Harakawa, et. al.,2002. The design of the proposed new type of charging station is, therefore, on renewable energy sources and direct current (DC) for the charging of electric vehicles so as to avoid the problems that may arise with regular charging stations due to harmonics, fluctuations, and voltage outages that affect the stability of the electrical grid. Thus, making it easier and more reliable in charging (Etezadi-Amoli et. al.,2010). Advances in renewable energy harvesting for EV charging infrastructures are regularly compared with other near approaches towards measuring effectiveness and aptness of these approaches. Now let's explore a few of these approaches and compare them with renewable energy harvesting:

The impacts of fossil fuel usage, greenhouse gas emissions, and air pollution due to the transportation sector. In response to these issues, a global trend towards electric vehicles powered by renewable energy sources has been made. However, in order to integrate them into the transportation system effectively, creating a reliable and sustainable charging infrastructure is necessary. EV charging infrastructure primarily depends on electricity grids and mainly run by fossil fuels. Using renewable energy sources in EV charging has been the fore-front of minimizing environmental impact and promoting the transportation service. Charging stations incorporating RES can further reduce the impact that environmental will have on the overall usage of EVs. New inventions on the integration of smart grids, storage, and renewable energy facilities have provided a route for innovative infrastructure in EV charging.

In the aggregate, it could stabilize the grid, improve the availability of clean and cheap energy for EVs, and contribute to a decentralized, more sustainable energy system. Such advancement is as important for policymakers and researchers as for industry stakeholders to know the current state of the art of renewable energy harvesting to power EV charging infrastructure. This literature review attempts to illustrate the state of the art, difficulties, and prospects in this sector by analyzing available literature and technical advancement. The analysis of available information and research gaps through the literature analysis might be helpful for future research orientations and policy decisions in the quest for sustainable transport and integration of renewable energy sources.

A few articles only superficially studied EV charging infrastructure research, but it analyzed unspecified technical parameters without keeping renewable energy sources in mind (H. Das et al., 2020). In the one study exploring RCI experiments, it centered the consumer. Interactions and preferences with EVs to our knowledge, no study has investigated RCI. In-depth studies by covering all relevant study areas (S. Hardman et. al., 2016). Uncontrolled charging of electric vehicles (EVs) can result in problems like overloaded feeders, inappropriate transformer placement, and strain on the power supply. Researchers have recommended using stationary energy storage and quick-charging technologies to address this issue. By providing power to EVs during moments of high demand, energy storage helps minimize the infrastructure needed for charging and lowers operational expenses. Energy storage can also deliver enough power during emergencies to go to charging stations, improving the overall stability of EV operations. Much research has been done (A. Hussain et al.,2019; A. Mohammad et al., 2020) done to show how using stationary energy storage with quick charging techniques is advantageous. The energy storage system must be the right size, considering things like the rate of energy supply, EV load patterns, and anticipated levels of EV penetration, to fully realize these benefits.

1.1 Motivations

Electric vehicles are becoming increasingly popular in several countries throughout the world. However, the absence of charging stations precludes EVs from being widely used by users worldwide. EV charging stations are being deployed in more public areas as EV usage increases. The urgent need to transition to sustainable transportation solutions will be explored due to the detrimental effects of fossil fuel consumption and greenhouse gas emissions. The motivations also include reducing air pollution, justifying climate change, and achieving energy independence by utilizing renewable energy sources. We will discuss the environmental benefits of renewable energy, such as lower carbon emissions, improved air quality, and the preservation of natural resources. Additionally, we will highlight the limitations and challenges associated with conventional energy generation, including the depletion of fossil fuel reserves, price volatility, and geopolitical tensions. Understanding these motivations will provide a comprehensive context for the advancements in renewable energy harvesting for EV charging infrastructure (F. Mwasilu et. al., 2014).

When balancing the electrical grid, energy storage technologies are no longer a ground-breaking idea. Other technologies use flywheels, dedicated batteries, concentrated solar energy systems, and hydro-pumped storage systems. Innovations in the sector include wireless sensors, smart meters, power converters, and improved communication. The most current developments in RCI technology and new findings in this important area are examined in this paper. The concept of RCI is covered in the article from a variety of angles, including the selection of suitable renewable energy sources for RCI, sitting, optimal planning, sizing, control and energy management, and RCI-related problems.

1.2 Contributions

This study emphasizes the positive impact of advancements in renewable energy harvesting on EV charging infrastructure. We will discuss how these advancements contribute to reducing carbon emissions, one of the major contributors to climate change. By harnessing clean and renewable energy sources such as solar power, wind, and power, EV charging infrastructure becomes more sustainable, aligning with global efforts to mitigate climate change. We will highlight charging networks' improved efficiency and reliability by integrating storage systems, renewable energy generation, and smart grid technologies. Additionally, we will address the positive socio-economic impacts, such as job creation and energy cost savings that arise from the growth and development of renewable energy harvesting for EV charging infrastructure.

1.3 Structure of the chapter

The chapter is organized into different Sections. Section 2 discussed the detailed, comprehensive survey on Renewable energy harvesting for EVs. Section 3 elaborates on the Renewable Energy Sources for EV Charging Stations. Section 4 discusses the Energy Storage System, which includes various approaches for energy storage and advancement in this field. Section 5, Integration of Renewable Energy and Energy Storage. Section 6, Benefits of Renewable energy harvesting for EV charging Infrastructure, Section 7. A real-world case study of a solar-powered EV charging station will be presented to demonstrate the practical implementation of RES in the EV charging infrastructure. Section 8, the conclusion drawn from the chapter. The last section contains a list of references.

2. LITERATURE REVIEW

(Mwasilu, et. al. 2014) offers an extensive review of the integration of EVs to the smart grid infrastructure. It outlines the nature of V2G and its manifestation with the VPP phenomenon, along with the interaction of EVs with smart grids that focuses on smart charging and advanced metering and communication infrastructures. The paper evaluates the possibility of RES integration using EVs and evaluates the case of EV integration into the infrastructure of the smart grid. In summary, it concludes that advanced control, communication, and metering technologies will permit the integration of EVs into the smart grid system; they also promote EV interoperability for grid support. The paper lists several important benefits of integrating electric vehicles into the smart grid, including improving air quality, reducing greenhouse gas emissions, and providing financial savings to the owners of electric vehicles.

(H. Das et. al, 2020) analyzed the scenario of electric cars (EVs) concerning the charging infrastructure as well as techniques of grid integration at that time. It covered crucial international standards, infrastructure configurations, and optimization algorithms for electric vehicle grid integration management. The article also evaluates the future impact of EV developments on the grid and offers recommendations for future infrastructure development. The main advantages of an electric vehicle in terms of grid integration involve improvement in power quality, support to renewable energy regulations, and augmentation of stability in the power grid. The paper also discusses supporting technologies that are relevant to the future of EVs and their integration into the grid. Overall, this paper provides a significant resource for researchers and engineers working in the area of EV charging and grid integration.

(T. Harakawa, et. al., 2002) Propose new technologies to improve the efficiency of solar power generation systems for electric vehicles. The authors aim to achieve this by introducing a new method for instantaneously extracting maximum power from solar cells under all kinds of conditions changes, as well as two efficiency tracking systems for electric vehicles. The ultimate goal is to improve energy conversion efficiency for charging EVs and contribute to solving global environmental problems. The authors also discuss the feasibility of solar power generation systems for various fixed portions like building roofs, the land of vacant space, walls of schools and public spaces, monuments and signals along the side of the roads, etc.

It has been talked about the prospects of electric cars to replace the conventional gasoline-powered automobiles. The problems and challenges that must be addressed in that are also brought up as part of the discussion. The authors' perspective is that at a reasonable price, the EVs already on the market can be provided to satisfy the driving needs of most drivers in several regions in the US, and larger batteries, which will surely come in the near future, would really make them more usable. Other recommendations to be derived from the report are calls for range-appropriate battery vehicles and charging rates, multiple policies customized to a range of driving needs and a greater understanding of the diverging needs of different market segments as part of a much broader approach to stepping up adoption of electric vehicles.

In the in-depth analysis of all aspects of modeling a grid-connected EV-PV system, (Mohammad A. et al., 2020) discuss control structures, charging algorithms, and uncertainty analysis. The paper draws a general framework for developing the grid-connected EV-PV system and aims to present the researchers with the tools they need to model an EV-PV system connected to the grid. It also assesses different components of system modeling.

The primary objective of the study by M Khalid et al. (2019) was to have a comprehensive investigation of EV charging facilities along with their impacts on the power quality of the utility grid. Through this paper, the author identifies difficulties in integrating EV charging infrastructure into the power grid, examines and analyses the impact of different types of EV charging stations on power quality, and, therefore, provides potential remedies with the aim of reducing the impact of EV charging on the grid.

Authors (Y Bayo et al., 2018) are introducing a new bi-level optimization approach to the rapid charging of electric vehicles. The strategy reduces the electricity prices, achieves peak shaving, and limits fluctuation in the range of the state-of-charge. The paper features the results of day-ahead as well as real-time rolling optimizations and includes a case study of a real rapid charging station in Beijing, China. Good performance with respect to electricity cost reduction and strengthening of the stability of the charging station is manifested by the proposed approach. The paper presents the problems and motivations for the research, the proposed BESS-based bi-level optimization method, and models of day-ahead and real-time optimizations in separate chapters. Discussion on the results obtained and their implications is done within the concluding section of the paper.

For the ESS for fast electric vehicle charging stations, Hussain, A. et al., 2019 discusses the batter and converter size at one time. In the course of that research, simulations of the proposed method are done relative to the existing size methodologies. The work establishes a mathematical model for ESS sizing and carries out sensitivity analyses regarding the size of the converter and battery. All these factors-the number of electric vehicles registered at the charging station, the fraction of the useful energy range of the ESS, the unpredictability of the EVs about when they would return to the charging station, and the commercial-to-private ratio of the fleet-are discussed. As conclusion to the article, future research axes are presented.

(Münderlein, J. et. al. 2020), The paper discusses how the techniques of operation and optimization of the hybrid storage system can be assessed. In the presented paper, authors review how BSS stabilizes the grid and whether they are fit for the market for primary control reserves (PCR). The paper includes a literature review on hybrid battery storage system operating strategies, the impacts of hybridizing the HSS with a power-to-heat system and optimizing the technology sizes. The authors further provide an optimization algorithm to calculate the most beneficial dimensioning of the hybrid storage system. It is crucial that in their study, (Logan et al., 2021) investigate the introduction of EVs and what that may mean for attaining net-zero emissions.

The study compares a simple operating emissions model with a life cycle analysis. Battery electric vehicles can result in a dramatic reduction of emissions, relative to conventional internal combustion engine vehicles, according to the findings comparing the two. This is dependent on several factors which range from the origin of electricity utilized to charge EVs to the types of batteries. The report continues by stating that since a vehicle will eventually be a waste, it is essential to consider the entire lifecycle of a vehicle from production through use and eventually become waste in order to determine how effectively the vehicle affects the environment. The paper also discusses impacts of the research findings on the UK's goal to become net zero. This includes ensuring the environmental impacts of battery production and recycling are considered with legislation in place to encourage the path toward electric vehicles. Table 1. Generally, discuss all methods for collection of renewable energy used at EV charging facilities.

Table 1. Literature Review Renewable Energy Harvesting for EV Charging Infrastructure

Author	Objective	Technology/Methodology	Strength
Mwasilu, et. al. (2014)	EVs Integration in a smart grid system and to evaluate the potential benefits of such integration. infrastructures, challenges associated with the integration of EVs into the smart grid	• advanced metering infrastructure (AMI), • EV management systems (EVMS).	• improved grid stability • Increased efficiency of energy and Improved air quality • Reduced the emission of greenhouse gases.
H. Das et. al. (2020)	Evaluate the existing EV standards, and grid integration techniques charging infrastructure.	• EV grid integration management by using Game-based algorithms	• improved power management • Improvement in Power Quality • Increased stability of the power grid
T. Harakawa, et. al. (2002)	Improve energy conversion efficiency for charging electric vehicles and contribute to solving global environmental problems	• conventional method and control system for power conversion circuits. • Feed forward method and climbing method.	• Improving the efficiency of solar power generation systems for electric vehicles. • Reducing the burden of external electrical power stations. •
W. Kempton et. al. (2016)	The potential of electric vehicles (EVs) to replace traditional gas-powered cars and the challenges that need to be overcome to make this transition.	• An approach to increasing EV adoption includes multiple policies, an improved understanding of the differing range needs of different market segments, and creating vehicles with batteries and charging rates appropriate for diverse driving requirements.	• the potential of reducing dependence on petroleum-based fuels, mitigating climate change • human health improvement in urban areas.
A. Mohammad et. al. (2020)	To present a framework for designing a grid-connected electric vehicle-PV system. Evaluate different modelling components of the system, including charging algorithms, control architectures, and uncertainty analysis.	• grid-connected Electric Vehicle-PV system	• increased use of renewable energy sources • Improved power quality and reliability • Reduced energy costs for EV owners

continued on following page

Table 1. Continued

Author	Objective	Technology/Methodology	Strength
M Khalid et. al. (2019)	provide a comprehensive review of EV charging infrastructures and their effect on the power quality of the utility grid	• AC/DC/DC converters, common AC link. multiple DC ports, and conductive and inductive charging structures	• High vehicle-to-wheel efficiency due to electric motors. • High well-to-wheel efficiency.
Y Bayo et. al. (2018)	Propose a new approach known as the bi-level optimization approach to charging load regulation of EV quick charging stations	• linear programming and predictive control-based real-time rolling optimization strategies to get the control objectives and maintain battery life.	• reducing electricity costs • Minimizing change in state-of-charge range • Real-time rolling optimization
A. Hussain et. al. (2019)	Propose a simultaneous sizing method for battery and converter in Stationary Energy Storage System (ESS) for Fast Electric Vehicle (EV) Charging Stations.	• current sizing method for battery and converter in Stationary Energy Storage System (ESS) for quick EV Charging Stations.	• Avoiding oversizing converters by sizing them together with batteries. • Shaving peak loads, which can help to reduce the price of purchasing power by the grid.
J. Münderlein et. al. (2020)	Optimize a hybrid storage system consisting of a lithium-ion battery, a lead-acid battery, and a power-to-heat system for participation in the primary control reserve (PCR) market	• surrogate optimization algorithm • Operation strategies for the hybrid storage system.	• Presents an optimization algorithms to determine the most profitable dimensioning of the hybrid storage system. • Evaluates operation strategies for the hybrid storage system
Logan et. al. (2021)	Investigate the effect of BEVs (battery electric vehicles) on reducing operating emissions and achieving net zero emissions in the UK.	• simple OPEM (operation emissions model) against an LCA (life cycle analysis) approach to assess the carbon dioxide emissions from internal combustion engine vehicles (ICEVs) and BEVs	• Reduced operating emissions • Potential to achieve net zero emissions • Reduced dependence on fossil fuels

3. RENEWABLE ENERGY SOURCES FOR ELECTRIC VEHICLE CHARGING INFRASTRUCTURE

The rise in the popularity of EVs has resulted in a marked increase in the demand for electric energy worldwide. It is challenging to ignore the role of environmentally friendly energy sources and the considerable influence of the combination of EVs and systems while examining the effects of EV grid integration. In this paper (Knezovic et al. 2017), the analysts first started from the standpoint of controlling greenhouse gas emissions before examining the feasibility and challenges of coordinating limitless energy sources. The power structure becomes less stable when adaptable loads are employed due to the misuse of sustainable energy, which has not yet been effectively remedied. PEVs can charge EVs when renewable energy is available or during off-peak hours. Table 2. provides a comparison of different renewable energy sources.

3.1 Solar Energy Harvesting

Collecting and storing solar energy is known as solar energy harvesting. The energy from the heat and light is then appropriately transformed into electrical energy. Considering energy tariffs, an ideal energy storage system size is needed for solar energy harvesting. Solar-powered batteries can meet unreliable grid electricity demands, such as powerful charging and discharge rates and irregular full charging periods. A variety of battery kinds meets these exact requirements. and flow batteries are the three main battery storage types examined for solar energy (Ibrahim et al., 2010). A central controller is necessary to route the generated electricity to the battery to conserve the extra energy generated by photovoltaics, as shown in Fig. 1. The photovoltaic sequence of controllers has been the subject of extensive research. They emphasized the necessity of using pulse width modulated (PWM) and MPPT technology (maximum power point tracker) to increase the productivity of solar energy generation (Singh R. et al., 2024). For EV charging (during the day), solar energy is connected to the home's power source, which is referred to as a sun-to-vehicle (S2V) power source. In the evening, the charging process will be reversed when the EV discharges home electricity using the vehicle-to-home (V2H) and vehicle-to-grid (V2G) processes.

Figure 1. Infrastructure for Solar PV Charger for EV Charging.

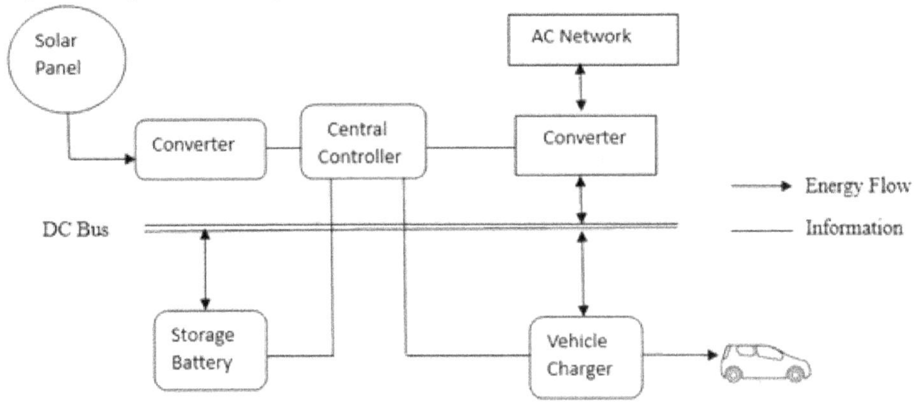

In PV systems, photovoltaic arrays are linked to a DC/DC converter, allowing accurate power point tracking control for solar converters. In addition, an AC/DC converter is used to convert power between AC and DC forms in both directions. Electric car charging and power transmission between the grid and electricity networks are made possible by DC conversion because most of the grid's power is AC. As a result, converters are essential in PV systems since they rely on balanced energy conversion (Singh R. et al., 2024). A wide range of converter types and specifications have been carefully evaluated, including string inverters that install panels using microinverters, central inverters that use separate inverters, and micro-inverter power optimizers that demand more monitoring.

3.2 Wind Energy Harvesting

The process of transforming the wind's kinetic energy into usable power is called wind energy harvesting. This renewable energy source has drawn much interest because of its potential to support sustainable power production and lessen dependency on fossil fuels. Wind turbines capture wind energy and turn the wind's mechanical energy into electrical energy. Several important factors and considerations are involved in the process of harvesting wind energy, including:

3.2.1. Wind Turbines: Wind turbines are the main technique for capturing wind energy. They comprise substantial rotor blades mounted on a tower that turn when the wind blows. The blades' rotation powers a generator to generate electricity. Vertical-axis wind turbines (VAWTs) and. horizontal-axis wind turbines (HAWTs) are two types of wind turbines, each with unique benefits and applications.

3.2.2. Wind Resource Assessment: A detailed evaluation of the wind resource is required before the installation of wind turbines. Wind direction and speed must be continuously measured at the intended location. The potential for wind resources and their seasonal fluctuation is calculated using information from meteorological stations, satellite data, or on-site anemometers.

3.2.3. Wind Turbine Siting: Wind turbines must be appropriately positioned to maximize energy production. Wind speed, wind direction, turbulence, geography, and the closeness of obstructions must all be considered. Computer simulations and models are frequently used to determine wind turbines' best location and configuration to maximize energy extraction.

3.2.4. Grid Integration: To provide power, wind farms made up of numerous wind turbines are connected to the electrical grid. To maintain a consistent and dependable electrical supply, proper integration with the grid involves careful coordination and management procedures (M Khalid et al., 2019). Specific criteria for wind farm connections, such as voltage and frequency management, power quality standards, and communication protocols, may be imposed by grid rules and laws.

3.2.5. Offshore Wind Energy: Harvesting offshore wind energy has become more popular recently as a result of the availability of stronger and more reliable winds over bodies of water. In addition to the usual obstacles of installation and maintenance logistics, environmental impact studies, and subsea cable connections to onshore grids, offshore wind farms are frequently situated in coastal regions.

Table 2. Comparison of Renewable energy sources

Approach	Description	Advantages	Limitations
Wind energy	Refers to harnessing wind's kinetic energy and converting it into electricity through wind turbines. This renewable energy source provides a sustainable, eco-friendly solution to meet electricity demands.	• No greenhouse gas emissions, • Reducing dependence on fossil fuels. • Contributing to a more sustainable and environmentally friendly energy mix	• Relies on the availability of wind to generate electricity, which can result in fluctuations in power output. • Noise pollution from wind turbines can be a concern for wildlife.
Hydroelectric Power	Hydroelectric power renewable energy technology generates electricity by using moving water as energy. It uses dams or flowing water to turn turbines.	• produces clean electricity without emitting greenhouse gases, • Helps in water management by regulating river flow and controlling floods.	• building large dams can lead to environmental impacts. • Droughts or changes in water availability can affect power generation.
Grid-Based Charging	Relies on the existing electrical grid infrastructure to supply electricity to EV charging stations, predominantly powered by conventional sources like fossil fuels and nuclear energy.	• convenient and widely available. • Cost-effective in many areas. • Established infrastructure and reliability.	• contributes to carbon emissions and environmental pollution. • Reliance on fossil fuels and non-renewable resources. • Limited long-term sustainability
Conventional Renewable Energy	Utilizes RES, like solar and wind power, for general electricity generation without targeting EV charging infrastructure.	• established technologies and infrastructure. • Supply electricity to various applications beyond EV charging.	• Not optimized or customized for EV charging demands. • Not fully integrated with the charging network or prioritise grid management.
Fossil Fuel Generators	Utilises on-site fossil fuel generators, such as diesel or gasoline, to produce electricity for EV charging stations.	• provide power in areas with limited access to the grid. • Immediate availability and independent operation.	• Increase carbon emissions and air pollution. • Relies on non-renewable resources. • Costly fuel requirements and ongoing maintenance.
Renewable Energy Harvesting	Utilises clean and sustainable energy sources, like solar and wind power, to generate electricity for EV charging infrastructure.	• Environment friendly and reduces carbon emissions. • Promote long-term sustainability. • Can be integrated with the grid and optimised for charging demands.	• Initial setup costs can be high. • Depending on weather conditions and renewable energy availability. • Requires energy storage systems for stable power supply.

3.3 Hydroelectric Power Harvesting

The act of turning the kinetic energy flowing into electricity is known as hydroelectric power harvesting, often known as hydroelectricity or hydropower. It offers a dependable and clean energy option and the oldest, most popular renewable energy sources. Large-scale electricity production is accomplished via hydroelectric power plants, which use the energy of water (Sbordone, D. et al. 2015).

Several important elements and concerns are involved in the process of harvesting hydroelectric power, including:

- Dam: Water flow is controlled by dams; a diversion structure is built across a river or other water source. The dam builds a pond or reservoir that retains water, allowing for a controlled release to power turbines and producing electricity. Without building a reservoir, diversion structures change the direction of water flow.
- Penstock: The penstock is a sizable pipe or conduit used to transport water from a dam or other structure with a diversion system to the turbines. The turbines receive the high-pressure water in the penstock, which transforms the water's potential energy into mechanical energy.
- Devices called turbines use the mechanical energy of moving water to create rotational motion. Depending on the unique properties of the water flow, many types of turbines, such as Francis, Pelton, and Kaplan turbines, are employed in hydroelectric power plants.
- Generators: To run generators, which turn mechanical energy into electrical energy, turbines must rotate. Generators in hydroelectric power plants frequently use electromagnetic induction to generate energy.
- Transmission and distribution: To reach distribution networks, where it is then distributed to consumers, hydroelectric power plants must transport the electricity they produce across power lines. Transformers are used to increase or decrease the voltage to facilitate effective transmission and distribution.
- Pumped storage hydropower is that special type of hydroelectricity which has two reservoirs in different elevations. When the energy demand is low, it draws down water from the lower reservoir and pumps up to the upper storage unit by utilizing surplus power available in the grid. When the demand for energy is high, the water from the higher reservoir flows through and reaches the power turbines generating electricity.
- Environmental Considerations: Hydroelectric harvesting benefits and negative effects on the environment following the construction of dams. Though hydroelectric power is one of the cleanest renewable sources of energy, its installation might cause damage to habitats, river ecosystems modification, as

well as interrupt the routes of fish migration in rivers. Thus, the use of these power sources should also involve, apart from offering preliminary investigations, carrying out environmental impact assessment and implementing mitigative measures in their applications.

Some advantages hydroelectric power holds include being renewable, having high energy conversion efficiency, and lasting for a long time when the facilities are put in place. It further provides grid stability by supplying constant and reliable electricity (Ehsan et al., 2020). However, one requirement for the feasibility of hydroelectric power is having access to sufficient water resources such as rivers or waterfalls. Hydroelectric power harvesting is still undergoing study and innovation, emphasizing improving environmental impact assessments, optimizing turbine designs, and investigating cutting-edge technologies, including underwater turbines and run-of-river systems. In adapting to a low-carbon energy future, these initiatives seek to improve hydroelectric power's effectiveness, sustainability, and compatibility with other renewable energy sources.

4. ENERGY STORAGE SYSTEM

Energy storage devices are essential for gathering renewable energy for EV charging infrastructure. By storing extra energy during high generation and releasing it when demand is high or renewable energy production is low, they facilitate the effective use of renewable energy (Ding et al., 2015). Overview of energy storage technologies in EV charging infrastructure

Energy storage technologies used in EV charging infrastructure include:

- Lithium-ion Batteries. In the past, lithium-ion batteries were the most popular type of energy storage technology for EV charging infrastructures due to their good cycle life, high energy density, and fast charging abilities. At peak demand, excess renewable energy can be stored in these batteries for later distribution to EVs, thereby giving an assured and reliable charging experience.
- Flow batteries Flow batteries rely on the liquid electrolytes stored in external tanks to store energy. They are especially useful at grid scale because they offer longer-duration energy storage. Along with flow batteries, intermittent sources of renewable energy may be supplemented by steady power for charging EVs, according to Y Bayo et al., 2018.
- Supercapacitors: Also referred to as ultracapacitors, supercapacitors contain quick energy storage and release characteristics. Supercapacitors are combined with other storage devices in many cases in EV charging infrastructure

as components handle sudden power fluctuations and regenerative breaking very well.

- Energy Storage in Hydrogen Fuel Cells: Hydrogen fuel cells can be exploited to store energy, for instance, by electrolysis of excess renewable electricity to form hydrogen gas. The stored hydrogen is then used for on-demand energy production, which can either be electricity generated in a fuel cell or direct charging of EVs.

Benefits from EV Charging Infrastructure

- Grid Stabilization: Energy storage systems actually relieve some of the pressure on the grid through balancing intermit application of renewable energy sources. They can make extra supply energy during high-demand times and absorb excess energy during the generation time thereby providing constant and reliable power supply for EV charging.
- The provision of peak load management: The energy storage systems provide stored energy at times of high electrical demand, thereby aiding in the management of peak loads. This then reduces pressure on the grid, and the utilization of renewable energy is maximized such that the infrastructure concerned with EV charging is operated both efficiently and economically (M Khalid et al., 2019).
- Time Shifting: Renewable energy sources can be time-shifted by means of energy storage technologies. The surplus generated can be stored and used later when there is a need for it. More usage of RES would be made at peak periods when the demand for electricity for charging EVs at those times will also be matched by supply.
- Additional Services: Energy storage systems can provide additional services to the grid such as voltage support and frequency regulation. These services are ensured to further achieve better efficiency and dependability of the charging infrastructure of the EVs and the stability of the grid.

An overview of energy storage technologies in EV charging infrastructure is provided below:

- Lithium-Ion Batteries: Lithium-ion batteries are among the most popular energy storage technologies for EV charging infrastructure. They have a significant cycle life, high energy density, and quick charging ability. When demand is at its highest, these batteries can store excess renewable energy and provide it to EVs, delivering a dependable and consistent charging experience.

- Flow batteries: Flow batteries use liquid electrolytes housed in external tanks to store energy. They are useful for grid-scale applications because they provide longer-duration energy storage. Intermittent renewable energy sources can be supplemented with flow batteries, offering steady power for charging EVs (Y Bayo et al., 2018).
- Supercapacitors: Also referred to as ultracapacitors, supercapacitors have quick energy storage and discharge characteristics. They are frequently employed in conjunction with other storage devices to manage sudden power fluctuations and regenerative braking in EV charging infrastructure.
- Hydrogen Fuel Cells: By electrolyzing excess renewable electricity to produce hydrogen gas, hydrogen fuel cells can be utilized to store energy. The stored hydrogen can then be used to produce energy on demand, either for fuel cell power generation or for directly charging EVs.

Benefits of EV Charging Infrastructure:

- Grid Stabilization: Energy storage systems reduce grid stress and enhance stability by balancing the intermittent nature of renewable energy sources. They may supply energy at times of high demand and absorb excess energy during high generation times, delivering a steady and dependable power supply for EV charging.
- Peak load management: Energy storage systems provide stored energy at times of high electrical demand, which aids in managing peak loads. As a result, the strain on the grid is lessened, and the use of renewable energy is maximized, ensuring the efficient and economical operation of the EV charging infrastructure (M Khalid et al., 2019).
- Time Shifting: Renewable energy can be time-shifted using energy storage technologies. Excess energy produced during off-peak hours can be stored and used to enhance the use of RES during periods of huge demand. This will match the supply of renewable energy with the demand for EV charging.
- Additional Services: Energy storage systems can offer the grid additional services like voltage support and frequency regulation. These services enhance the overall efficiency and dependability of the EV charging infrastructure while preserving the grid's stability.

Various ways of energy conservation can be connected to the infrastructure for EV charging., including the following:

- Station-Level Storage: By installing energy storage systems at individual EV charging stations, it is possible to maintain a steady and dependable power supply even when the production of renewable energy is low.
- Community-Level Storage: The technologies for storing energy can be installed at the neighbourhood or community level, supplying several charging stations and balancing the demand and supply of renewable energy in the neighbourhood.
- Grid-Scale Storage: By storing extra renewable energy from diverse sources and supplying it to the charging stations as needed, large-scale energy storage systems integrated with the grid can assist EV charging infrastructure.

Ongoing energy storage research and development aims to increase the quantity of energy that can be stored, cutting prices, increasing charging efficiency, and investigating novel technologies, including solid-state batteries and sophisticated hydrogen storage techniques Improved smart grid technology and demand response have enabled better coordination and integration of energy storage systems with the charging infrastructure of electric vehicles. The technique has optimized charging patterns for maximum use of renewable energy sources. These, therefore, are energy storage techniques that are necessary for gathering renewable energy for the charging infrastructure of EVs. They enable a good amount of renewable energy to be used efficiently, ensuring grid stability and controlling peak demands, while also providing ancillary services. Lithium-ion batteries are often used as these have the ability to be charged at a high energy density and very quickly, whereas flow batteries provide scalability and longer-term storage. Supercapacitors and hydrogen fuel cells have different applications and benefits (Hardman et al., 2018). At various levels, energy storage systems can be combined with EV charging infrastructure. These include station-level, community-level, and grid-scale storage. Station-level storage reliably powers individual charging stations, while community-level storage balances supply and demand for renewable energy in a certain region. Extra renewable energy is controlled by grid-scale storage, offering broader assistance for charging infrastructure. 4.1 Advanced Batteries

Advanced batteries used in the charging infrastructure for EVs are very important for battery capture and storage of renewable energy. Compared to traditional approaches, advanced batteries provide superior safety, longer lifetimes, and higher charging rates. They have higher energy density and improved performance. Some of the most relevant modern battery technologies are listed below:

4.1.1. Lithium Ion (Li-ion) Batteries: These batteries are very well used in many areas of applications, starting from electric cars and other transportation means to renewable sources. They have high energy density, long cycling life, and lightweight designs. Therefore, even further research is going on to enhance the specific energy and power density, reduce cost, and solve the issues related to resource availability and environmental impacts.

4.1.2. Solid-State Batteries: This is an emerging technology that uses a solid-state electrolyte instead of the liquid electrolyte found in the conventional Li-ion battery. The benefits of this technology include higher energy density, better safety, and faster charging rates.

4.1.3. Flow batteries: Flow batteries make use of liquid electrolytes of which are kept in tanks outside the system. They provide the potential benefit of decoupling power and energy capacity and scalability. Flow batteries are especially appropriate for large-scale applications involving long-duration and high-capacity systems to balance supply due to intermittency of renewable energy resources and to satisfy changing energy consumption.

4.1.4. Sodium-Ion Batteries: Sodium-ion batteries are also being considered as a replacement for Li-ion batteries. Sodium is an attractive alternative in this regard to provide widespread energy storage due to its availability and lower cost compared to lithium. Even though the energy density and specific power of sodium-ion batteries are now low compared with those of Li-ion batteries, research is ongoing with developments that greatly enhance their functionality and feasibility in a commercial product.

4.1.5 Advanced lead-acid batteries. They are mature technologies but are still widely used for many applications from the storage of renewable energy sources. Modern lead acids, such as AGM and gel batteries, also offer longer cycle lives, faster charging times, and higher partial state of charge tolerance. All of these improvements enhance their application off grid and for renewable energy.

4.1.6. Metal-Air Batteries: These batteries use ambient air as a cathode and a reactive metal as anode. They offer the potential for long-term storage and high energy density. For instance, zinc-air batteries have drawn interest due to their high energy density, affordability, and environmental friendliness. However, there are still concerns about electrolyte stability and restricted rechargeability.

Better battery technologies must be developed and adopted to advance renewable energy harvesting and the infrastructure for charging electric vehicles. Research is still being done to increase energy density, charging speed, cycle life, and cost-effectiveness while guaranteeing sustainability and considering environmental issues. A cleaner and more dependable energy future will be made possible by ongoing improvements in sophisticated battery technology, supporting renewable energy source integration and sustainable transportation.

4.2 Hydrogen Fuel Cells

Modern hydrogen fuel cells allow for the electrochemical direct conversion of hydrogen gas into power. Hydrogen is fuel for fuel cells, providing a clean and effective method of generating power. They have attracted much interest in various applications, like portable electronics, stationary power generation, and transportation. The following are some essential features of hydrogen fuel cells:

4.2.1. Working Principle: A cathode, an anode, and an electrolyte make a hydrogen fuel cell. The anode receives hydrogen gas, which then undergoes an oxidation reaction to liberate electrons and create positively charged hydrogen ions (protons). An electrical current is produced as the electrons go via an external circuit. The only result is water when airborne oxygen from the cathode interacts with electrons and hydrogen ions. Fuel cells use an electrochemical technique that is extremely efficient because combustion processes are not involved.

4.2.2. PEM (Proton Exchange Membrane) Alkaline Fuel Cells (AFCs), fuel cells, Solid Oxide Fuel Cells (SOFCs), Phosphoric Acid Fuel Cells (PAFCs), and Molten Carbonate Fuel Cells (MCFCs) are a few examples of the several types of fuel cells that are available. Because each type has unique operating temperatures, electrode materials, and electrolytes, they can each be used for various purposes.

4.2.3. Advantages of hydrogen fuel cells include: Hydrogen fuel cells produce energy by mixing hydrogen and oxygen, leaving only water vapour as a byproduct. They are clean and emit no emissions. As a result, fuel cells are a clean and eco-friendly energy technology because they don't emit any pollutants or greenhouse gases when in use.

- Excellent Conversion Efficiency: Fuel cells offer excellent conversion efficiencies that frequently outperform conventional technologies based on combustion. Fuel cells are a potential solution for generating sustainable electricity since they directly convert chemical energy to electrical energy. This results in improved energy efficiency.
- Versatile Uses: Hydrogen fuel cells have a variety of uses, including grid-scale energy storage, portable devices, transportation (such as fuel cell cars), and home and business power generation. They are useful for stationary and mobile power applications because of their adaptability.
- Scalability: To accommodate different power needs, fuel cells may be simply scaled up or down. Hydrogen fuel cells offer versatility and adaptability in terms of power output and application, from tiny handheld devices to massive power production systems.

4.2.4. Challenges and things to think About:

- Infrastructure for hydrogen: Building an infrastructure for hydrogen, including facilities for production, storage, distribution, and refuelling, is a difficult task. A strong and accessible hydrogen supply network is necessary for the wide-scale implementation of hydrogen fuel cells.
- Cost and Commercial Viability: Now, the cost of hydrogen fuel cells is much higher than traditional power production methods. The commercial feasibility of hydrogen fuel cells depends on lowering prices through improvements in components, production techniques, and economies of scale.
- The safe storage and handling of hydrogen is a crucial component of fuel cell technology. Because of its great energy density, hydrogen must be safely stored and used
- Fuel cells have great conversion efficiency, but further research and development are required to increase their performance, sturdiness, and dependability. Improvements are being made to catalyst materials, membrane technology, and system architecture to maximise fuel cell performance.

5. INTEGRATION OF ENERGY STORAGE AND RENEWABLE ENERGY

Techniques for energy storage are crucial for ensuring the reliable operation of power systems and making renewable energy sources more widely adopted. Recent developments in materials, knowledge management, energy, control systems, and artificial intelligence have made significant discoveries and accomplishments in this field possible. This study examines the various facets of storage technologies for energy, including their functions in power systems, design optimization methods, and applications in various industries, including transportation, heating, cooling, and utility power generation. The study begins by discussing various energy storage technologies and their applications. It then provides a brief overview of energy storage systems' different states, processes, and storage capacities. The third part examines the suitability of different energy storage technologies, optimal system configurations, and energy control strategies, with a particular emphasis on the application of artificial intelligence (Waag et al., 2014).

Lastly, the study presents a range of subjects and perspectives, offering new suggestions and future research ideas on integrated energy storage systems.

5.1 Maximizing the Use of Renewable Energy

Due to growing concerns over energy use and environmental laws, the need for energy systems has increased in modern times. Several crucial elements, including cost control, infrastructure development, power system reliability, and increased flexibility, have evolved to overcome these difficulties. Additionally, it has become crucial to reduce CO_2 emissions, mitigate climate change, and give remote places with growing energy requirements access to reliable energy (Münderlein, J. et. al. 2020).

By utilizing renewable energy sources, the quick development of energy storage systems (ESS) has significantly contributed to addressing energy conservation and carbon reduction demands. However, ESSs encounter difficulties because of their unpredictable and intermittent nature, which can cause voltage and frequency changes, whether they are running independently or connected to the grid. With the growing integration of renewable energy sources (RESs) like solar and wind power, these difficulties have grown even more serious. The stability of the power system is influenced by these sources' varied and unpredictable output (Wong L. et al., 2019; Xu B. et al., 2015). Despite these obstacles, the capability of distributed energy storage and renewable energy sources has increased. However, the integration of renewable energy, frequency control, and peak demand regulation on the power grid has been made more challenging by the advent of new technologies.

Utilizing various storage devices and allowing quick and flexible changes to the power supply is made possible by energy storage technology (Zahraee et al., 2016; Sebastian et al., 2012). The issues created by the rising use of RES are partly addressed by this technology, which also contributes to the effective running of the power grid.

5.2 Energy Storage and Capture

Storage refers to engineering solutions that can capture solar energy. It can then be converted into mechanical, thermal, or chemical energy, stored, and released as necessary. Energy storage makes it possible to utilize energy at times other than when it was created. It's crucial to remember that energy storage is never 100 percent effective because energy conversion and retrieval always result in losses. However, energy storage can improve power quality by balancing the supply and demand for energy. It also strengthens the energy system's durability and effectiveness (Alsayegh et al, 2010).

5.2.1. Types of Energy Storage

In electrical systems, Pumped hydropower is the most efficient kind of power storage. However, thermal and electrochemical storage are the technologies most typically paired with solar power facilities. Other forms of storage, including flywheels and compressed air storage, have unique qualities like quick discharge and very large capacity. We discuss various options for energy storage (H. Garg. et al., 2022).

5.2.1.1. Pumped-Storage Hydropower: Hydropower is a water-based energy storage method that uses electrical energy to pump water uphill into a reservoir during low demand. When there is a greater need for electricity, the water that has been held is released, running downhill and powering a turbine to produce electricity. Since 1929, pumped hydro has been a reliable and cutting-edge storage technique in the US. But to put it into practice, you need the right landscapes and reservoirs, which you may get by building dams or using existing lakes. It's important to remember that the construction process could take a while and that the initial setup expenditure could be considerable. Pumped hydro remains a dependable and effective energy storage option.

5.2.1.2. Storage of electrical energy: Electrochemical batteries, such as those frequently used in mobile phones, tablets, and laptops. When electricity is applied to a battery, the chemical process causes it to store energy. This chemical reaction is reversed when a battery is drained, causing current to leave the battery and voltage between two electrical contacts. Lithium-ion batteries are the most popular, however lead-acid, sodium and nickel-based batteries are also often used.

5.2.1.3. Thermal Energy Storage: The term "thermal energy storage" refers to a group of techniques for storing heat in fluids like molten salt or water. In insulated tanks, this thermal energy is preserved. Later, electricity can be produced by heating water to create steam, which drives a turbine similar to those seen in conventional power plants. As an alternative, the thermal energy that has been stored can be used right away to heat and cool things. Thermal energy storage is particularly useful in Concentrated Solar Power (CSP) facilities because it enables the captured sunlight to heat a working fluid, maintaining a constant power supply even when natural sunlight is not present (Singh, M. et. al. 2024).

5.3 Consistent and Reliable Power Supply

It is vital to comprehend how this technology operates and its advantages in a day where a constant power supply is necessary for the efficient operation of both large and small organizations. Here, we'll go over some of the important characteristics of the continuous power supply and explain why it's so crucial to contemporary business and industry.

Reliable Electricity: Achieving high power reliability is crucial for any organisation or industry. A continuous power supply system is essential to achieve power system reliability. A straightforward definition of power system dependability is the likelihood that the system will carry out its intended function within predetermined parameters for a predetermined period. Typically, a percentage is used to express it. If a power system, for instance, has an annual reliability of 99%, it is anticipated to operate properly 99% of the time during that year. Various factors, including environmental circumstances, natural catastrophes, operational blunders, maintenance concerns, and equipment malfunctions can impact power system reliability. By offering an uninterruptible power source that can take over in the event of a power outage, a continuous power supply solution, also known as an uninterruptible power supply (UPS), can help reduce some of these hazards. Even if there are fluctuations in the main power source, it can help regulate the power supply to maintain a constant power level (Syed et al., 2006).

6. RENEWABLE ENERGY HARVESTING FOR EV CHARGING INFRASTRUCTURE

Wind and solar power are various Renewable energy sources with many advantages over traditional fossil fuels. They are much more environmentally friendly, emitting fewer harmful emissions. Solar and wind power will not deplete as quickly as fossil fuels. They are less expensive and more dependable in the long run because they are not subject to price fluctuations generated by market changes or geopolitical events. Integrating smart energy management systems into renewable energy allows for more efficient energy use, resulting in even greater cost savings. Decentralized energy production from renewable resources can reduce dependency on centralized power networks by enabling people and communities to produce their own electricity. Because they lower GHG emissions and other air pollution emissions, EVs are especially helpful for reducing environmental impact in the transportation sector. Fuel costs can be eliminated by employing RESs for charging or onboard energy production (M. Ceraolo et. al., 2008).

6.1 Reduced Reliance on the Grid

An energy grid's main function is to transmit electricity more effectively across the country to commercial, residential, and industrial premises. An energy grid comprises intricate electrical systems that produce energy from numerous sources to rapidly restore electricity following energy disturbances. An energy grid can also lower customer electricity bills and the costs associated with utility management and operations. Extreme weather, high demand that results in blackouts, and other dangers can all cause disruptions to an energy grid. An energy grid's ageing infrastructure may lead to frequent service interruptions. Hence, reliance on the energy grid should be lowered by investigating alternate ways to produce and use electricity (Miao. R et al.,2014).

6.1.1. Spend money on solar energy: Even if you still use the energy grid, investing in solar energy is a great way to lessen your reliance on conventional electricity sources. Solar energy transforms solar energy into usable electricity that can be used to run your house or place of business. By adding solar panels to your home or place of business, solar energy can be harvested and stored for later use. The energy emanating from the sun is estimated to be about 1000 times more powerful than traditional grid electricity. Solar energy generally costs 25% less than grid power, though prices vary greatly depending on the market.

6.1.2. Put in a green roof: An existing roof extension with top-notch waterproofing, a drainage system, a root-repellent system, filter cloth, and vegetation or plants is called a "green roof system." The cooling effects of green roofs decrease power plant demand. This roof is a practical way to lessen reliance on the energy grid for cooling or heating buildings. A green roof system provides evapotranspiration or water movement from the land to the atmosphere via evaporation and a wind barrier, insulation, and shadow for the roof layer. Most heat is lost through roofs in the winter and is generated in the summer. Green roofs are more insulating than other types of roofing, which lowers the amount of electricity needed to control a building's temperature (Protected, A. et al., 2023).

6.1.3. Smart Technologies Uses: Technology is an effective instrument for promoting energy efficiency. Utilizing the most recent technologies to manage your electricity will help you consume less energy and rely less on the grid. Nowadays, many electrical equipment, devices, and appliances have smart technology to help consumers rely less on the power grid. Appliances and building systems can be equipped with temperature sensors, light sensors, remote management, artificial intelligence, and warning functions to help monitor power use.

6.1.4. Use good energy practices: Your electricity prices rise with rising inflation. Your energy habits directly impact your electricity usage and utility costs. Even though you wish to save time and effort, putting the following strategies into practice can lessen your reliance on the energy grid:

- During the summer, air-dry laundry outside in the sun rather than using a fan or a dryer.

- Washing dishes and clothes by hand when you have lots of free time will help you use fewer appliances.
- Sweep and mop the floor rather than using a Hoover.
- When not in use, turn off electrical appliances and fixtures.
- Turn off your main power source before a lengthy vacation.
- Cook using natural gas rather than electricity.
- Replace your electricity lighting with solar-powered options.

6.2 Minimal Effects on the Environment

Because EVs run on electricity instead of fossil fuels, they do not have all of the same parts as ICE cars. For instance, fuel tanks, fuel lines, and exhaust pipes. EVs reduce air pollution because they don't carbon dioxide (CO2). EVs are a significant development in sustainable mobility as a cleaner substitute. The top four ways EVs can help the environment are listed below.

6.2.1. EVs can have emissions-free tailpipes: Electric vehicles (EVs) are different from cars that run on gas or petrol because they don't make any exhaust, so they don't need a tailpipe. Instead, EVs store and use energy through batteries. A lithium-ion battery is the most popular battery used in electric vehicles. You can charge and drain these batteries many times without causing air pollution. Even if the electricity used to charge an EV comes from fossil fuels, it still produces less pollution than a standard car with an internal combustion engine. Some EV charging sites use clean and renewable energy sources, while others use less eco-friendly sources like coal-burning power plants. This could be higher in areas where coal, oil, or natural gas happens to be the primary source of energy during the charging of EVs. However, overall, EVs create less pollution compared to normal cars. For instance, the use of EVs has lowered greenhouse gas pollution by 20% in China, which still relies more on coal as their source of energy.

Fully electric vehicles do not emit exhaust gases and use batteries as a means of energy storage. However, even if it will be charged through fossil fuels, EVs still emit less than conventional vehicles. The use of electric vehicles contributes toward the reduction in greenhouse gas emissions, especially to countries that rely much on coal (Logan et al., 2021).

6.2.2. Production of EV Batteries Can Be Green Too: Electric vehicles do not produce air pollutants; however, much has been said regarding the negative external environmental impacts related to the production of EVs. The processes of battery manufacturing consume energy, thereby producing greenhouse gases. The energy sources utilized in EV technology, which is still developing, may not be well enough in terms of efficiency levels and can cause a level of carbon footprint than necessary. There has been some progress in this area, however. The carbon footprint for EV batteries is a fraction of what it was just two years ago and continues to be reduced. EV manufacturers are actively working with their battery suppliers to set standards for battery production. Today, they require them to source clean power, like solar and wind power, for their productions. Renewable energies might provide such vast quantities of energy as will be needed for the production of these batteries free from harmful polluters. One example is Tesla, an electric vehicle manufacturer that aims to produce all of its batteries using green energy. This commitment demonstrates a shift towards more sustainable practices in the EV industry.

6.2.3. ICE cars constantly emit pollution: Once electric cars are made, they don't add to air pollution. They only do this when they are charged at coal-powered stations. The majority of pollutants, though, come from making batteries. So, even before an EV is used for the first time, its general effect on the environment can be measured. On the other hand, every time an internal combustion engine (ICE) runs, it gives off CO2. A car running on petrol puts out about 5 to 6 metric tons of CO2 each year. Surprisingly, research by the Union of Concerned Scientists shows that ICE car emissions exceed EV emissions from "well to wheel" (the whole life cycle) after only 6 to 18 months of use. Since ICE cars are so common, they produce a lot of emissions. On the other hand, electric cars that run on green energy sources leave no carbon footprint over the course of their lives. This makes EVs much better for the earth, especially in places where the power comes from cleaner sources (Anadón et al., 2023).

6.2.4. EV makers use green materials: Manufacturers face a huge difficulty in designing a useful and lightweight electric vehicle (EV). When EVs are made with standard materials, making them go farther and use less energy is hard. But there is hope because organic and recycled materials can now work as well as traditional materials. These green choices are strong and light, making them a good choice. Even though many traditional makers use recycled materials for small parts, cars are still mostly made of materials that are not recycled. Still, EV makers are working hard to use and improve eco-friendly materials to make lighter cars and use less fuel (Singh, M. et. al. 2024)

6.3 Effective and Sustainable Charging

Electric vehicles (EVs) are like cars that run on electric motors powered by batteries. They can be charged using plug-in outlets. EVs are considered environmentally friendly and are becoming a popular choice for transportation. Charging stations for EVs can be categorized based on accessibility and charging levels. Private charging stations are located at residences and are accessible only to one household. Semi-public stations are available to a group of people, such as those at workplaces. Public stations are open to all users and are usually located in easily accessible areas (Zhao, P. et. al. 2014)

There are three charging levels: slow chargers, which take 10-13 hours to fully charge; quick chargers, which take 1-3 hours; and rapid chargers, which take 30-100 minutes. Private stations are usually slow chargers, while fast and rapid chargers are commonly found at public stations. Most EV owners use private charging stations to charge their vehicles at home, especially at night. This can fulfil over 90% of their daily travel needs. However, many EV owners rely on public charging stations in urban areas with limited parking facilities, especially for long-distance trips. The number of public charging stations worldwide is relatively low, and their growth rate and usage per car are declining. High costs and limited demand have hindered the widespread deployment of public stations. This creates the "chicken and egg" problem whereby low adoption of EVs is discouraged by the unavailability of re-charging structures and vice versa: although the current low number of EVs seems not enough to justify investing in more recharge stations.

Public agencies are promoting the adoption of EVs by introducing laws that give special privileges to the drivers of such vehicles, including access to special lanes, reduced fees and free parking, and unrestricted access to city centers. They are electrifying their fleets and public transport and installing public charging station networks. Strategic placement of efficient public charging station networks should be in semi-private and public spaces. Various stakeholders such as owners of EVs, government bodies, electro-mobility service providers, and charging service operators are required in the running management process of charging stations (Koohi-Kamali et al., 2013). Business models can be used by public authorities to determine the commercial feasibility of public charging networks. Business models come in handy by helping to understand how public value, including the social and environmental benefits and profitability, could be created and persisted. However, most studies focus on the technical implications of charging networks without involving business models as the source of funding for such infrastructure (Sun et al., 2019). Summary, Electric vehicles form a choice for greener transportation, and the supporting infrastructure, namely charging stations, private and public, plays a crucial role in supporting the practice of EV adoption. Public agencies are also working on establishing efficient public charging networks, but so far, it seems that there is a lack of both funding and viable business models within the entire equation.

7. CASE STUDY: SOLAR-POWERED EV CHARGING STATION

A demonstration of how RES can be easily integrated into the EV charging infrastructure is given through this case study, where we describe an operational solar-powered EV charging station within a city. The project was commissioned to promote environmentally friendly travel and minimize carbon emissions. It was positioned strategically near areas with commercial activities and busy throughfares for easy access by the owners of EV. Proper solar viability analysis that determined the capability of solar panels with an optimal capacity was conducted. Working with renewable energy specialists resulted in an efficient integration strategy being created.

The 100-kW solar array took advantage of a canopy as the mounting structure for the solar-powered EV charging station. High efficiency solar panels were used to garner maximum sunlight. A DC Fast and Level 2 charger was made available for fast charge during peak hours. A 150-kWh battery storage system was also added to ensure that charging occurred continuously even with little sunlight. The station was seamlessly connected into a smart grid for effective energy management. Public awareness campaigns focused on its dependence on renewable resources and actively advertised the station's eco-friendly features while encouraging people to adopt EVs. It mitigated virtually close to 200 tons of CO2 emissions annually, thus, thus propelling the city's sustainability goals by a mile. Lastly, the refurbishment would keep electricity costs down, thus leading to long-term cost savings.

Integration of solar panels, battery storage, and smart grid technologies formed the backbone of an efficient and sustainable charging infrastructure. Public involvement initiatives in promotion and awareness contributed significantly to the adoption of EVs. The success demonstrated by these stations was replicated and scaled up further to increase the RES-based network of charging infrastructure. In this case, an installation that is getting closer to becoming an ideal solar-powered EV charging station that promotes sustainable transportation, thereby reducing environmental effects, could be useful and inspirational. This is how the project was trying to pursue goals set out for promoting green transportation, reducing carbon emissions, and utilizing advanced technologies. It is an inspiring model for cities and organizations interested in developing green and effective solutions for urban mobility.

8. CONCLUSION

Incorporation of renewable energy harvesting at charging stations for EV is a major step toward sustainability and reduced carbon emissions as well as cleaner transportation. Electric vehicle charging may therefore be considered less fossil fuel dependent and more energy-efficient when implemented from renewable energy sources such as solar, wind, or hydroelectric power. Coupled with energy storage and smart grid technology, charging networks become much more efficient, reliable in terms of power supply, and constant. All these developments pave the way toward a greener and sustainable future and reflect on the socioeconomic impacts by creating jobs and reducing energy costs. Even though there are issues of intermittency, grid integration, and energy density, continuous research and development of technologies actively tackle these problems. Ongoing advancements in renewable energy harvesting for EV charging infrastructure promise a robust, effective, and sustainable ecology for electric vehicles. The use of renewable energy for EV charging is one key change toward achieving sustainability in transport. In this, solar energy especially has to be credited. Solar powered EV charging stations are getting progressively cheaper and increasing in availability because of improvements in the technology of solar panel systems. In effect, these charging stations take advantage of the sunlight to generate electricity by installing solar panels on the rooftops or surrounding places, lessening reliance on conventional grids and hastening the change to cleaner sources of renewable energy. Another thing is the renewable energy harvesting for EV charging stations, transformed by smart grid technologies. Smart grids can divert power flow to charging stations effectively whenever they know that renewable energy is in periods of peak production because smart grids are always working constantly in monitoring energy production, distribution, and consumption. This smart strategy ensures effective and efficient use of electricity at the charging station while maximizing the utilization of clean energy. Together, these efforts work to create a transportation ecosystem that shall prove more ecologically friendly and sustainable. Renewable energy harvesting may help in making the future of transportation better and greener by decreasing the usage of fossil fuel, lessening greenhouse emissions, and fostering cleaner energy habits.

REFERENCES

Alsayegh, O. A., Alhajraf, S., & AlBusairi, H. (2010). Grid-connected renewable energy source systems: Challenges and proposed management schemes. *Energy Conversion and Management*, 51(8), 1690–1693. DOI: 10.1016/j.enconman.2009.11.042

Anadón Martínez, V., & Sumper, A. (2023). Planning and Operation Objectives of Public Electric Vehicle Charging Infrastructures: A Review. *Energies*, 16(14), 5431. DOI: 10.3390/en16145431

Bayo, Y., Luo, Y., Zhang, W., Huang, M., Wang, L. Y., & Jiang, J. (2018). A Bi-Level Optimization Approach to Charging Load Regulation of Electric Vehicle Fast Charging Stations Based on a Battery Energy Storage System. *Energies*, 11(1), 229. DOI: 10.3390/en11010229

Benato, A., & Stoppato, A. (2018). Pumped Thermal Electricity Storage: A technology overview. *Thermal Science and Engineering Progress*, 6, 301–315. DOI: 10.1016/j.tsep.2018.01.017

Ceraolo, M., di Donato, A., & Franceschi, G. (2008). A General Approach to Energy Optimization of Hybrid Electric Vehicles. *IEEE Transactions on Vehicular Technology*, 57(3), 1433–1441. DOI: 10.1109/TVT.2007.909268

Das, H., Rahman, M. M., Li, S., & Tan, C. H. (2020). Electric vehicles standards, charging infrastructure, and impact on grid integration: A technological review. *Renewable & Sustainable Energy Reviews*, 120, 109618. DOI: 10.1016/j.rser.2019.109618

Ding, H., Hu, Z., & Song, Y. (2015). Value of the energy storage system in an electric bus fast charging station. *Applied Energy*, 157, 630–639. DOI: 10.1016/j.apenergy.2015.01.058

Ehsan, A., & Yang, Q. (2020). Active Distribution System Reinforcement Planning With EV Charging Stations—Part I: Uncertainty Modeling and Problem Formulation. *IEEE Transactions on Sustainable Energy*, 11(2), 970–978. DOI: 10.1109/TSTE.2019.2915338

Etezadi-Amoli, M., Choma, K., & Stefani, J. (2010). Rapid-Charge Electric-Vehicle Stations. *IEEE Transactions on Power Delivery*, 25(3), 1883–1887. DOI: 10.1109/TPWRD.2010.2047874

Garg, H., Singh, M., Sharma, V., & Agarwal, M. (2022). Decentralized Application (DAPP) to enable E-voting system using Blockchain Technology. *2022 Second International Conference on Computer Science, Engineering and Applications (ICCSEA)*. DOI: 10.1109/ICCSEA54677.2022.9936413

Gupta, M., Singh, M., Sharma, A., Sukhija, N., & Aggarwal, P. K. (2023). Unification of machine learning and blockchain technology in healthcare industry. *Innovations in Healthcare Informatics*, 185–206. DOI: 10.1049/PBHE041E_ch6

Harakawa, T., & Tujimoto, T. (2002). Efficient solar power equipment for electric vehicles: improvement of energy conversion efficiency for charging electric vehicles. https://doi.org/DOI: 10.1109/IVEC.2001.961718

Hardman, S., Jenn, A., Tal, G., Axsen, J., Beard, G., Daina, N., Figenbaum, E., Jakobsson, N., Jochem, P., Kinnear, N., Plötz, P., Pontes, J., Refa, N., Sprei, F., Turrentine, T., & Witkamp, B. (2018). A review of consumer preferences and interactions with electric vehicle charging infrastructure. *Transportation Research Part D, Transport and Environment*, 62, 508–523. DOI: 10.1016/j.trd.2018.04.002

Hardman, S., Shiu, E., & Steinberger-Wilckens, R. (2016). Comparing high-end and low-end early adopters of battery electric vehicles. *Transportation Research Part A, Policy and Practice*, 88, 40–57. DOI: 10.1016/j.tra.2016.03.010

Hussain, A., Kim, H., Baek, J., & Kim, H. (2019). Stationary Energy Storage System for Fast EV Charging Stations: Simultaneous Sizing of Battery and Converter. *Energies*, 12(23), 4516. DOI: 10.3390/en12234516

Ibrahim, H., Younes, R., Ilinca, A., Dimitrova, M., & Perron, J. (2010). Study and design of a hybrid wind–diesel-compressed air energy storage system for remote areas. *Applied Energy*, 87(5), 1749–1762. DOI: 10.1016/j.apenergy.2009.10.017

Kempton, W. (2016). Electric vehicles: Driving range. *Nature Energy*, 1(9), 16131. Advance online publication. DOI: 10.1038/nenergy.2016.131

Khalid, M., Alam, M. K., Sarwar, A., & Asghar, M. I. (2019). A Comprehensive Review of electric vehicles charging infrastructures and their impacts on power-quality of the utility grid. *eTransportation*, 1, 100006. DOI: 10.1016/j.etran.2019.100006

Knezovic, K., Marinelli, M., Zecchino, A., Andersen, P., & Træholt, C. (2017). Supporting involvement of electric vehicles in distribution grids. *Energy*, 134, 458–468. DOI: 10.1016/j.energy.2017.06.075

Koohi-Kamali, S., Tyagi, V., Rahim, N. A., Panwar, N. L., & Mokhlis, H. (2013). Emergence of energy storage technologies as the solution for reliable operation of smart power systems. *Renewable & Sustainable Energy Reviews*, 25, 135–165. DOI: 10.1016/j.rser.2013.03.056

Logan, K. G., Nelson, J. D., Brand, C., & Hastings, A. (2021). Phasing in electric vehicles: Does policy focusing on operating emission achieve net zero emissions reduction objectives? *Transportation Research Part A, Policy and Practice*, 152, 100–114. DOI: 10.1016/j.tra.2021.08.001

Miao, R., Yang, J., Feng, X., Jia, H., Wang, J., & Nuli, Y. (2014). Novel dual-salts electrolyte solution for dendrite-free lithium-metal based rechargeable batteries with high cycle reversibility. *Journal of Power Sources*, 271, 291–297. DOI: 10.1016/j.jpowsour.2014.08.011

Mohammad, A., Zamora, R., & Lie, T. T. (2020). Integration of Electric Vehicles in the Distribution Network. *Energies*, 13(17), 4541. DOI: 10.3390/en13174541

Mohammad, A., Zamora, R., & Lie, T. T. (2020). Integration of Electric Vehicles in the Distribution Network: A Review of PV-Based Electric Vehicle Modelling. *Energies*, 13(17), 4541. DOI: 10.3390/en13174541

Münderlein, J., Ipers, G., Steinhoff, M., Zurmühlen, S., & Sauer, D. U. (2020). Optimization of a hybrid storage system and evaluation of operation strategies. *International Journal of Electrical Power & Energy Systems*, 119, 105887. DOI: 10.1016/j.ijepes.2020.105887

Mwasilu, F., Justo, J. J., Kim, E. K., Duc, T., & Jung, J. (2014). Electric vehicles and smart grid interaction. *Renewable & Sustainable Energy Reviews*, 34, 501–516. DOI: 10.1016/j.rser.2014.03.031

Mwasilu, F., Justo, J. J., Kim, E. K., Duc, T., & Jung, J. (2014). Electric vehicles and smart grid interaction: A review on vehicle-to-grid and renewable energy sources integration. *Renewable & Sustainable Energy Reviews*, 34, 501–516. DOI: 10.1016/j.rser.2014.03.031

Protected, A. (2023). 4 Ways to Reduce Energy Grid Reliance! *Supply Chain Game ChangerTM*.https://supplychaingamechanger.com/4-ways-to-reduce-energy-grid-reliance/

Sbordone, D., Bertini, I., Di Pietra, B., Falvo, M. C., Genovese, A., & Martirano, L. (2015). EV fast charging stations and energy storage technologies: A real implementation in the smart microgrid paradigm. *Electric Power Systems Research*, 120, 96–108. DOI: 10.1016/j.epsr.2014.07.033

Sebastian, R., & Alzola, R. P. (2012). Flywheel energy storage systems: Review and simulation for an isolated wind power system. *Renewable & Sustainable Energy Reviews*, 16(9), 6803–6813. DOI: 10.1016/j.rser.2012.08.008

Sharma, A., Singh, M., Gupta, M., Sukhija, N., & Aggarwal, P. K. (2022). IoT and blockchain technology in 5G smart healthcare. In *Blockchain Applications for Healthcare Informatics*, 137–161. https://doi.org/DOI: 10.1016/B978-0-323-90615-9.00004-9

Singh, M., & Malik, A. (2024). Multi-hop routing protocol in SDN-based wireless sensor networks: A comprehensive survey. In *Software-Defined Network Frameworks* (pp. 121–141). CRC Press.

Singh, M., Sharma, R., Kumar, K., Singh, M., & Vajpayee, P. (2024). Securing lives and assets: IoT-based earthquake and fire detection for real-time monitoring and safety. In *Communications in Computer and Information Science* (pp. 15–25). https://doi.org/DOI: 10.1007/978-3-031-56703-2_2

Sun, L., Jin, Y., Pan, L., Shen, J., & Lee, K. Y. (2019). Efficiency analysis and control of a grid-connected PEM fuel cell in distributed generation. *Energy Conversion and Management*, 195, 587–596. DOI: 10.1016/j.enconman.2019.04.041

Syed, F. U., Kuang, M. L., Czubay, J., & Ying, H. (2006). Derivation and Experimental Validation of a Power-Split Hybrid Electric Vehicle Model. *IEEE Transactions on Vehicular Technology*, 55(6), 1731–1747. DOI: 10.1109/TVT.2006.878563

Von Jouanne, A., Husain, I., Wallace, A., & Yokochi, A. (2005). Gone with the wind: Innovative hydrogen/fuel cell electric vehicle infrastructure based on wind energy sources. *IEEE Industry Applications Magazine*, 11(4), 12–19. DOI: 10.1109/MIA.2005.1458270

Waag, W., Fleischer, C., & Sauer, D. U. (2014). A critical review of the methods for monitoring of lithium-ion batteries in electric and hybrid vehicles. *Journal of Power Sources*, 258, 321–339. DOI: 10.1016/j.jpowsour.2014.02.064

Wong, L. T., Ramachandaramurthy, V. K., Taylor, P., Ekanayake, J., Walker, S. I., & Padmanaban, S. (2019). Review on the optimal placement, sizing, and control of an energy storage system in the distribution network. *Journal of Energy Storage*, 21, 489–504. DOI: 10.1016/j.est.2018.12.015

Xu, B., Li, P., & Chan, C. L. (2015). Application of phase change materials for thermal energy storage in concentrated solar thermal power plants: A review. *Applied Energy*, 160, 286–307. DOI: 10.1016/j.apenergy.2015.09.016

Zahraee, S. M., Assadi, M. K., & Saidur, R. (2016). Application of Artificial Intelligence Methods for Hybrid Energy System Optimization. *Renewable & Sustainable Energy Reviews*, 66, 617–630. DOI: 10.1016/j.rser.2016.08.028

Chapter 6
Sustainable Energy Solutions for Charging Electric Vehicles by Integrating Renewable Sources

Anupriya Sharma Ghai
https://orcid.org/0000-0002-9725-6939
Graphic Era Hill University, India

Ramesh Chandra Sharma
https://orcid.org/0000-0002-1371-1157
Graphic Era Hill University, India

Kapil Ghai
Graphic Era Hill University, India

Shwetank Avikal
Graphic Era University, India

Vishan kumar Gupta
Amity University, Mohali, India

Gulsun Kurubacak Cakir
Ankara Haci Bayram Veli University, Turkey

Sanjay Jasola
Graphic Era Hill University, India

ABSTRACT

Integrating sustainable renewable energy sources into electric vehicle (EV) charging is critical to reducing carbon emissions and promoting environmental sustainability. This integration holds tremendous potential for enhancing affordability, sustainability, and grid resilience. By incorporating renewable energy sources like solar, wind, and hydroelectric power into EV charging stations, it becomes possible to produce clean and sustainable energy on-site, reducing the carbon footprint associated with transportation. This book chapter explores the optimisation of renewable energy

DOI: 10.4018/979-8-3693-6844-2.ch006

integration in EV charging stations, focusing on key challenges, strategies, and future directions. The chapter highlights the need for effective energy storage solutions and smart charging algorithms to balance supply and demand by examining the intermittent nature of renewable sources like solar and wind power. The importance of the strategic placement of charging stations near renewable energy sources and the role of vehicle-to-grid (V2G) technologies in enhancing energy flexibility.

INTRODUCTION

In this global call for climate change mitigation and reduced emissions of greenhouse gases, the transportation sector has been targeted for intervention strategies. Electric Vehicles are generally accepted as a revolutionary approach towards less dependence on fossil fuel and minimized ecological impact. The environmental benefits of EVs depend on source of electricity used to charge them. Renewable energy sources like solar and wind power should be integrated into EV charging to realize the true potential of EVs as sustainable transportation solutions. It is through integrating renewable energy into EV charging systems that realize total sustainability. It can greatly decrease the carbon footprint in the generation of electricity used in recharging electric vehicles through the exploitation of clean and unlimited sunlight and wind energy.

The direct emissions that are associated with renewable sources are not present as they do not harm the environment. In addition, electric cars are putting immense pressure on the electrical grid, running mostly and having no capacity to meet increasing demands. Distributed energy sources may provide relief by ensuring stability and reliability in grids by including renewable energy. The installation of solar panels and wind turbines at or close to charging stations would provide charge to the EV from generated local clean energy, which reduces the burden of a grid and encourages both energy independence and security. Economic benefits that involve integrating renewable energy come up with it into the EV charger infrastructure. The renewable energy sectors are great job enforcers, hence leading and encouraging economic growth and technological innovation. Renewable energy sources are in line with global efforts toward mitigating climate change and transitioning into a low-carbon economy. Governments and organizations have set ambitious targets to reduce carbon emissions and promote sustainable development. This brings renewable energy into the electric vehicle systems. This is a promise for the care of the environment, thus showing sustainability. It is an important step to make a future that will be strong and carry less carbon emissions. This just goes hand in hand with the right use of benefits from electric vehicles because nations are fighting climate change all over the world. The infrastructure will harness the renewable sources like solar and wind energy for a seamless charging of the electric vehicles to avoid the environmental impacts.

Javed et al. (2023) further elaborates on the use of renewable sources in charging EVs with the use of solar panels and wind turbines. Renewable energy will help in sustaining a cleaner, more resilient system that is beneficial for both the environment and consumers in terms of energy consumption. Renewable energy will be beneficial for delivering a cleaner and more reliable energy system; this will also protect the environment and the average consumer from all ends. The most significant change in the transport sector in greener ways is known as EVs. For EVs' environmental and emissions benefits, the cars can only ever be as clean as the electricity that powers them. They can not replace or eliminate the demand for fossil fuels without being charged from renewable sources. Our chapter findings suggest that it is vital to introduce renewable energies, including solar and wind power, in EV charger infrastructure. The combined approach offsets the carbon impact and eases the load on the electrical grid, given the higher adoption rates of EVs and renewable charging. Hopefully, there will be a green future with our fossil fuel reliance reduced even further by integrating renewable energy for EV charging. The World Resources Institute (2019) explores emerging ways to pair electric vehicles with renewable energy, including on-site solar energy systems and managed charging with batteries.

The main reason for including renewable energy sources at the charging stations of EVs is to prevent environmental effects. Currently, electricity is primarily generated using fossil fuels, which has been one of the key sources of emissions of greenhouse gases and pollutants. These renewable sources, like solar and wind, are clean and highly available, and they do not cause any direct emissions. In any case, these sources of energy will significantly reduce the carbon footprint of transportation. Panda et al. (2023) discuss solar energy as a primary source for DC charging in electric vehicles with ESS integrated to offset intermittencies and PV mismatch losses. This will significantly cut the carbon footprint of the EVs. This is one way of enhancing the environmental impacts associated with the use of solar energy to produce electricity. It also aims to resolve some of the disadvantages facing solar energy. It offers power to the vehicle for charging, which is reliable and continuous. As it appears, ESS helps determine the solution to the erratic occurrence of solar energy, charging the battery efficiently and stably.

Renewable Energy

The main categories of renewable energy are solar, wind, hydro, geothermal, and biomass, which are all derived from renewable, naturally replenished resources. It has benefits such as energy security, environmental sustainability, and economic growth. In electric vehicles, renewable energy is essential for sustainable transport that minimizes the emission of greenhouse gases as well as the use of fossil fuels. Mohanty et al. (2024) did insightful research to link renewable energy and electric vehicles in a system striving towards 100% sustainability. The paper analyzed different charging models for EVs at home, public, and community-based infrastructures. This research work was also centered on examining various tariff structures set to motivate the adoption of electric vehicles with the promotion of utilizing renewable sources of energy. It is a major finding that storage capacity is very limited in achieving a sustainable energy system. Information about the challenges and solutions that would be faced and solved when integrating renewable energy with an electric vehicle system in a regional context will be obtained.

According to Gaertner et al. (2022), IEA and NREL focus on electric vehicle and power system integration, focusing on the integration with the grid, charging infrastructure, and recommendations for policy. Such collaboration will be of utmost importance for policymakers, industry, and researchers. The report discusses the integration of EVs into power systems and the required infrastructure. If interested, for more information, refer to IEA or NREL. Kobashi et al. (2020) discuss the role of EVs and photovoltaics in decarbonization of transportation systems. This study shows that for both sustainability and climate change factors, EVs are essentially required with renewable electricity. Mutually, the authors come up with cooperation between the EVs and RE, mainly solar power, to reduce emissions from the transport sectors. Their study presents an opportunity to integrate renewable energy with electric vehicles and build a sustainable energy and transport system. Figure 1 shows that multiple forms of renewable energy might be used to charge an electric vehicle.

Figure 1. Types of Renewable Energy

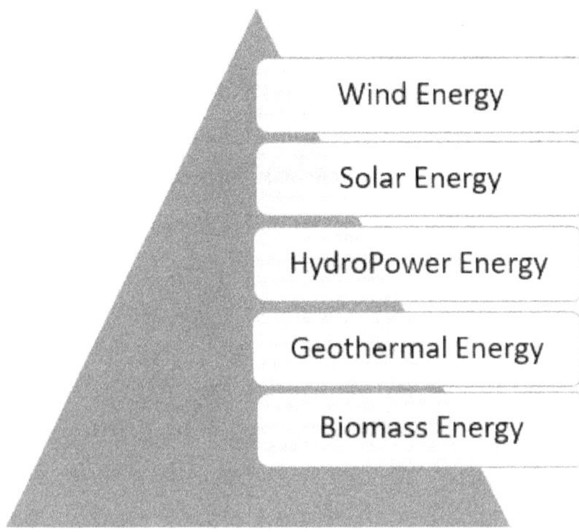

Figure 1. Types of Renewable Energy

Solar Power

Solar Power is a particularly viable option for EV charging stations. Solar panels can be installed on charging station parking lots rooftops or even integrated into the stations' design. During daylight hours, these panels can generate substantial electricity, which can be used directly for charging vehicles or stored in batteries for later use. This reduces the need for electricity from the grid, especially during peak sunlight hours. By harnessing solar energy, EV charging infrastructure can become more self-sufficient and environmentally friendly, significantly reducing carbon emissions. Hutchinson (2023), in his book, says, "Campbell, California-based company Paired power, has developed the PairTree, a modular, off-grid EV charger powered by a solar canopy. It can be installed in just four hours and provides Level 2 charging. Ach unit employs ten bifacial solar panels that provide up to 75 miles of the daily range. Optionally, with lithium iron phosphate batteries, the daily range can go up to 230 miles. Pair Tree's rapid installation and substantial range of capabilities make it flexible and efficient for various charging needs. Another practical approach to follow, as Tran (2018) mentioned, is installing a home solar photovoltaic (PV) system and connecting a Level 1 or 2 EV charger that runs off your home electricity supply. This method will charge the EVs by making use of the energy that is produced on their premises, thus enhancing independence in energy and sustainability. Using a home solar PV system cuts the cost of electricity as a large part of transportation-related energy is available free daily in the form of solar energy. Many solar PV public EV charging stations are also coming online, featuring on-site photovoltaics that power a portion or all of their electric-vehicle charging operation from generated electricity without burning fossil fuels. Stations can be found through an app or a navigation feature. Solar-powered public charging stations offer a sustainable and eco-friendly option for EV owners, supporting the broader adoption of renewable energy in the transportation sector.

Wind Power

Wind energy is complementary to solar energy because wind patterns often peak when solar energy generation is low, such as during the night or in overcast conditions. This complementarity can help provide a more consistent and reliable renewable energy supply for EV charging stations, enhancing the overall stability and sustainability of the energy grid. Roga et al. (2022) note “Change Wind Corporation,” which has developed the Wind & Solar Tower, which ingeniously combines wind and solar energy at a single site. Each 82-foot tower features a six-helical-blade wind generator with a self-cleaning solar panel. These towers can be paired with a 1,000-kWh energy storage array, effectively harnessing wind and solar power to generate up to 52.5 kW, with an annual output of 169,000 kWh. This output can provide up to 8,455 EVs annually with a 20-kWh road-trip top-off. These towers can generate power, even with wind speeds as low as five mph, significantly lower than the 26 mph minimum required for traditional propeller-based designs. The estimated installation cost is approximately $484,000 before tax credits, with an average electricity cost of 6.5 cents per kWh over 30 years. Urban Green Energy has introduced the Sanya Skypump, a 4 kW vertical axis wind turbine mounted on a 42-foot tower. This net-zero energy solution allows EV drivers to recharge their vehicles using wind power, promoting sustainable and renewable energy usage for transportation. The first wind-powered EV charger in the United States was established in Chicago in 2010, aptly named due to Chicago's “Windy City” nickname. This initiative marked a significant step towards integrating renewable energy with EV charging. Additionally, Denmark has tested a vehicle-to-grid system that leverages vehicle battery packs to store excess power generated by the country's wind farms, demonstrating the potential for large-scale renewable energy integration into the transportation sector, Le Fouest & Mulleners (2024).

RENEWABLE ENERGY USAGE INTO EV CHARGING INFRASTRUCTURE

The increase in demand for a number of EVs also creates a peak in electricity usage, which can be too much of a load on an existing electrical grid, especially at peak charging times. One of the ways of reducing the strain from the above is by introducing renewable energy sources into the charging infrastructure of EVs.. This decentralization helps ease the load on the grid by generating electricity locally, thereby minimizing transmission losses. In addition to that, its integration would further support peak shaving activities. Electricity is stored when electricity demand is low and used when the demand is at a high peak hour. This way, there will be ample supply of stable and reliable electricity for the EVs and avoid loading the grid. Advanced EV charging stations having renewable energy sources will support the provision of additional services like frequency regulation and demand response. This increases stability in the grid by giving stored energy back into the grid when the demand is peak and absorbing surplus energies when the demand is less. Proper energy storage applications like lithium-ion are seen as very important tools for addressing intermittency and maintaining a constant energy supply from all sources of renewable energies. It will contribute to reduced dependence on fossil fuels, ensure sustainability through low greenhouse gas emissions.

Distributed Generation: Renewable energy systems can be installed nearly anywhere, including homes, businesses, and public charging stations. Distributed generation has the advantage of reducing demand on large power plants and long-distance transmission lines because fewer of them are necessary to serve the same loads. With a renewable source at every charging station-including solar panels or wind turbines-more can be powered locally, thus reducing demand on the grid. For instance, photovoltaic cells mounted on charging stations or car park rooftops can generate power at local levels and thus avoid energy transmission loss and losses at long distances. In that regard, Hussain et al. (2021) discuss opportunities and challenges in relation to EVs as an energy storage device with regard to distributed generation as a solution for stress at the grid level. According to Gür (2018), hybrid energy systems could combine solar and wind power in distributed generation, therefore reducing reliance on top-down centralized power plants.

Peak Shaving: The integration of renewable energy provides the ability to take part in peak shaving activities, which store electricity that was produced during low periods and can then be stored for later full use down the road. What Are Some Common examples? However, there are plenty of ways to integrate renewable energy, with Gehrung picking out an example: A grid-connected solar photovoltaic array might generate excess power in the early afternoon, while those same vehicles can store it and then charge overnight when demand is highest. The result is lower peak demand on the grid and stable, reliable power passing through those above-ground boxes to charge EVs. One is that peak shaving reduces grid load to decrease the chance of blackouts or brownouts during especially high-demand times for electricity. Rahulkumar et al. (2022) describe how using renewable energy more efficiently to integrate with EV chargers achieves a peak reduction, which benefits the grid and aids in further progress in clean energy

Reducing Dependency on Fossil Fuels: Integrating renewable energy sources into EV charging infrastructure is critical in reducing fossil fuel dependency. The use of renewables increases work capability and decreases greenhouse gases and cleaner air; thus, this fosters efforts globally to fight climate change; it is within the national as well as international goals pertaining to the carbon footprint and sustainable development. In the same way, the exploitation of local renewable sources increases energy security through other energy sources besides reducing the risks stemming from fossil fuel price shocks. Mwasilu et al. (2014) discuss the motivations and alternatives for using hybrid systems, combining generators and batteries, that point towards the cost-saving and environmental benefits of combining different sources of renewable energy as a means of strengthening the grid and reducing dependence on fossil fuels. Liu et al. (2015) offer the status quo and alternatives available for different alternative energy sources. It is this element that such sources are put to use to reduce the usage of fossil fuels, mainly towards recharging electric car stations. Renewable energy used in charging stations also lightens the load on electric utilities while increasing energy and environmental safety. Renewable energy improves the efficiency and strength of charging systems for EVs by vastly improving peak shaving, grid services, energy storage solutions, and less dependence on fossil fuels. The method provides a holistic approach to fulfilling the ever-growing demand for electric vehicles with sustainable and reliable sources of energy.

ENERGY MANAGEMENT STRATEGIES FOR EV CHARGING STATIONS

With electric vehicles becoming the norm, energy management in charging stations is a necessity. Many benefits accompany the integration of renewable energy into the charging station, but proper methods are necessitated to ensure that power and energy are optimally utilized and within supply requirements. This chapter discusses several methods for load control, demand control, and problems associated with renewable energy up and downs. It also discusses energy storage systems and how they should be placed and designed to charge optimally for renewable energies. Strategies on smart load management are required to really use renewable energy at its best when utilized in EVs for charging. The techniques ensure that the energy supply is well matched with energy demand so that the renewable source is utilized appropriately and properly.

Demand-side management (DSM) is also aimed at shifting the load of EV charging away from the main demand periods, which are called peak periods. Renewable DSM is a specific technique that falls under the umbrella of DSM; its goal is to shift the highest electricity demand to lower scales during periods of availability of renewable energy or when the aggregate load is low. Therefore, DSM and Peak shaving planning help charging stations adjust to renewable energy resources and decrease the load pressure and costs. For instance, during periods of solar or wind generation, charging of EVs guarantees maximum utilization of renewable power, thus decreasing the use of conventional power and flattening the load curve. It also assists in the avoidance of power overloads, especially during customer's periods of high usage, which results in blackouts or having to use highly costly and less eco-friendly peaking power plants. Based on the studies by Mohanty et al. (2022), through increasing EV charging to correlate with periods of renewable power generation in addition to minimizing grid demand, DSM and peak shaving significantly improve the effectiveness of renewable energy and reduce grid stability and costs.

Time-of-use (TOU) pricing will make the owners of electric vehicles charge their cars during times when electricity is relatively cheap and abundant, meaning at a time when there is an oversupply of renewable energy. Most TOU pricing requires charging different prices for electricity at busy and less busy times. Charging stations can exploit TOU pricing to force users to charge at those best times that enable maximum use of renewable energy. For instance, the low rate of electricity at night or midday would make the owner of an electric car charge his car. In such a case, the energy system works better, balancing high and low demand times and saving the consumer money on his electricity bill. Alsokhiry et al. 2022 on sustainable energy policy work state that "Time-of-use" pricing is an important mode through which people can use energy when it is available from renewable sources. It helps in making the energy system better and more sustainable. Both consumers and the larger energy system benefit by saving costs and encouraging the use of clean energy.

CHALLENGES WITH RENEWABLE ENERGY

The fast pace of adoption of electric vehicles is generating demand in the charging infrastructure for renewable sources of energy. As far as the clean source is concerned, the output produced from sunlight and winds are cleaner and much more sustainable than fossil fuels as those have proven their destructive trail to the environment as well. Sources like those would be very challenging for charging systems and integration within the existing energy grid. That's why proper analysis related to such issues is critical and must be made to ensure the reliability, efficiency, and resilience of any network of EV charging. Issues on strategic issues dealt with in this section include intermittency of renewable sources, energy storage, strategic placement and design of EV charging stations, and important issues that include intermittency of renewable energy sources.

Solar and wind are the natural sources of intermittent natures. Their electricity is varied because of fluctuations of the weather and changes of day and night times. There is a massive task for the alignment between demand and supply of the electricity, which will be provided in order to charge an EV. For example, generated from the solar source, peak of generation occurs during the daytime; hence, the output that generates from the wind power source is very unsmoothed. All this mandates the management of the power which would ensure smooth generation at the charging stations of EV. Energy storage solutions are absolutely necessary to solve this very critical problem. The systems would act as good buffers as they would absorb the surplus generated energy during peak production periods. Such systems would release the amassed energy during low production rate periods, hence controlling the fluctuations in the supply of energy. The process would establish a stable and predictable inflow of energy to ensure an uninterrupted supply of energy and to meet the demand at all points in time.

Advanced ESS technologies include lithium-ion batteries, thus ensuring the necessary capacity and efficiency for the uninterrupted supply of energy for EVs. Solid-state and flow batteries have further benefits of extended lifespan and higher safety. With ESS, charging stations can ensure a constant and predictable supply of energy even at times when the renewable sources have less generation. With good use of renewable energy and reliable charging infrastructures, cars no longer rely so much on the normal power grid. The locations and design of EV charging stations, therefore, depend on where and how they can be nearer to renewable sources and connection to the electrical grid. Charging stations should as close as possible to the renewable energy sites, for instance, solar farms or wind turbines, to efficiently achieve a lower transmission loss. Besides, renewable energy systems, energy storage, and smart grid technologies should complement charging station design in order to ensure sustainability and improvement in performance. The method of combining in this process ensures that charging stations will be self-supportive to support the whole energy grid towards stability and strength, Laimon & Yusaf (2024).

Renewable energy installations and energy storage systems at charging stations are critical for maximizing their utilization. The capacity of renewable energy systems should be carefully aligned with the demand patterns of EV charging to ensure that sufficient energy is available when needed. For instance, the solar panels' capacity should correspond to peak daylight hours when EVs are most likely to be charged. Similarly, wind turbines should be sized according to local wind patterns to optimize energy capture. Strategic placement of these renewable energy installations is equally essential. Locating charging stations near renewable energy sources, such as solar farms or wind turbine clusters, minimizes transmission losses and enhances the overall efficiency of the energy system. Kumar et al. (2023) stated that this placement reduces the distance that electricity must travel from generation to consumption, thereby reducing energy losses and improving the reliability of the power supply.

Integrated energy management systems have to control renewable energy generation, storage, and charging of vehicles. EMS can alter the power supplied by considering the current situation and the future use and supply of energy. For example, if renewable production is at a higher level, then EMS must charge the vehicle directly by renewable sources, and all unused energy will be stored in the battery so that it can charge later on. On the other hand, EMS can utilize stored energy at times when renewable energy production is low to help satisfy fluctuating demands. Dynamic optimization enhances the effectiveness and sustainability of charging stations for EVs while making the grid stable through balance between supply and demand. Complex energy management strategies are needed to effectively integrate renewable energy into the charging infrastructure of EVs. Proper placement and sizing of renewable energy systems and storage ensure proper utilization with the minimum loss in the energy transportation process. Collaborating energy management systems make the charging setup more efficient and more eco-friendly through organization of renewable energy, storage, and electric vehicle charging work together.

Using these optimization techniques and advanced technologies, it can be possible to serve the increasing demand for electric vehicles through charging infrastructure which would be helpful in environment protection and energy saving. Renewable sources of energy are not always available; hence, through the proper location of the charging stations and advanced systems for energy storage, it would be possible to supply energy in a reliable manner in the future transportation context.

SMART CHARGING ALGORITHMS

The rapid expansion of EVs calls for smart charging infrastructure that could maximize efficiency and sustainability. In other words, it has to support smart charging algorithms, advanced infrastructure solutions for supporting renewable energy sources and grid stability, and rely on the reliability of charging services. Smart charging algorithms use data of current to improve about when and how fast they should charge electric vehicles. It means using renewable energy at the least cost. Additionally, the Vehicle-to-Grid (V2G) feature allows EVs to feed energy back into the grid. This is good for renewable energy and strength of the grid. Comparing the grid-connected and off-grid charging stations reveals various benefits and challenges. Thus, careful planning is required for the EV charging infrastructure. With these technologies, we can establish a strong and long-lasting charging network supporting the growing market of EVs and fulfilling environmental goals. Smart charging algorithms are of extreme importance to use renewable energy to its maximum extent for charging the EVs. All use real-time data streaming-in from multiple sources-which include weather forecasts, actual grid conditions, renewable sources of energy production-with each other to plan and determine how best to control/charge an EV. Section Summary This section describes the crux of smart charging algorithms and will illustrate why compositing real-time information, demand response, user's preferences, and analytics using predictive techniques is such an imperative.

Smart charging algorithms include real-time data that guide the charge rates and scheduling decisions. They can dynamically change the charge rates based on the availability of renewable energy, demand from the grid, and prices of electricity in order to maximize usage of available renewable energy and also cut down costs. For example, if the period is marked with high production of solar or wind energy, smart charging algorithms can preferentially charge EVs in order to take full advantage of the surplus of renewable energy. Contrarily, when the supply of renewable energy is low, these algorithms may decrease charging rates or charge when more renewable energy can likely be generated. Mohamed et al. (2013) said that real-time data are useful in making smart charging algorithms such that energy use becomes better at lower costs. These algorithms charge EVs with the lowest-cost sources of renewable energy and provide sustainability, based on inputs of renewable energy availability and demand in the grid plus electricity prices. Dynamic rates help balance supply against demand in electricity, thus making the system as efficient as possible. They respond to grid signals and change loads dynamically based on prevailing conditions in the grid. This one it will charge at speed while all the excess at maximum production of renewable energy are utilized.

Albogamy et al. (2022), the preservation of demand response is crucial such that smart charging algorithms capable of ensuring grid stability besides renewables are maximally optimized. These algorithms balance supply and demand energy through charging loads based on signals derived from the grid. This makes the grid less susceptible to overloads and therefore increases its reliability at a highly critical point when massive amounts of renewable energy are being generated, making it easy to utilize and store the excess energy.

Smart charging algorithms are pretty critical to optimize the usage of renewable energy in EV charging. They allow for efficient, real-time data EV charging with grid conditions and balance the desires of the users with predictive analytics. Advanced technologies make a balance between supply and demand for energy by aligning the shift in the charging rates and schedules on the basis of real-time data and grid signals, so such an energy system becomes stabilizing and efficient. Using such smart charging algorithms, furthering environment sustainability, along with increasing electric vehicle demands, these technologies can fulfill various applications.

CASE STUDIES AND EXAMPLES

Here are some examples of eco-friendly charging stations for EVs that harness renewable energy plus smart charging. In any case, in terms of function, price, and rewards, there is no doubt whether the combined system works. The following are some examples of key cases that reflect exactly how renewable and smart-charging technologies are put into action.

1. Amsterdam: Solar-Powered EV Charging Stations

Some of the EV charging stations in Amsterdam are solar panel-equipped. This allows them to charge electric vehicles using solar energy. The smart systems of these stations adjust their charging speeds based on current information such as weather, the amount of energy needed by the grid, and the amount of available solar energy. These stations reduce the use of energy from the grid and carbon emissions through the efficient use of solar energy. Sørensen et al., (2018). Amsterdam's purely solar electric vehicle charging station really made renewable energy the real deal. The possibility of smarter algorithms for recharging would open up the doors for realizing the benefits of solar energy, both for the environment and for one pocket at the same time. This is how cities can afford to accommodate more users driving electric vehicles.

2. California's ChargePoint Network

In California, ChargePoint is the largest deployment of renewable energy-based EV charging stations. Most of its stations use solar and wind energy; besides, it has onsite energy storage. These smart charges with technologies for on-site management of supply and demand for optimum use of renewable energy. An example that Sinai et al. (2020) give is in California, where it grows and spreads the ChargePoint network as a model to expand and define renewable energy usage in electric vehicle charging. A good model to replicate is that of solar and wind energy, smart charging, and energy storage. The case study reveals the economic and environmental benefits of renewable energy in mass-scale EV charging.

3. London's Source London Network

Source London has renewable energy and smart grid technologies in its electric vehicle charging stations. Its stations are close to renewable sources of energy, such as solar farms and wind turbines, so the energy lost when it is transmitted is reduced. Smart charging methods control how grid conditions and user needs influence the way energy will be shared. Barman et al. (2023) illustrate how practical examples work to prove that integrated renewable energy and smart charging systems perform well. This way, readers can find out how such systems would perform, how much they would save money, and what the environmental advantages would be. An evidence-based approach will help to confirm the advantage of using renewable energy for EV charging systems.

REGULATORY AND POLICY CONSIDERATIONS

Highly necessary for government policies and incentives to increase the usage of renewable energy in the charging infrastructure of EVs. Useful regulations will expedite the usage of green practices in transport. For example, the installation aid for renewable energy, tax relief for owners of EVs, and the regulation of the use of renewable energy in public charging stations will considerably increase the number of integrated systems. According to Esmaeili Et al. (2022), government policies and incentives are significant in using renewable energy in the charging system of EVs. Such policies and incentives provide financial aid and regulations that encourage green practices and technologies. This support would be necessary to overcome the high initial costs and to widely employ renewable energy solutions in transport.

Public awareness and opinion will call for eco-friendly electric vehicle charging solutions. Teaching more people about how renewable energy will be used will make other consumers more aware of the said solutions. Public campaigns on enlightenment, demonstrations, and education on benefits will allow increasing confidence to transition into alternative sustainable energy. According to Singh et al. (2022), public awareness is critical in making positive use of the eco-friendly charging solutions of EVs. Stakeholders' embracing of renewable energy can be enhanced by educating consumers on the economic and environmental benefits. Good communication strategies will help in building public trust that allows several people to embrace sustainable practices.

FUTURE OUTLOOK AND RESEARCH DIRECTIONS

Discussion on new technologies and areas of research related to the betterment of the utilization of renewable energy in electric vehicle charging. V2X technologies will provide a bright prospect to make the system more connected and better. The improvement in energy storage technologies, smart grid systems, and predictive analytics will work better and be more reliable for the integrated systems. Mwasilu et al., 2014, further suggest considering new technologies and research work towards better utilization of renewable energy in EV charging. Such an investment in V2X technologies, better energy storage options, and smart grid systems can create more viable and reliable systems. Hence, continuous research and new concepts are essential to solve ongoing problems and fully realize the benefits of renewable energy in replacing the traditional networks for EVs.

The above chapter can be said to provide an idea regarding the use of renewable energy at charging stations for EVs, as that addresses some critical points about key factors, plans, and further research ideas pertaining to using sustainable energy together with an EV charging system.

REFERENCES

Albogamy, F. R., Paracha, M. Y. I., Hafeez, G., Khan, I., Murawwat, S., Rukh, G., Khan, S., & Khan, M. U. A. (2022). Real-time scheduling for optimal energy optimization in smart grid integrated with renewable energy sources. *IEEE Access : Practical Innovations, Open Solutions*, 10, 35498–35520. DOI: 10.1109/ACCESS.2022.3161845

Alsokhiry, F., Siano, P., Annuk, A., & Mohamed, M. A. (2022). Novel time-of-use pricing based energy management system for smart home appliances: Cost-effective method. *Sustainability (Basel)*, 14(21), 14556. DOI: 10.3390/su142114556

Barman, P., Dutta, L., Bordoloi, S., Kalita, A., Buragohain, P., Bharali, S., & Azzopardi, B. (2023). Renewable energy integration with electric vehicle technology: A review of the existing smart charging approaches. *Renewable & Sustainable Energy Reviews*, 183, 113518. DOI: 10.1016/j.rser.2023.113518

Esmaeili, M., Anvari-Moghaddam, A., Muyeen, S. M., & Perić, V. S. (2022). On the role of renewable energy policies and electric vehicle deployment incentives for a greener sector coupling. *IEEE Access : Practical Innovations, Open Solutions*, 10, 53873–53893. DOI: 10.1109/ACCESS.2022.3176012

Gaertner, E., Rinker, J., Sethuraman, L., Zahle, F., Anderson, B., Barter, G. E., & Viselli, A. (2020). *IEA wind TCP task 37: Definition of the IEA 15-megawatt offshore reference wind turbine (No. NREL/TP-5000-75698). National Renewable Energy Lab*. NREL.

Gür, T. M. (2018). Review of electrical energy storage technologies, materials and systems: Challenges and prospects for large-scale grid storage. *Energy & Environmental Science*, 11(10), 2696–2767. DOI: 10.1039/C8EE01419A

Hussain, M. Z., Anbalagan, R., Jayabalakrishnan, D., Muruga, D. N., Prabhahar, M., Bhaskar, K., & Sendilvelan, S. (2021). Charging of car battery in electric vehicle by using wind energy. *Materials Today: Proceedings*, 45, 5873–5877. DOI: 10.1016/j.matpr.2020.08.341

Hutchinson, L. (2023). *Design and simulation of photovoltaic system for reduced demand from electric bus charging*. University of Delaware.

Javed, S., Raza, S., & Riaz, M. (2023). An energy management approach for isolated power system using low load diesel. *Sukkur IBA Journal of Emerging Technologies*, 6(1), 1–10. DOI: 10.30537/sjet.v6i1.1268

Kobashi, T., Yoshida, T., Yamagata, Y., Naito, K., Pfenninger, S., Say, K., Takeda, Y., Ahl, A., Yarime, M., & Hara, K. (2020). On the potential of "Photovoltaics+ Electric vehicles" for deep decarbonisation of Kyoto's power systems: Techno-economic-social considerations. *Applied Energy*, 275, 115419. DOI: 10.1016/j.apenergy.2020.115419

Kumar, B. A., Jyothi, B., Rathore, R. S., Singh, A. R., Kumar, B. H., & Bajaj, M. (2023). A novel framework for enhancing the power quality of electrical vehicle battery charging based on a modified Ferdowsi Converter. *Energy Reports*, 10, 2394–2416. DOI: 10.1016/j.egyr.2023.09.070

Laimon, M., & Yusaf, T. (2024). Towards energy freedom: Exploring sustainable solutions for energy independence and self-sufficiency using integrated renewable energy-driven hydrogen system. *Renewable Energy*, 222, 119948. DOI: 10.1016/j.renene.2024.119948

Le Fouest, S., & Mulleners, K. (2024). Optimal blade pitch control for enhanced vertical-axis wind turbine performance. *Nature Communications*, 15(1), 2770. DOI: 10.1038/s41467-024-46988-0 PMID: 38553502

Liu, L., Kong, F., Liu, X., Peng, Y., & Wang, Q. (2015). A review on electric vehicles interacting with renewable energy in smart grid. *Renewable & Sustainable Energy Reviews*, 51, 648–661. DOI: 10.1016/j.rser.2015.06.036

Mohamed, A., Salehi, V., Ma, T., & Mohammed, O. (2013). Real-time energy management algorithm for plug-in hybrid electric vehicle charging parks involving sustainable energy. *IEEE Transactions on Sustainable Energy*, 5(2), 577–586. DOI: 10.1109/TSTE.2013.2278544

Mohanty, A., Ramasamy, A. K., Verayiah, R., & Mohanty, S. (2024). Neighborhood and community battery projects: A systematic analysis of their current state and future prospects. *Journal of Energy Storage*, 95, 112525. DOI: 10.1016/j.est.2024.112525

Mohanty, S., Panda, S., Parida, S. M., Rout, P. K., Sahu, B. K., Bajaj, M., Zawbaa, H. M., Kumar, N. M., & Kamel, S. (2022). Demand side management of electric vehicles in smart grids: A survey on strategies, challenges, modeling, and optimization. *Energy Reports*, 8, 12466–12490. DOI: 10.1016/j.egyr.2022.09.023

Mwasilu, F., Justo, J. J., Kim, E. K., Do, T. D., & Jung, J. W. (2014). Electric vehicles and smart grid interaction: A review on vehicle-to-grid and renewable energy sources integration. *Renewable & Sustainable Energy Reviews*, 34, 501–516. DOI: 10.1016/j.rser.2014.03.031

Panda, S., Mohanty, S., Rout, P. K., Sahu, B. K., Parida, S. M., Samanta, I. S., Bajaj, M., Piecha, M., Blazek, V., & Prokop, L. (2023). A comprehensive review on demand side management and market design for renewable energy support and integration. *Energy Reports*, 10, 2228–2250. DOI: 10.1016/j.egyr.2023.09.049

Rahulkumar, J., Narayanamoorthi, R., Vishnuram, P., Bajaj, M., Blazek, V., Prokop, L., & Misak, S. (2022). An empirical survey on wireless inductive power pad and resonant magnetic field coupling for in-motion EV charging system. *IEEE Access : Practical Innovations, Open Solutions*, 11, 4660–4693. DOI: 10.1109/ACCESS.2022.3232852

Roga, S., Bardhan, S., Kumar, Y., & Dubey, S. K. (2022). Recent technology and challenges of wind energy generation: A review. *Sustainable Energy Technologies and Assessments*, 52, 102239. DOI: 10.1016/j.seta.2022.102239

Singh, S. K., Park, L., & Park, J. H. (2022). Blockchain-based federated approach for privacy-preserved IoT-enabled smart vehicular networks. In *The 13th International Conference of ICT Convergence (ICTC-2022)*, Jeju, Korea, Oct. 19-21, pp. 1-5. DOI: 10.1109/ICTC55196.2022.9952835

Sørensen, Å. L., Jiang, S., Torsæter, B. N., & Völler, S. (2018). Smart EV charging systems for zero emission neighbourhoods.

Szinai, J. K., Sheppard, C. J., Abhyankar, N., & Gopal, A. R. (2020). Reduced grid operating costs and renewable energy curtailment with electric vehicle charge management. *Energy Policy*, 136, 111051. DOI: 10.1016/j.enpol.2019.111051

Section 3

Energy Management Strategies and Optimization Techniques

Chapter 7
An Efficient Energy Management Strategy to Reduce the Comprehensive Cost of Electric Vehicles by Recent Advances

Suriya Ponnambalam
https://orcid.org/0000-0002-8775-0474
Bharath Institute of Higher Education and Research, India

M. K. Ilampoornan
https://orcid.org/0000-0002-6233-8141
Bharath Institute of Higher Education and Research, India

ABSTRACT

The Energy Management Strategy (EMS) to reduce comprehensive cost of EVs can be approached based on the efficiency improvement using a model, considering various parameters in co-ordination. The main parameters considered are vehicle mass with various resistances such as rolling resistance and average resistance. The energy desired from primary source of battery and emergency source of super-capacitor is consider in the model. Further the model considers energy conversion via converter, bus to motor. The mechanical output of motor is managed by power train/drive train, Hence, overall efficiency improvements is achieved by reducing losses at various stages, such as switching losses of converter, transmission loss of bus, frictional and magnetic loss of motor and energy loss of drive train. The

DOI: 10.4018/979-8-3693-6844-2.ch007

internal considerations for the above loss reduction are discussed in detail along with supercapacitor support in case of emergency and also reverse generation by means of regenerative braking.

1. INTRODUCTION

Electric Vehicles (EVs) are the way of the future for environmentally friendly transportation. Compared to conventional gasoline-powered cars, EVs are more affordable and environmentally benign due to their low energy usage and straightforward power system design. As the globe moves closer to a more sustainable future, electric vehicles (EVs) are becoming the obvious choice for anyone looking for a more environmentally friendly mode of transportation. The performance of power batteries will deteriorate, their life cycles will be shortened, and the cost of using EVs will rise due to the sudden elevated battery temperature, high charge/discharge rate, and extended periods of heavy workload (Martel et al., 2016; Wu et al., 2017; Ji et al., 2020). Supercapacitors have a long service life and high power density, however they are not a good option for EVs as their only power source (Dubal et al., 2014). Hybrid Energy Storage Systems (HESS) for electric vehicles (EVs) has been established to overcome this constraint. These systems combine power batteries with super capacitors (Huan et al., 2019; Song et al., 2017a; Zhao et al., 2012). Combining the high specific energy of power batteries with the high specific power of super capacitors, HESSs take advantage of the advantages of both parts. They also effectively retain the kinetic energy used in braking. According to Dubal et al. (2014), these benefits and features are essential for extending a vehicle's range, protecting power batteries, and lowering total running expenses.

Supercapacitors and batteries can be optimally distributed in power thanks to an efficient energy management system (EMS) used by HESSs in order to increase the vehicles' overall power and efficiency. This strategy makes sure that electricity is distributed in a way that optimizes performance everywhere, which helps to increase the system's efficacy and cost efficiency (Yao et al., 2019). (Haleema et al., 2024) assesses the efficiency of several methods in terms of maximizing energy utilization, cutting expenses, cutting emissions, and enhancing sustainability and reliability. A model predictive control (MPC)-based multi-time-scale energy management method is introduced in (Ding et al., 2023) in order to create energy efficient 5G networks as well as lower the operational costs related to 5G base stations. As stated in (Lu et al., 2024), EMS might coordinate control of various power supply types to improve the electricity systems' overall operational effectiveness, dependability, and financial sustainability while helping to realize dual carbon targets. It lessens the strain on the equipment, lessens fuel cell deterioration and lithium battery capacity loss, and improves the hybrid power system's efficiency as described in (QU et al., 2024). In order to create energy-efficient data centers for GPU-intensive workloads, (Smith et al.,2024) present a unique machine learning-based technique that tries to reduce energy usage through thermal-aware task distribution. Power batteries age more rapidly the longer they are used, and their cost makes up a significant amount of the total cost of the vehicle. Rule based and optimization based EMSs are the two most often utilized forms of EMSs at the moment (Panday and Bansal, 2014). While optimization-based EMSs make use of convex optimization, genetic algorithms, and dynamic planning algorithms, rule based EMSs employ methods using fuzzy control and logic thresholds.

Nevertheless, the effect of degree of battery life on power distribution in HESSs using power batteries & supercapacitors has not received much attention in study. (Wang, 2018) researched EVs using HESSs and developed a logic sill regulator approach when taking into account the motor power, power batteries, and supercapacitors' state of charge (SOC). These results suggest that the strategy can improve vehicle performance and lessen the maximum battery power requirement. Hu et al. suggested a combined fuzzy logic control technique to get the best power output from batteries while avoiding high currents throughout the charging as well as discharging processes (Hu et al., 2018). (Siskat et al., 2016) created a fuzzy control EMS that takes into account HESS performance traits. The best release power for supercapacitors and power batteries was determined using fuzzy rules, which decreased the current variation in the batteries (Sisakat and Barakati, 2016). In order to obtain the best energy distribution for HESSs, (Santueei et al., 2014) presented the battery capacity loss modelling as well as dynamic programming technique. This can maximize battery life and lower HESS costs. (Santucci et al., 2014). Drawing from the model for battery life, the EMS was designed with cost of integrated energy utilization and cost of a corresponding battery life loss. It was determined that incorporating battery life into the strategy resulted in a significant increase in battery life cycle when compared with the scheme that did not take battery life into account (Xie et al., 2019). Zhou et al. (2017) presented a genetic algorithm for optimizing fuzzy control rules and fuzzy membership function. Furthermore, (Song et al., 2017b) presented an EMS using Convex Optimization (CO), that avoided problems with the large computational weight of dynamic programming and unsatisfactory search algorithm results by finding a worldwide best solution that was near to dynamic programming and greatly increased the computational effectiveness.

In summary, the fuzzy control strategy's control principles are based solely on straightforward algorithms and expert expertise. But according to Zhang et al., (2015) and Xu et al., (2017), there is a big difference between the method's optimization effect and the globally optimal outcome, and the approach is also arbitrary. Although dynamic programming can yield the best solution for a given issue, its large computation time overhead makes it unsuitable for real-time applications. Convex optimization is difficult to accomplish because it requires previous convex processing before its use. As an example of an instantaneous optimization technique, instantaneous optimization control (IOC) offers real time benefits and is very useful in various uses. As a result, industrial control is paying more and more attention to algorithms.

This chapter developed a thorough cost function for running EVs with HESSs made up of power batteries and supercapacitors, taking into account the cost of the energy use of HESSs and the decreasing cost of the power batteries, provided that the previously mentioned EMSs don't include the expense of battery aging into account and have issues like meagre optimization things, challenges in real-world applications and long computation times. The function considered the cost of battery aging, utilized the instantaneous power battery SOC and supercapacitor as the state variables, and solved the coefficient of optimal power combination of a supercapacitor that fulfilled the constraints at any given time using the Hierarchical Instantaneous Optimization Control (HIOC) algorithm in order to minimize the instantaneous comprehensive cost. A unique EMS through good real-time optimization effects was created based on the previously discussed technique.

1.1 Motivation

To get the optimum fuel efficiency, EVs' new degree of freedom the power distribution between the engine as well as the electric motor must be carefully utilized. Energy management solutions are frequently used in the literature to describe tactics that regulate this power distribution inside the power train. These tactics are usually separated into two categories: optimum strategies and heuristic techniques.

Real-time implementations frequently use heuristic techniques. Although it is possible to obtain low fuel usage, the performance is highly dependent on how the rules are tuned. When the tuning is done for a typical driving cycle and the driving patterns that actually happen are very different from the cycle used for the tuning, this sensitivity becomes crucial. This causes the fuel usage to be higher than what might be accomplished. Since these rules are frequently not model-based, such heuristic control systems are generally not scalable.

On the other hand, optimal strategies are usually developed utilizing a suitable physically based model and techniques from optimal control theory. In order to determine the control signal, such algorithms frequently include minimizing a local cost function at each time step. As a result, reduced fuel usage was achieved. Usually, this type of local cost function is an equivalent fuel utilization, which is made up of the weighted electric consumption and the actual fuel consumption. As a result, these local minimization techniques are readily scalable and necessitate a power train model. However, because an online optimization must be performed in real-time, such solutions usually need more processing power than rule-based approaches.

The information at hand has a significant impact on how well any energy management plan performs. Only when the total driving cycle is known in progress can truly optimal fuel economy be attained. Although it establishes a standard against which every causal technique can be evaluated, the resulting approach is not causative. For the task at hand, dynamic programming (DP) is ideally suited to identifying such an ideal solution. Observe that even in real-world scenarios, causal solutions may be helpful in some circumstances.

2. OVERALL VEHICLE MODEL OF PURE EVS

A pure Electric Vehicle (EV) is powered entirely by electricity, typically using one or more electric motors and a large battery pack instead of a conventional internal combustion engine. An overview of the vehicle model for pure EVs as follows.

2.1 Overall Vehicle Parameters and Power Train Configuration

Table 1 enumerates the primary EV parameters. A drive motor, a supercapacitor pack, and a ternary lithium battery pack make up its principal parts. Signal connections exist between the BMS, EMS, vehicle controller, and DC-DC converter, while the remaining parts are electrically connected. The mechanical connections include the motor, primary gear reducer, and ties. The model's HESS structure is semi-active, with a bus formed by the connections of the motor controller, BMS, EMS, and power battery in series. The DC-DC converter and supercapacitor are connected to the bus line first in series and subsequently in parallel. With this HESS system, the power battery can meet the vehicle's kindnecessities, while the supercapacitor can provide a brief burst of high power to meet the vehicle's power needs.

Table 1. Electric Vehicle Simulation Model Parameters

Parameter	Value
Body mass of electric vehicle /kg	1,7000
Coefficient of rolling resistance	0.0100
Area of windward /m^2	1.6000
Air resistance coefficient	0.3000
Air density (kg/m^3)	1.1800

continued on following page

Table 1. Continued

Parameter	Value
Rotational inertia for conversion factor	1.1000
Mechanical drivetrain efficiency	0.9500

2.2 Whole Vehicle Longitudinal Dynamics Model

Electric vehicle EMSs that have supercapacitors within their HESSs was the primary focus of this chapter, hence only the kinematic features In the direction of the longitudinal drive were taken into account. The longitudinal force state equation through driving was used for the entire power model needed for the vehicle (Liu et al., 2019). This learning estimated the power required based on the pattern of acceleration for automotive propulsion and velocity throughout driving cycles by using the backward simulation method. The following is the resultant equation:

$$\textit{Required Power (Preq)} = (\textit{Rolling resistance} + \textit{Wind flow resistance} + \textit{Change in the mass} + \textit{Gravity acting along the slope}) \quad (1)$$

3.MODELLING THE POWER COMPONENTS AND STRUCTURE OF HYBRID ENERGY STORAGE SYSTEMS (HESS)

Hybrid Energy Storage Systems (HESS) is designed to combine multiple types of energy storage technologies to optimize the performance of an electric vehicle (EV) or another application. The primary goal is to control the strengths of dissimilar storage systems, such as batteries and supercapacitors, to meet varying power and energy demands efficiently. Here's how the power components and structure of HESS can be modelled.

3.1 HESS Configuration

A supercapacitor and a battery pack are connected in parallel to form a HESS. A bi-directional DC-DC converter is connected in series with this configuration, which has benefits including low cost, easy maintenance, and basic construction. By keeping an eye on variables like power consumption and the state of charge of the supercapacitor (SOCu) and battery (SOCb), a system is put in place to control the structure's energy flow. This makes it possible to calculate the power produced by the power battery and supercapacitor (Zhou et al., 2019).

3.2 Model of the Motor

Because of its remarkable qualities—including accurate control, steady torque production, high torque density, and low noise—the permanent magnet synchronous motor is commonly acknowledged as the best motor for electric vehicles (EVs) (Chen et al., 2021). A motor with 8,000 revolutions per minute, 105 killowatts of maximum power, and 1800 Nm of maximum torque is selected in this chapter. A phrase relating to the efficiency and power production of the motor can be used to express the electric power of the drive motor.

$$\textit{Electric motor power (PEM)} = \frac{\textit{Torque} * \textit{Rotational speed}}{\textit{Efficiency}} \tag{2}$$

Using the look-up table method, the speed and current motor torque were acquired, and from there, the motor efficiency was computed.

3.3 Model of Bus Power Balance

The following provides the bus power expression:

$$\textit{Drive Power (PD)} = (\textit{Power received and delivered by battery pack in KW}) + \textit{Super capacitor power} * \textit{DC} - \textit{DC conversion efficiency}) \tag{3}$$

$$\textit{Energy Recovery Power (PER)} = (\textit{Power received and delivered by battery pack}) + (\textit{Super capacitor power/ DC} - \textit{DC conversion efficiency}) \tag{4}$$

3.4 Model of HESS

Hybrid Energy Storage Systems (HESS) are designed to combine multiple types of energy storage technologies to optimize the performance of an electric vehicle (EV) or another application. The primary goal is to leverage the strengths of different storage systems, for example batteries and supercapacitors, to meet varying power and energy demands efficiently. Here's how the power components and structure of HESS can be modelled.

3.4.1 Supercapacitor Model

With a rated voltage of 399 V, the supercapacitor is made up consisting of 133 monoliths arranged in series to create a group. The internal resistance state fluctuation model for open-circuit voltage, which may be written as follows, serves as the establishment for the monolith model used in supercapacitors.

$$\text{Super capacitor output power } (Pu) = (\text{Super capacitor output voltage} * \text{Super capacitor current}) - (\text{Super capacitor current} * \text{Internal Resistance}) \quad (5)$$

3.4.2 Battery Model

(1) Model of Battery Circuit: For the battery cell model, the R_{int} model of internal resistance voltage was used (He et al., 2015). The battery pack's equivalent internal resistance and open-circuit voltage have the following expressions:

$$\text{Voltage of Battery } (Vb) = \text{Number of series connection in battery pack} * \text{Battery output Voltage} \quad (6)$$

SOC as well as the battery's ambient temperature (DU et al., 2020),

$$\text{Battery output voltage}(Vo) = \text{Open circuit voltage of Battery} + (\text{Battery current} * \text{Equivalent internal resistance}) \quad (7)$$

The integration approach was used to calculate the power battery's current state of charge (SOC) using the following expression.

$$\text{Resistance}(R) = (\text{Number of series connections in battery pack} * \text{Equ. Internal resistance}) \quad (8)$$

*(2) Battery thermal model:*The following phrase is the use of the Bernardis-Masoud and Tare Yelmek (2014) developed a heat production model in the power battery temperature-rise model.

$$\text{Temperature Rise } (TR) = \text{Heat Conduction} - \text{Radiation} \quad (9)$$

The battery temperature was controlled by the EV's thermal management system to keep it operating within a safe range. In this study, the upper temperature limit of the power battery was determined using an EV model to be 35° C, or 308.15

K. Heat was produced by the current passing through the internal resistance of the battery, which was subsequently lost as heat energy.

3.4.3 Parameter Matching of HESS

Supercapacitors and Lithium batteries, which have varying densities of energy and can be paired in numerous ways to suit range needs, make up the HESS. This is a multi-objective optimization issue where the goal of combined power supply parameter identical is to maximize battery life while minimizing total cycle energy consumption. The objective formulation for the HESS's parameter identical optimization is as follows.

$$\textit{Energy consumption of the HESS}\ (Ec) = (\textit{Energy consumption of the battery pack}\ (10) + \textit{Energy consumption of super capacitor})$$

Table 2. Power Parameters and Losses Concerned

S.No	Parameters	Efficiency	Losses
1	Electric motor power (P_{EM})	Efficiency improved	Friction, core, copper loss reduced
2	Drive Power (P_D)	Efficiency improved	Thermal, core loss reduced
3	Super capacitor output power (Pu),	Power Efficiency improved	Copper loss reduced
4	Temperature rise (T_R)	Thermal Efficiency improved	Core loss reduced

Table 2 shows that the energy recovery power (PER), drive power (PD), temperature rise (TR), voltage of battery (Vb), super capacitor output power (Pu), and electric motor power (PEM) are all maximized by lowering the loss taken into account at each stage of the electric vehicle's development.

4. EMS FOR EVS WITH HYBRID ENERGY STORAGESYSTEM

When using electric vehicles (EVs) with longer driving ranges over the course of their service lives, energy management is crucial, distribute power batteries' and supercapacitors' output power sensibly in order to prolong battery life, lessen frequent battery replacement requirements, and maintain the vehicle's economic viability.

4.1 Formulation for Objective Function

The total of the HESS's aging and electricity consumption costs is its comprehensive cost. The price of the power produced and utilized by the battery and supercapacitor is known as the power consumption cost. In addition, the battery aging cost is calculated by dividing the cost of purchasing the battery pack by the amount of capacity lost during the current trip.

$$J(u^*(t), t) = \min Cs(u(t), t) = Ce(u(t), t) + Cbr(u(t), t) \tag{11}$$

In this case, $u^*(t)$ is the ideal combination coefficient at time t;$u(t)$ is the power coupling coefficient at time t, $u(t)$ $[0,1]$; and $C_e(u(t),t)$,$C_{br}(u(t),t)$, and $C_s(u(t),t)$ are the electricity consumption cost (CNY), the comprehensive cost (CNY), and the battery aging cost (CNY) at time t, correspondingly.

$$Ce = (Pe\int(Pbat(t) + Pu(t))dt))/(3.6 \times 10^{\wedge}6) \tag{12}$$

$$Cbe = Pbe \cdot Ncell \cdot nb\int Eb \cdot \sigma\,(Ib(t)).(|Ireal(t)|/3600Imm)dt \tag{13}$$

The energy consumption cost (CNY) is represented by Ce, the battery aging cost (C_{br}), the purchasing cost of the battery pack (P_{br}), the unit price of power (P_e), which is 0.51 CNY/(kW·h), the total time the vehicle spends on a single trip (T), and the number of single cells associated in similar to the battery pack (N_{cell}).

5. THE CLASSIFICATION OF EMSS

Using two primary headlines from (Zhang et al., 2020), we offer a new organization method of EMSs are available for all kinds of Hybrid Electric Vehicles in this chapter.

1) Offline EMSs are distributed into two categories: rule based and globally optimized based, depending on the degree of information available for the circumstances while being used.
2) Three types of online emergency management systems (EMSs) are available: learning-based, predictive, and immediate optimization-based.

According to the EMS categorization, an EMS in flexible condition can use a variety of online as well as offline strategies to create an incorporated EMS that enhances performance and fuel economy. As a result, although offering a specific

EMS classification, various combinations with other approaches may be included in this chapter. Two classifications for offline. Both optimization-based and global rule-based EMSs are introduced. Achieving a global optimal power split for a given driving cycle and providing adjusted online EMSs are the primary objectives of worldwide optimization-based EMSs. Because of their computing complexity and presumption of knowledge of the full driving cycle, they are not immediately useful in real-time control. However, it can serve as a standard by which to modify the control parameters. Although conventional techniques, such dynamic programming, are capable of global optimization across current driving cycles, they are not readily transferable to actual automobiles. Therefore, this technique can be used to assess how well other optimization techniques extract the control rules. Rule-based energy management systems (EMSs) use pre-established control rules to determine how much electricity to divide, however their power allocation efficiency is inferior to that of offline globally optimized EMSs. However, online crisis management systems can be employed in real-time control and are dependent on local optimization. Among these strategies, instantaneous fuel usage at each instant which does not require information of the whole driving cycle beforehand is the only one that produces restricted optimal results. The three instant optimization-based EMSs are the Robust Control (RC), adaptive-ECMSs (A-ECMS), and equivalent energy minimization strategy (ECMS). The core method of ECMS can be executed in real time due to its flexibility, which depends on the Equivalent Factor (EF). It is found that there is a strong correlation between the equivalent factor and the performance of ECMS. Selecting an appropriate equivalent factor remains one of the main issues with ECMS. As a result, a number of methods including A-ECMS are proposed to divide the power according to ECMS and alter the EF online.

The main objective of the subsequent discussion on predictive EMSs is to optimize the power split based on projected rate over a certain time period. With the use of GPS and ITS traffic data, the future power demand is computed throughout the course of time. Predictive control is being implemented,It divides energy by optimizing fuel efficiency within a given time frame. This is made feasible by the helpful data from the prior car that was gathered through the cars' communication channels, since traffic management systems are using more and more intelligent transportation system technology. For predictive EMSs, the driving cycle forecast is therefore important. Predictive control for hybrid electric vehicles (HEVs) can be implemented by means ofModel Predictive Control (MPC), a common solution method that relies on the precision of aEV model for calculation. The primary method used by learning based EMSs to adjust the devicevariables in response to shifting road conditions is training data.

5.1 Offline EMSs

5.1.1 EMSs Based on Global Optimization

Since these methods needpast information of the normal driving cycle, they are non-causal and search for global optimal solutions. They cannot be used straight in real time situations due to the non-casual solution; on the other hand, non-casual For a given driving cycle, the best solutions can be identified offline. These can be used as benchmarks by different algorithms or updated online EMSs. Therefore, these techniques can be used as a baseline to achieve worldwideideal outcomes in a particular driving cycle. This section compares and illustrates typical techniques like convex optimization, game theory, robust control, genetic algorithms, stochastic dynamic programming, robust programming, pseudo spectral approach, and game theory. Table 1 presents a comparison of various ways in order to clearly demonstrate the disadvantages and advantages of each method. Low computational burden is necessary for "computational complexity" to score highly because it promotes efficiency and quick operation.

The term "adaptability" describes how easily the EMSs can be adjusted to various drive cycles. When control parameters are easily adjusted to varied driving cycles for fuel economy, IT performs well. When compared to other approaches, the SDP might offer the highest degree of versatility. The quantity of driving future information needed for formulation and calibration is indicated by the term "priori knowledge of driving cycle." In order to achieve the optimum fuel efficiency with these strategies, the DP needs the greatest amount of a prior information about the driving cycle's future information.

- Convex Optimization
- Genetic Algorithm (GA)
- Game Theory (GT)
- Pseudo Spectral Method
- Dynamic Programming (DP)

5.1.2 EMSs Rule Based Technique

- Deterministic Rule Based EMSs
    - On/Off EMSs
    - The power follower EMSs
- Fuzzy Logic –Based EMSs

5.1.3 Online EMSs

Since online EMSs typically don't require previous awareness of the complete driving cycle, they are connecting and local optimization based algorithms. EMSs can be developed in real-time with little processing overhead by converting the global optimization issue of offline EMSs into an instant optimization problem.. The implementation of online EMSs in real-time control problems is possible since they require less computational effort. There are three types covered: learning based EMSs, predictive based EMSs, and immediate optimization based EMSs. While predictive EMSs primarily use information from future driving cycles Instantaneous optimization based EMSs are used to optimize the power split use an ideal algorithm to estimate the power split. Moreover, the primary intention of the instantaneous optimization EMSs is to minimize the performance indices at each instant in order to establish the ideal power split. The relevant headlines are underlined and these EMSs are thoroughly explored in the part that follows.

- *Instantaneous Optimization Based EMSs*
  Robust Control
  Equivalent Consumption Minimization Strategy (ECMS)
  Adaptive Equivalent Consumption Minimization Strategy (A-ECMS)
- *Predictive EMSs*
  The Driving Cycle Prediction Approach
  1. Driving pattern recognition
  2. Traffic flow modelling
  3. Based on driving cycle calculation an Intelligent Transportation System (ITS)
  4. artificial intelligence based driving cycle prediction

  Learning-Based SMPC
  Stochastic Model Predictive Control (SMPC)
  Model Predictive Control (MPC)

6. CHALLENGES AND ISSUES OF ENERGY MANAGEMENT STRATEGIES IN EVS

Even while EVs have significantly improved in terms of improving fuel efficiency, lowering emissions, and vehicle performance, they still face several difficulties when it comes to EMSs. Durability of the batteries, processing load, and power sharing amongst various energy sources are the problems. Novel EMSs have been presented in order to address these problems.

6.1 Benefits Drawbacks of Existing Methods

Benefits drawbacks of existing methods is presented in Table 3

Table 3. Advantages and Disadvantages of Existing methods

Approaches	Advantages	Disadvantages
DP	•attains worldwide optimal outcomes; • sets a standard for other EMSs;	• is less flexible to modifications; • has the largest computational complexity (3-Level); • is aware of the full driving cycle beforehand
SDP	• Greater flexibility • Near-optimal fuel efficiency is attained	• The most difficult computation (3 Level) and the need for driving cycle database
GA	• more computational complexity (3-Level); • global optimality;	• strong global search performance; • reduced adaptability
GT	• trade-off between competing goals, • driver behaviour in EMSs,	• maximum computational complexity (3-Level), • and low adaptability

6.1. Battery Durability

Long driving distances require the EV system to have enormous energy reserves, and it also needs substantial power reserves for stopping, ascending, and accelerating. The battery must use a lot of energy and power to be constantly discharged and recharged. Rapid discharge-charge ratios, however, shorten battery life and impair fuel efficiency. Because of this, an effective for EVs, a power management system is necessary to ensure the battery's long-term viability.

The term "battery durability" describes how long a battery can continue to operate and hold its capacity. There are other aspects to take into account, even if the relationship between battery capacity, charging power, and discharging power indirectly affects battery durability. Numerous elements, such as a battery's chemistry, design, usage habits, operating environment, and charging and discharging profiles, affect how long it will last. The entire quantity of energy a battery is capable of storing is measured by its capacity. The pace at which energy is extracted from the battery during use is referred to as charging power. Elevated power consumption during charging or discharging may result in increased internal resistance, faster deterioration of the battery's components, and eventually a decrease in total capacity. As a result, exceeding the battery's recommended parameters when charging or discharging could potentially shorten its lifespan.

To address battery durability, a system-level design is required. A novel optimal control strategy was presented, which reduced the maximum current to dynamically improve battery longevity by using the off-line solution of an isoperimetric issue. The results showed how well the method worked to reduce significant charging and discharging current peaks and increase battery longevity. Hysteresis control is used for electric vehicles (EVs) with 3 fuel cell loads. In this technique, every fuel cell hoard maintains a consistent working position while the dynamic duration is decreased by switching (on-off) control. (Wang et al., 2019) created a management strategy based on bounded state machines and suggested an ideal oxygen surplus ratio control to exploit the fuel cell's net power while also taking into account its capabilities and state of charge. The simulation's findings showed that the plan guarantees the required power under operating conditions. Furthermore, by considering battery durability, the power management systems may achieve a nearly ideal outcome. The optimal power management strategy (Zhang et al., 2016) recommended based on the DP algorithm and tested by the varied battery SOH and battery SOC situations ensures a improved approach for system efficiency. (Robin et al., 2016) created aautomatic catalyst dissolving model to calculate the fuel cell's lifespan as well as estimate performance loss in order to optimize the fuel cell system's control.

6.2 Computational Load

Because of the “curse of dimensionality” the amount of information needed for verification a valid conclusion starts to expand exponentially as the number of variables increases. A massive amount of data burdens computers heavily or even prevents them from working. Therefore, a number of power management techniques cannot be used for the power management of EVs described in (Jiang et al., 2021) without realistic optimization for computational load. (Larsson et al., 2014) concentrated on local estimates at the suitable control inputs for the gridded cost to move in order to decrease the quantized interruption. The findings showed that only a gradual decline in simulated fuel efficiency was necessary to achieve orders of magnitude reductions in computing complexity. In addition, it is possible to significantly reduce the amount of memory storage needed by utilizing a spline approximated cost to go. A computationally intensive optimum control method like DP is used when the design search space is big. Therefore, a more rapid optimization strategy with improved computing efficiency and appropriate precision is required. The authors Yang et al. (2019) propose the use of PSO in combination with a rapid-

DP optimization technique to determine the ideal motor control state and improve the fuel efficiency of the vehicle.

The time it took to make decisions for ideal control was greatly reduced by this technique. The level-set DP idea described in (Elbert et al., 2012) resulted in a 300-fold reduction in DP computation time. The recommended method achieves comparable precision with a far lower amount of discretization as compared to the traditional DP. However, there were also other challenges, such the standards and Markov problems. Zhou et al. (2018) projected a unified solution approach to address these problems, which involved creating a single equation of state using the Markov features of DP. To reduce the amount of data used in the calculation, filters based on state variables as well as control variables were employed. This technique worked more quickly than both level-set DP and conventional DP due to the short computing complication of neural networks. Stochastic gradient drop and expertise replay were used in (Li et al., 2017) to demonstrate an EMS on reinforcement learning without having to worry about the curse of dimensionality in difficult stations. This ensured the accuracy and stability of the method.

6.3 Multiple Energy Sources

The choice of energy management systems is significantly predisposed by the various forms of energy storage. The energy density, power density, efficiency, cycle life, and cost of each form of energy storage technology vary. These qualities have an impact on which EMS is best suited for a certain application. Fuel cells, for instance, have a long operational life and high energy density, but they also usually have a poor power density and sluggish response times. However, compared to fuel cells, lithium ion batteries may have a lower energy density despite having a comparatively high power density and prompt response times. Meanwhile, UCs is identified for their elevated power density and quick response times; yet, when compared to fuel cells moreover lithium-ion batteries, they usually have a lower energy density (Foles et al., 2023).

The primary energy storage system (ESS), or batteries, has a major influence on the overall presentation of electric vehicles. The primary purpose of a hybrid energy storage system (HES) made up of different types of energy storage is to integrate different energy storage technologies into one more dependable and efficient energy storage and utilization solution. Through the integration of several energy storage technologies, including batteries, fuel cells, ultra-capacitors, and UCs, a hybrid system can optimize energy conversion and management, boost energy density, and improve overall storage system dependability and flexibility. This leads to enhanced grid resilience and stability as well as greater integration and use of renewable energy sources (Liu et al., 2022).

The drawbacks with batteries, such as their large size, high cost, short cycle life, and reduced power density, can be remedied by using a hybrid energy storage system (HES), which combines two or more ESS. Longer driving ranges are possible with several energy sources compared to typical EVs. Additionally, the engine mimics the condition in traditional EVs by allowing the vehicles to run while the battery SOC is low. Additionally, the engine is turned off and the vehicle moves only on its electric power system when the speed or else power need is low. Driving concert is therefore impacted by the way power is allocated among different energy sources, as shown in (Iris et al., 2019).

The particular needs of the application, such as the required energy capacity, power consumption, response time, and cost limits, determine which EMSs should be used. It will also be contingent upon the viability and accessibility of the many energy storage technologies that have been suggested. It will be necessary to take into account variables like longevity, efficiency, charging and discharging rates, and length of energy supply.

6.4 Regenerative Braking

Apart from the previously described difficulties, regenerative braking—also referred to as recovering the energy used during normal braking—is a crucial remedy for the EMS of electric vehicles. Research on innovative and contemporary control logic is growing in an effort to maximize energy recovery and attain the best possible force distribution during braking. RBS (Hamada et al., 2022 and Lin et al., 2019) currently includes the ABS and other safety systems. RB is a method of reclaiming kinetic energy during a vehicle's slowdown and storing it for later use. This aids in the recovery of some energy that would otherwise be wasted in the traditional vehicles described in (Xu et al., 2019 and Ganesh et al., 2023) as heat and frictional energy. In a hybrid or electric car, the electric motor serves two purposes: first, it drives the wheels and moves the car forward; second, it recharges the batteries (generated power). When the driver depresses the brake pedal rather than the accelerator pedal, power transmission from the motor to the wheels is sporadic. Because of the EMF the resolving wheels emit, the battery is now being charged by the motor, which is now acting as a generator. In adding to getting better generally vehicle performance, restoration may really extend the life of the braking system because the mechanical parts do not wear out as quickly. When it comes to safely stopping the automobile or reducing its speed rapidly, RB is insufficient on its own. Consequently. It is combined with an extra braking mechanism, like hydraulics or friction (Zhao et al., 2018; Yang et al., 2023) A appropriate control system for the distribution of braking force is necessary since both braking systems are worn depending on the driver's demand both during and after braking.

6.5 Uncertainties and Robustness

A power switching unit, converters, and several energy sources (like a motor, engine, and battery) make up an EV's power system. These parts are often affected by variations in parameters for example temperature, load change, and discharge charge rate. Consequently, variable power requirements, battery overheating, and low power quality are brought on by the uncertainties in the parameters. which together eventually cause the vehicle's power system to become unstable (Zhang et al., 2017). To solve the issue of system stability and guarantee stable system operation, the design of the EMSs should obtain the incurable cost and stability constraints into consideration (Zhang et al., 2020).

The robustness of the EMSs is the control system's capacity to ensure model stability and withstand noise and disturbances in the system. That being said, most EMSs for electric vehicles are evaluated under predetermined conditions, ignoring unknowns. Such stationary EMSs may cause the car to perform poorly. Conversely, certain EMSs, like Markov chains, rely on the data collected. Therefore, if the data from the recorded driving cycles differs from the actual driving situations, the algorithms cannot be optimized. However, as the system runs, parts like generators, capacitors, and batteries always deteriorate and age, creating uncertainty about the configuration and performance of the system as described in (Saeed et al., 2016).

7. CONCLUSION

The adaptive energy management of EVs using HESSs made up of batteries and supercapacitors was covered in this chapter. In order to increase effective power and decrease the loss taken into account at each stage of the electric vehicle's development, a hierarchical optimum searching EMS was developed. The optimization objective includes Electric Motor Power (P_{EM}), Drive Power (P_D), Energy Recovery Power (P_{ER}), Super Capacitor Output Power (P_u), Battery Voltage (V_b), and Temperature Rise (T_R), with the state variables being the supercapacitor and battery SOC. These demonstrated how EMS may effectively lower losses and increase the efficiency of electric vehicles equipped with HESSs.

8. REFERENCES

Azim Mohseni, N., Bayati, N., & Ebel, T. (2024). Energy management strategies of hybrid electric vehicles: A comparative review. *IET Smart Grid*, 7(3), 191–220. DOI: 10.1049/stg2.12133

Chen, L. M., Liao, Z. L., & Ma, X. J. (2021). Hierarchicalcontrol-basedreal-timeenergy management strategy for hybrid electric vehicles. *ActaArmamentar*, II(8), 1580–1591. DOI: 10.3969/j.issn.1000-1093.2021.08.002

Chen, Z. Y., Xiong, R., Wang, K. Y., & Jiao, B. (2015). Optimalenergymanage-mentstrategyofaplug-inhybridelectricvehiclebasedonaparticleswarmoptimization-algorithm. *Energies*, 8(08), 3661–3678. DOI: 10.3390/en8053661

Ding, T., Wang, J., Matas, J., Guerrero, J. M., Qiao, R., & Wang, C. (2023, June). Multi-Time Scale Energy Management Strategy based on MPC for 5G Base Stations Considering Backup Energy Storage and Air Conditioning. In *2023 IEEE International Conference on Predictive Control of Electrical Drives and Power Electronics (PRECEDE)* (pp. 1-6). IEEE. DOI: 10.1109/PRECEDE57319.2023.10174449

Dubal, D. P., Ayyad, O., Ruiz, V., & Gómez-Romero, P. (2014). Hybrid energystorage: The merging of battery and supercapacitor chemistries. *Chemical Society Reviews*, 44(7), 1777–1790. DOI: 10.1039/C4CS00266K PMID: 25623995

Elbert, P., Ebbesen, S., & Guzzella, L. (2012). Implementation of dynamic programming for $ n $-dimensional optimal control problems with final state constraints. *IEEE Transactions on Control Systems Technology*, 21(3), 924–931. https://ieeexplore.ieee.org/document/6178777. DOI: 10.1109/TCST.2012.2190935

Foles, A., Fialho, L., Horta, P., &Collares-Pereira, M. (2023).Economic and energetic assessment of a hybrid vanadium redox flow and lithium-ion batteries, considering power sharing strategies impact.Journal of Energy Storage, 71, (pp.108167). .DOI: 10.1016/j.est.2023.108167

Ganesh, S., Nivas, K. V., Kavya, A., Arunteja, S., & Guruvulunaidu, P., & KUMAR, B. (. (2023). DESIGNING OF ELECTRIC VEHICLE WITH REGENERATIVE BRAKING SYSTEM. *Journal on Electrical Engineering*, 16(3). Advance online publication. DOI: 10.26634/jee.16.3.19419

Haleema, K., Akhilesh, R., Shashank, G., Baig, M. A., & Krishnamoorthy, M. (2024, May).A Comprehensive Analysis of Load Management Strategies and Hybrid Renewable Solutions for Sustainable Rural Electrification System.In *2024 5th International Conference for Emerging Technology (INCET)* (pp. 1-6).IEEE. DOI: 10.1109/INCET61516.2024.10593192

Hamada, A. T., &Orhan, M. F. (2022).An overview of regenerative braking systems. Journal of Energy Storage, 52, (pp.105033). DOI: 10.1016/j.est.2022.105033

He, Z. C., Yang, G., & Lu, L. G. (2015). BatteryDCinternalresistancetestmethodbased on the constant current external characteristic and SOC. [SciTechnol]. *J Tsinghua Univ*, (5), 9–18.

Hu, J. J., Zheng, Y., & Hu, Z. H.. (2018). Parameter matching and controlstrategiesofhybridenergystoragesystemforpureelectricvehicle.*ChinaJ.Highways and Transportation*, (3), 142–150. DOI: 10.19721/j.cnki.1001-7372.2018.03.016

Huan, N., Yao, E. J., & Yang, Y.. (2019). Stochastic dynamic user equilibriumassignment model considering penetration of electric vehicle. *Journal of Traffic and Transportation Engineering*, (10), 150–160. DOI: 10.19818/j.cnki.1671-1637.2019.05.015

Iris, Ç., & Lam, J. S. L. (2019). A review of energy efficiency in ports: Operational strategies, technologies and energy management systems. *Renewable & Sustainable Energy Reviews*, 112, 170–182. DOI: 10.1016/j.rser.2019.04.069

Ji, C. W., Pan, S., & Wang, S.. (2020). Experimental study on effect factors ofagingrateforpowerlithium-ionbatteries.*J.BeijingUniv.Technol.*, (10), 1272–1282. DOI: 10.11936/bjutxb2019040020

Jiang, J., Jiang, Q., Chen, J., Zhou, X., Zhu, S., & Chen, T. (2021).Advanced power management and control for hybrid electric vehicles: a survey.Wireless Communications and Mobile Computing, 2021(1), (pp.6652038). .DOI: 10.1155/2021/6652038

Larsson, V., Johannesson, L., & Egardt, B. (2014). Analytic solutions to the dynamic programming subproblem in hybrid vehicle energy management. *IEEE Transactions on Vehicular Technology*, 64(4), 1458–1467. DOI: 10.1109/TVT.2014.2329864

Li, T., Zou, Y., & Liu, D. (2015). EnergymanagementstrategyforHEVsincludingbatterylifeoptimization. *IEEE Transactions on Transportation Electrification*, 70(1), 98–112. DOI: 10.1109/TTE.2015.2471180

Li, Y., He, H., Peng, J., & Zhang, H. (2017). Power management for a plug-in hybrid electric vehicle based on reinforcement learning with continuous state and action spaces. *Energy Procedia*, 142, 2270–2275. DOI: 10.1016/j.egypro.2017.12.629

Lin, C., Liang, S., Chen, J., & Gao, X. (2019). A multi-objective optimal torque distribution strategy for four in-wheel-motor drive electric vehicles. *IEEE Access : Practical Innovations, Open Solutions*, 7, 64627–64640. DOI: 10.1109/ACCESS.2019.2917313

Liu, X. D., Ma, J., Zhao, X., Zhang, Y., Zhang, K., and He, Y. (2019).Integrated-component optimization and energy management for plug-in hybrid electric buses. *Process*7(8),(pp.477).. https://doi.org/.DOI: 10.3390/pr7080477

Liu, Z., Li, Y., Fan, G., Wu, D., Guo, J., Jin, G., & Yang, X. (2022). Co-optimization of a novel distributed energy system integrated with hybrid energy storage in different nearly zero energy community scenarios. *Energy*, 247, 123553. DOI: 10.1016/j.energy.2022.123553

Lu, J., Zheng, W., Yu, Z., Xu, Z., Jiang, H., & Zeng, M. (2024). *Optimizing Grid-Connected Multi-Microgrid Systems With Shared Energy Storage for Enhanced Local Energy Consumption*. IEEE., DOI: 10.1109/ACCESS.2024.3351855

Martel, F., Dube, Y., Kelouwani, S., Jaguemont, J., & Agbossou, K. (2016). Long-term assessment of economic plug-in hybrid electric vehicle battery lifetime degradationmanagement through near optimal fuel cell load sharing. *Journal of Power Sources*, 318(30), 270–282. DOI: 10.1016/j.jpowsour.2016.04.029

Masoud, S., & Tare Yelmek, K. (2014). Multi-objective optimal design of hybridrenewable energy systems using PSO-simulation based approach. *Renewable Energy*, 68, 67–79. http://hdl.handle.net/1993/31040. DOI: 10.1016/j.renene.2014.01.011

Panday, A., & Bansal, H. O. (2014). A review of optimal energy managementstrategiesforhybridelectricvehicle. *International Journal of Vehicular Technology*, 2014(2), 1–19. DOI: 10.1155/2014/160510

Qu, J., Wang, H., Zou, L., Zhang, L., Zhang, T., Zhou, J., & Zhang, B. (2024). A Two-Layer Energy Management Strategy of Fuel Cell Hybrid System in Electric Ships. *IEEE Transactions on Industry Applications*, 1–11. Advance online publication. DOI: 10.1109/TIA.2024.3447628

Robin, C., Gérard, M., Quinaud, M., d'Arbigny, J., & Bultel, Y. (2016). Proton exchange membrane fuel cell model for aging predictions: Simulated equivalent active surface area loss and comparisons with durability tests. *Journal of Power Sources*, 326, 417–427. DOI: 10.1016/j.jpowsour.2016.07.018

Saeed, M. A., Ahmed, N., Hussain, M., & Jafar, A. (2016, January).A comparative study of controllers for optimal speed control of hybrid electric vehicle.In *2016 International Conference on Intelligent Systems Engineering (ICISE)* (pp. 1-4).IEEE. DOI: 10.1109/INTELSE.2016.7475142

Santucci, A., & Sorniotti, A. andLekakou,C.(2014).Powersplitstrategiesforhybridenergystoragesystemsforvehicularapplications.*J.PowerSources*258,(pp.395–407).. https://doi.org/DOI: 10.1016/j.jpowsour.2014.01.118

Sisakat, S. T. andBarakati,S.M.(2016).Energymanagementusingfuzzycontroller-for hybrid electrical vehicles.*J. Intelligent Fuzzy Syst.* 30, (pp.1411–1420), DOI: 10.3233/IFS-152054

Smith, M., Zhao, L., Cordova, J., Jiang, X., & Ebrahimi, M. (2024, January).Machine Learning-Based Energy-efficient Workload Management for Data Centers. In *2024 IEEE 21st Consumer Communications & Networking Conference (CCNC)* (pp. 799-802). IEEE. DOI: 10.1109/CCNC51664.2024.10454842

Song, C. X., Zhou, F., & Xiao, F. (2017a). Energy management optimization of-hybrid energy storage system based on dynamic programming. *J. Jilin Univ. Eng. Technol.Ed.*, (1), 8–14. DOI: 10.13229/j.cnki.jdxbgxb201701002

Song, C. X., Zhou, F., & Xiao, F.. (2017b). Parameter matching of on-board hybridenergy storage system based on convex optimization method. *Jixie Gongcheng Xuebao*, 53(08), 44–51. DOI: 10.3901/JME.2017.16.044

Wang, B. W. (2018). *Studyonparametermatchingandcontrolstrategyofthepowersystemofnewenergyvehicle*. HunanUniv., DOI: 10.1177/0954407020936781

Wang, Y., Sun, Z., & Chen, Z. (2019). Energy management strategy for battery/supercapacitor/fuel cell hybrid source vehicles based on finite state machine. *Applied Energy*, 254, 113707. DOI: 10.1016/j.apenergy.2019.113707

Wu, C., Zhu, C., Sun, J., & Ge, Y. (2017). A synthesized diagnosis approach forlithium-ion battery in hybrid electric vehicle. *IEEE Transactions on Vehicular Technology*, 66(7), 5595–5603. DOI: 10.1109/TVT.2016.2634125

Xie, S., Hu, X., Qi, S., Kun, L., Zongke, X., & James, B.. (2019a). Modelpredictiveenergymanagementforplug-inhybridelectricvehiclesconsideringoptimalbatterydepthofdischarge. *Energy*, 173, 667–678. DOI: 10.1016/j.energy.2019.02.074

Xu, W., Chen, H., Zhao, H., & Ren, B. (2019). Torque optimization control for electric vehicles with four in-wheel motors equipped with regenerative braking system. *Mechatronics*, 57, 95–108. DOI: 10.1016/j.mechatronics.2018.11.006

Yang, J., Liu, B., Zhang, T., Hong, J., & Zhang, H. (2023). Multi-parameter controlled mechatronics-electro-hydraulic power coupling electric vehicle based on active energy regulation. *Energy*, 263, 125877. DOI: 10.1016/j.energy.2022.125877

Yang, Y., Pei, H., Hu, X., Liu, Y., Hou, C., & Cao, D. (2019). Fuel economy optimization of power split hybrid vehicles: A rapid dynamic programming approach. *Energy*, 166, 929–938. DOI: 10.1016/j.energy.2018.10.149

Yao, T. Z., Xie, C. J., & Zeng, T.. (2019). Multi-Fuzzy control based energymanagementstrategyofbattery/supercapacitorhybridenergysystemofelectricvehicles. *Automotive Engineer*, (6), 615–640. DOI: 10.19562/j.chinasae.qcgc.2019.06.002

Zhang, F., Wang, L., Coskun, S., Pang, H., Cui, Y., & Xi, J. (2020). Energy management strategies for hybrid electric vehicles: Review, classification, comparison, and outlook. *Energies*, 13(13), 3352. DOI: 10.3390/en13133352

Zhang,L.,Hu,X.S.,Wang,Z.B.,etal.(2017).Multi-objectiveoptimalsizingofhybridenergy storage system for electric vehicles.*IEEE Trans. Veh. Technol.*, 1.. https://doi .org/DOI: 10.19562/j.chinasae.qcgc.2017.12.003

Zhang, P., Yan, F. W., & Du, C. Q. (2015). A comprehensive analysis of energymanagement strategies for hybrid electric vehicles based on bibliometrics. *Renewable & Sustainable Energy Reviews*, 48(48), 88–104. DOI: 10.1016/j.rser.2015.03.093

Zhang, Q., Deng, W., & Li, G. (2017). Stochastic control of predictive power management for battery/supercapacitor hybrid energy storage systems of electric vehicles. *IEEE Transactions on Industrial Informatics*, 14(7), 3023–3030. DOI: 10.1109/TII.2017.2766095

Zhang, S., Xiong, R., & Cao, J. (2016). Battery durability and longevity based power management for plug-in hybrid electric vehicle with hybrid energy storage system. *Applied Energy*, 179, 316–328. DOI: 10.1016/j.apenergy.2016.06.153

Zhang, Z., Guan, C., & Liu, Z. (2020). Real-time optimization energy management strategy for fuel cell hybrid ships considering power sources degradation. *IEEE Access : Practical Innovations, Open Solutions*, 8, 87046–87059. DOI: 10.1109/ACCESS.2020.2991519

Zhao,B.,Song,Q.,andLiu,W.(2012).Powercharacterizationofisolatedbidirectional dual-active-bridge DC–DC converter with dual-phase-shiftcontrol.*PowerElectron. IEEETransaction*27(9), (pp.4172–4176)..DOI: 10.1109/TPEL.2012.2189586

Zhao, D., Chu, L., Xu, N., Sun, C., & Xu, Y. (2018).Development of a cooperative braking system for front-wheel drive electric vehicles.Energies, 11(2), (pp.378). DOI: 10.3390/en11020378

Zhou, F., Song, C. X., & Liang, T. W.. (2017). Parameter matching of on-boardhybrid energy storage system using NSGA-II algorithm. *J. Jilin Univ. Eng. Technol. Ed.*, (9), 1336–1343. DOI: 10.1109/TVT.2017.2762368

Zhou, M. L., Feng, J. F., & Zhang, Y. (2019). Composite energystorage system anditsenergycontrolstrategyforelectricvehicles. *Electr.Mach.Control*, (5), 51–59.

Zhou, W., Yang, L., Cai, Y., & Ying, T. (2018). Dynamic programming for new energy vehicles based on their work modes Part II: Fuel cell electric vehicles. *Journal of Power Sources*, 407, 92–104. DOI: 10.1016/j.jpowsour.2018.10.048

Chapter 8
A Smart Traffic Management System for Improving EV Range Using Machine Learning Algorithms

Pawan Kumar
Ajay Kumar Garg Engineering College, Ghaziabad, India

Ramveer Singh
https://orcid.org/0009-0003-8773-3863
Raj Kumar Goel Institute of Technology, Ghaziabad, India

Anu Chaudhary
Ajay Kumar Garg Engineering College, Ghaziabad, India

Shashank Sahu
https://orcid.org/0000-0001-6720-4284
Ajay Kumar Garg Engineering College, Ghaziabad, India

Deepika Yadav
Bharati Vidyapeeth's College of Engineering, New Delhi, India

ABSTRACT

Nowadays, Hybrid Electric Vehicles (HEVs) have become the most demanded technology for reducing pollution emissions and fuel saving in the transport sector. The demand growth rate for electric vehicles is rapidly increasing due to zero emissions

DOI: 10.4018/979-8-3693-6844-2.ch008

of carbon dioxide from electric vehicles. The main challenges with HEVs are their limited battery capacity and difficulty in deciding when and where to charge EVs. The Concerned EVs charging problem and range anxieties still make customers uneasy about purchasing an electric vehicle but new advancements in EVs try to resolve the above problems and improve EVs range. The effectiveness of HEVs relies on energy management strategies (EMSs). A big challenge for the researcher is to develop suitable EMSs for HEVs, to enhance the range of EVs and increase their performance and battery life.

INTRODUCTION

The current population of India stood at about 1.44 billion in January 2024, increasing with a growth rate of 0.9% every year. It is expected to surge to 1.7 billion in 2047, and 51% of the total population resides in urban areas. As a country, India has the third-largest road network in the world, with around 60 per cent of the population using personal or shared vehicles for commuting, with 326.3 million registered vehicles in India up to the year 2020. The Growth rate of vehicles is also increasing rapidly every year, and according to the survey report of SIAM (Society of Indian Automobile Manufacturers), the vehicle population increased by 8.4 per cent in the last financial year (April- March) 2023-24. From various survey projects, it is estimated that the total two-wheeler population will increase by 420 million to 562 million, and similarly, population of cars will also increase from 262 million to 432 million in 2047 (M. Zichichi et al.,2020) As we all know the transport sector also plays a major role in contributors to the GDP of India and it is expected to contribute around 20% of GDP by 2026 from the current 6.7%. Both vehicles and populations increase rapidly, and it is not easy to manage traffic loads with existing outdated transportation networks that cause more traffic congestion, road accidents, etc. It has become important for cities to look for a Smart Transportation System to tackle these issues (A. Pundir et al.,2022). The smart Intelligent Transportation System operates, evaluates and manages transportation systems across the city by implementing advanced information and seamless technology that helps to enable reliable and more personalized travel experiences around the cities. The data is collected and analyzed according to the sectors. The system focuses mainly on smart mobility vehicle infrastructure and stakeholders from a holistic point of view and rectify the errors related to them. The research proposal helps to overcome the problems of existing transport management systems, helps minimize battery consumption of electric vehicles, and also reduces traffic congestion and air pollution.

The travelling time and probability of road accidents are also reduced by smart transport systems (Z. Khan et al.,2023).

The transport sector has a very high impact on the economic growth of any country; 17 per cent of the total economic growth of India is directly dependent upon the transport sector. India spends almost a fifth part of their economy on purchasing fuels from other countries. Growth of the Transport sector is increasing rapidly due to the increasing number of vehicles day by day. As we know, the Government of India has been more focused in the last few years on the revolution in the transportation sector and many new policies have been designed for improving the transport sector. The first step is the replacement of existing vehicles with electric vehicles, which indicates positive signs towards green development in India in 2047. The government of India provides many incentives to customers for the purchase of new electric vehicles, such as free vehicle registration fees. The main objective of the government is to make India free from pollution (Shanthamallu et al., 2017). Electric vehicles emit zero pollution, and this is the reason the government decided to increase the sales of personal vehicles by 30 per cent, commercial vehicles by 70 per cent and 80 per cent with two and three-wheelers by 2030 and is projected in 2047, 91% of three-wheelers,90% of two-wheelers79% of passenger vehicles and buses are shifted to electric vehicles (N. Choudhury et al.,2018). However, despite the efforts, the EV industry in India is still facing numerous challenges. The biggest challenge for implementing EVs is to travel long distances without recharging because current batteries are very small in size and have low voltage capacities, so they can't travel long distances. It has been observed that 10% of battery capacity is consumed due to traffic congestion because the existing traffic management systems work on traffic light timings, and even if there is no traffic congestion, traffic has to wait until the light is not green (N. Choudhury et al.,2021).

Existing traffic systems suffer from various drawbacks such as economic losses, more time waste, and more fuel or battery consumption. Battery consumption is a crucial factor in EVs, and therefore, a smart traffic management system is needed to help increase the battery life of EVs. ML and Deep Learning traffic management systems dynamically adjust traffic lights by existing traffic, helping in congestion reduction, battery consumption and saving of money. After the research survey, it has been identified that the existing transport management system has several drawbacks and needs to be improved (A. B. M. Adam et al.,2023). In the existing traffic management system, there is no proper planning to manage the traffic load, and due to this, the traffic congestion is increasing day by day. The existing traffic system still uses a timing approach to manage traffic lights whether we have traffic or not. If traffic congestion is very less, the driver has to wait till the traffic light is green, which results in more fuel or battery consumption. In India, no advanced traffic management system provides information about route optimization. Suppose the driver is well aware of traffic congestion in advance. In that case, route optimization plays important role in reducing traffic congestion or help in finding the ultimate route that helps in saving money and time (I.García-Magariño et al., 2020). In the existing transport system, there is no tracking of vehicles but in some other countries, they are using a n AI-based transportation system that manages vehicle records. Vehicle manufacturing companies are inserting IoT sensors in their vehicle that can help to track the position of vehicles and that can help the users to identify the real-time location of their vehicles (S. Kaleem et al., 2023) .

The involvement of AI in the transportation sector is very helpful for reducing road accidents, providing safer lives and also helping in the growth and development of the country. Presently, many low-income countries are suffering from a lack of traffic resources or infrastructure, and the frequency of road accidents is very high., The safety issues of drivers and pedestrians are a big challenge, and sometimes human error, poor roads, and outdated vehicles cause big problems. The other big challenge in the existing transport system is climate change. The emission of CO2 from vehicles highly affects our environment, and 2% of emissions are caused by existing transport systems (Teoh et al.,2022). Climate change again rethinks the people's or researchers' way of living. The AI-based transport system provides a new approach to solving global problems, and researchers are developing AI applications in their existing transportation systems. The results help reduce carbon emissions, fuel usage, and operational costs.

RELATED WORK

Shwetha R.J proposed a traffic signal intelligent model that worked on cloud vision API by using Machine learning. Traffic images are captured from the junction point and sent to the cloud, and all images are updated by Cloud Vision API. It helps to detect the position of vehicles and also keeps a record of the return's previous signal status. Previous and present signals check the next signal status, and based on the status of the signal, it performs a further operation (Shwetha R.J et al., 2018). Aditi Yadav proposed a new approach based on adaptive traffic management that worked with the help of IOT devices and machine learning. It uses Camera sensors to capture the image and two controllers to keep records of several counts of vehicles from two different lanes, and the traffic signal light is adjusted accordingly. If the difference between counts of both lanes is high, then it applies a min-max approach, and a low waiting average time will be given priority. If the difference is not too high, then it applies the Round Robin approach, and low congestion traffic has to be given high priority (Aditi Yadav et al.,2019). Alharahsheh proposed a smart city traffic management system that used a Reinforcement learning approach. Compound optimisation error is easily handled by reinforcement learning successfully and deep learning methods have drawn enormous fascination (Alharahsheh Y et al, 2010).G. Sathya, et[19]. al used "AARS using GPRS 3G TECHNOLOGY" and provided a safe and smooth flow of ambulances by adjusting traffic lights. It adjusts the traffic light according to the real position of the location of the ambulance so that it can reach the hospital in time. GPS are used to find the location of the ambulance. Salama et al. proposed a model for monitoring and managing traffic lights intelligently with the help of various sensors that are deployed on both sides of traffic lights. The complete work of the proposed model is divided into four subparts. First is the control system unit that was installed in place of the traffic control unit, which is used for switching traffic lights, generating emergency signals from sensors, and sending information to a central database. The second component is Sensors and in the proposed model, two sensors are used, the first sensor is applied to check whether extensions of roadways are clear or not. Second-type sensors are deployed on both sides of traffic lights to keep records of vehicles. The third component is an emergency alarm. It is used to provide free space to police, hospital ambulances and fire vehicles, and the traffic light is open in case of emergency alarm. The last component of the proposed model was Traffic lights that were connected to the control unit and that were handled by the central control system (G.Sathya et al.,2018). Collotta proposed dynamic-based methods for traffic light systems that were based on WSN. The main emphasis of the paper is to minimize vehicle waiting time and light time. RLR is a daily problem that is faced by the driver when it decides to cross the road, the traffic light turns green to red. To solve this problem, the designed model applies two techniques. First, it tries to collect information related to real traffic with the help of WSN and then after it operates dynamically traffic lights and asks the controller to manage traffic light cycles depending on several vehicles that are in the queue. Reduced Function Devices (RFD) are used to monitor road junction points that are placed every 12 meters, and vehicles were detected by detected by magnetic sensors provided that are deployed at each junction. After collection of the data through RFD, it is supplied to a Full Function Device (FFD) which is deployed every 60 meters and finally sent to First Pan Controller (FPC) that is placed at the end side of the road (Collotta M et al.,2014). For calculation points in his research, the length of the road was taken at 1.5 km, and the traffic light cycle was fixed at the 30s, 50s, 75s, and 90s-120s. It was concluded a result that 80 per cent of vehicles from the queue when traffic lights operated dynamically and 43.35% passed when the traffic light was fixed. Ghazal designed a new model that evaluates the density of traffic density with the help of infrared (IR) sensors and manages periods of timing dynamically based on different levels of the Programmable Intelligent Controller (PIC) microcontroller. It also tries to solve emergency traffic congestion problems through the mobile controller. The proposed model used numerous modules that comprise various components such as PIC 16F877A microcontroller, Liquid Crystal Display (that was used to send alert message to manager about the emergency case), XBee transceivers that provides a secure support and simple full-duplex communication between microcontrollers over serial port data transmission (Ghazal B et al.,2016). (Islam M, el. Al 2016) proposed a new method for the adaptive approach for traffic light systems that were based on image processing. The proposed work is divided into two sub-components: the first one is used for data acquisition and the second delivery of data. They used 22 videos and in which each video size consisted of (300 to 500) frames. The methodology of research work depends on object recognition, which detects the moving object of street distance vector and is classified into three object types, i.e. walkers, cars, and vehicles of other categories. The accuracy of the model is found to be around 98% for the correct identification of an objects.

BACKGROUND

Machine learning is a sub-division of artificial intelligence that helps to predict output-based values on previous data values by using creation of an intelligent algorithm. Machine learning works on mathematical functions for making predictive decisions with the help of previous information. For designing a predictive model – a first requirement of appropriate information (data) and before supplied data to the designed model, data need to be filtered with the help of data preprocessing process of machine learning. A third important step is the training and testing of data, for which data is distributed into two sub-parts. The fourth step is the selection of appropriate features and models for designing a model and the last step is to check the effectiveness of the model with good appropriate results and effectiveness output of machine learning algorithms, now it is used in various fields such as education, medicine, transport, banking, finance etc (Mahadevan, et al.,2009). It provides accurate results without error and handles complex and large datasets easily. In the last few years, machine learning has performed a tremendous output in the transport sector and automatically identifies patterns and their relationships in traffic data and uses that for to make future predictions for traffic conditions. The smart transport system manages transport intelligently and reduces traffic congestion, saving money, time, fuel and battery consumption. Machine learning is further subdivided into three subparts –Supervised, Unsupervised and reinforcement learning (Andreoletti,D et al.,2019). The flow diagram of ML model Is given in figure1.

Figure 1. ML Model flow diagram.

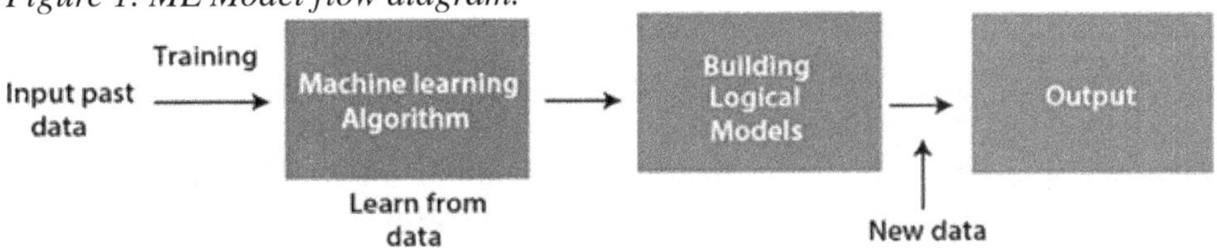

Supervised Learning works on labelled datasets and the model is trained on available datasets and predicts output. This model is generally used for mapping input and output data. Supervised algorithms are mostly used for to predict both discrete and continuous output. In the case of continuous output prediction, the model is designed or trained for the prediction of continuous output values. The value of the output target feature is not fixed; it decreases and increases with a change in the value of the input variable, for example, home price prediction, salary, age, etc. In continuous, all input values are labelled with some feature, and correct output data is predicted. The model is prepared on trained data and, with the help of an algorithm and machine, predicts the accurate output correctly. In another case, the model predicts discrete output, the value of the target feature is fixed and does not change with input variables. The value of the target(output) variable is discrete, and it is classified as whether the output value is true or false. Supervised learning is further divided into two categories based on output feature values - regression (continuous output) and classification (discrete output) (Rau, F et al.,2021).

Regression algorithms are supervised algorithms that work on labelled data and predict the continuous output. It is generally used to find the relation between dependent and independent variables. The size of independent variables depends upon the dataset and can vary from one feature to a number of features. The size of feature variables can be reduced with the help of feature reduction techniques like PCA, ANOVA, and Pearson Correlation coefficient. A large selection of features sometimes creates the problem of overfitting (Liu, Z. et al.,2021). Dependent variable values depend upon independent variables and change accordingly with a change in the value of independent variables. The regression model generally creates a best-fit line for all data points and predicts output values. The accuracy of the model is increased if more numbers of data points lie on the best-fit line and decreases if fewer data points lie on the best-fit line (Singh, R et al.,2105). These algorithms are mostly used for Weather forecasting, prediction of house prices, Market Trends, etc. Linear Regression and Multiple Regression algorithms are some popular regression algorithms. Regression models use previous historical traffic data for the prediction of future traffic conditions. Classification algorithms are also supervised algorithms that all work on labelled datasets and are generally used for predicting discrete or categorical output. This algorithm classified input data into appropriate different output classes. The output values may be binary classifications or multilevel classifications. In binary classification, the output class is distributed into two output values like No-Yes, false-true, 1 or 0, etc. Spam Filtering and disease prediction are some examples of classification algorithms. Random Forests, Decision Trees, LR, and SVM are some more demanded algorithms of classification (Wu, J et al.,2012). In multilevel classification, input values are classified into multiple classes (more than two classes). In multilevel classification, output classification may be 0,1 or 2, low, medium, high etc.

Unsupervised learning: It is also a sub-division of machine learning that uses unlabeled datasets, data are not arranged in order. The unsupervised learning maps the input data according to similarities of the available data sets and predicts the output. It is of two types: clustering and Association. Clustering is an unsupervised algorithm that maps the input into a group of similar data values. The groups are represented by the cluster objects of the same similarity are mapped into one cluster, and different similarities are mapped into another cluster; there is no similarity between a group of clusters and objects of groups. Cluster generally finds commonalities among data objects and categorizes them into respective clusters on basis of similar commonalities (Mao, Q et al.,2018).

The second method is generally applied in large datasets to find relationships among items and try to find a list of items that occur simultaneously in the dataset and apply association policy. Such as a person buys A items (suppose a written book)) are supposed to purchase B (Pens) items. some more frequently used algorithms are K-means, KNN, and Hierarchal model, and for large datasets, the most demanded methods are Anomaly detection and Neural Networks; the Book chapter emphasizes the uses of various classification algorithms for machine learning for the prediction of real-time traffic and minimizing battery consumption of EVs. As mentioned above the classification algorithms are used for predicting the categorical outputs either in yes or no. The traffic light is red or green, depending upon the available traffic data inputs (Hou, Y et al., 2022).

Reinforcement Learning: It is worked on agent system and feedback-based learning approaches, and on each right decision, a positive reward is given to the agent. In this approach, the agent learns or understands the merits or demerits by interacting with the nearby environment and, on the basis of the learning experience, makes the decision. The agent updates itself automatically from the environment and improves its score and efficiency through feedback, and the main task of the agent is to improve its efficiency by acquiring a maximum number of reward points. Robot systems work on reinforcement learning and improve efficiency by automatically learning from nearby environments (Huang, S.C et al., 2018).

PROPOSED WORK

Objectives

As we see the existing traffic system works on the old methodology of traffic management for managing the traffic load. The traffic light timings are fixed and more time-consuming. If there is no traffic congestion, traffic has to still wait until the light is not green. The more fuel or battery consumption in the existing system and in the case of EVs battery consumption is a more crucial factor. As we know, populations and vehicles are increasing at a very fast growth rate, and it is very difficult to manage the increasing traffic load with the existing traffic management system. The proposed transport traffic management system overcomes all the problems of existing traffic light management. Battery consumption is a more crucial factor in EVs and therefore, need a smart traffic management system that helps in increasing the battery life of EVs. Smart transportation systems use ICT (information and communication technology) in new transformation systems that help provide safety to drivers and increase efficiency with the help of advanced information technology systems and communication sensors(Xiao, C.,2018).

The Proposed smart transport management system manages the traffic light by new advanced approach that manages traffic light timing based on the actual traffic data in place of fixed timing. If the traffic load is less, the driver does not have to wait more time. The Smart transport management system also manages the traffic light according to the time of day. In India, the traffic load is very high in the morning as compared to the evening time, and in the proposed system, traffic lights operate accordingly. Sometimes, the conditions o f weather have a high impact on traffic management Systems. It has been observed in India, especially in the Summer season, that temperature is very high between 12: 00 AM and 3:00 PM and traffic load is observed very less at that particular time due to the hot temperature(Zhang, X. et al., 2021). The research focus is to provide a better solution to improve the traffic management system and reduce fuel consumption, and battery consumption, saving money and time. The smart traffic management system improves existing transportation systems by minimizing traffic congestion and providing safety to drivers. The objective of the book chapter is to reduce the battery consumption of EVs by handling transport traffic and designing a machine learning algorithm that manages the traffic light by analyzing traffic patterns and congestion hotspots and optimizing traffic light timings dynamically based on real traffic. This significantly helps in improving traffic flow, reducing congestion, and saving battery consumption of EVs (Impedovo, D et al., 2019).

The key objectives of the proposed traffic management system for managing the traffic light are:

Improve battery consumption of EVs by reducing traffic congestion

Manage traffic lights based on real-time traffic in place of fixed traffic time.

Manage the traffic light with timing of day.

Manage the traffic light with days of week.

Manage Traffic Light with road conditions.

To design an intelligent traffic management system, the first step is the requirement of collection of appropriate traffic data for the specified location where we want to implement the proposed traffic management system. The proposed model used a machine learning classification algorithm for implementing the intelligent traffic management method that accurately predicts output. The traffic light is operated with the help of the classification technique. Working flow of designed model is given below in figure 2(Sha, S et al., 2020).

Figure 2. Working diagram

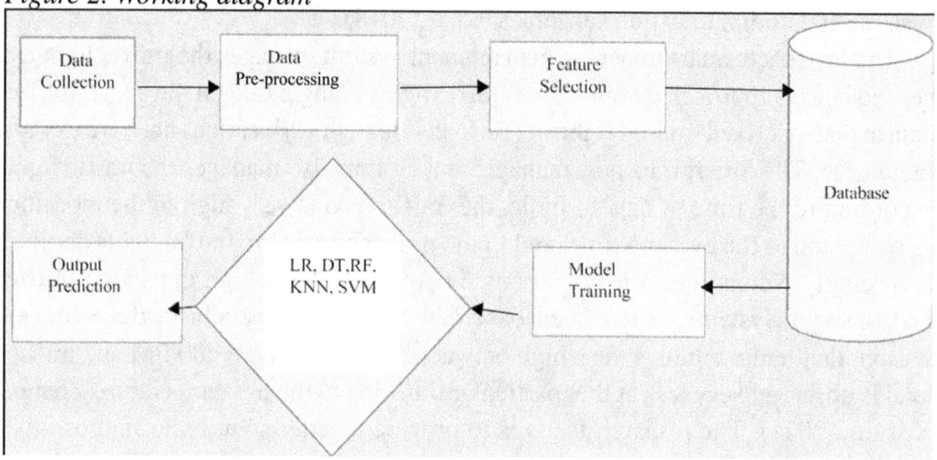

Data Collection

The collection of data is the first primary step for designing an efficient working model. Authenticity and accuracy of data play a very crucial role in achieving good accuracy or results from designed predicted models. The data set of Traffic is collected from the Kaggle online repository and consists of a data size of 2975 rows and 9 columns. The dataset contains nine feature columns, and out of these nine columns features, eight are input key features and one 'Traffic Situation' is output (target) key feature. The data set contains the hourly base counts of four vehicles (bike, bus, car and truck) along with the time and days of the week that are passed from junction points. On the basis of the total number of vehicles, the traffic load is distributed into four categories: Low, Normal, Heavy, and High(Zeng, C et al., 2022). The Description of the dataset with all features is given in Figure 3.

Figure 3. Description of Data Set

```
a.info()

<class 'pandas.core.frame.DataFrame'>
RangeIndex: 2976 entries, 0 to 2975
Data columns (total 9 columns):
 #   Column               Non-Null Count  Dtype
---  ------               --------------  -----
 0   Time                 2976 non-null   object
 1   Date                 2976 non-null   int64
 2   Day of the week      2976 non-null   object
 3   CarCount             2976 non-null   int64
 4   BikeCount            2976 non-null   int64
 5   BusCount             2976 non-null   int64
 6   TruckCount           2976 non-null   int64
 7   Total                2976 non-null   int64
 8   Traffic Situation    2976 non-null   object
dtypes: int64(6), object(3)
memory usage: 209.4+ KB
```

Out of eight input columns and one output column, two features values are in categorical form and remaining seven features values are in numerical form. The target output contains the categorical data with four variables value (low, normal, heavy and high) and input variable 'days of week' has seven categorical values. Columns variables (Features description) values information is given in Figure 4(Khan, Z et al., 2019).

Figure 4. Feature Description

```
[1]: import numpy as np
     import pandas as pd
     a=pd.read_csv("desktop/Traffic1.csv")
     a.head(10)
```

[1]:

	Time	Date	Day of the week	CarCount	BikeCount	BusCount	TruckCount	Total	Traffic Situation
0	12:00:00 AM	10	Tuesday	31	0	4	4	39	low
1	12:15:00 AM	10	Tuesday	49	0	3	3	55	low
2	12:30:00 AM	10	Tuesday	46	0	3	6	55	low
3	12:45:00 AM	10	Tuesday	51	0	2	5	58	low
4	1:00:00 AM	10	Tuesday	57	6	15	16	94	normal

Feature Engineering and Data Exploratory Analysis

EDA is the second most important step for designing a model and it helps data scientists to look at the data before making any assumptions. It is generally used to discover patterns from data, provide a better understanding of data variables and their relationship with each other and also help for identifying obvious errors, and detecting outliers or anomalous events. The main components of EDA are feature engineering and data encoding (Chui, K.T et al .,2021).

Feature Engineering

Feature Engineering is a technique that is used to improve the effectiveness of a model using creating new features from available features by selecting relevant information from given raw data. The main purpose of feature engineering is to enhance model accuracy by providing more meaningful and relevant information. As we know, success of any machine learning model highly depends upon the quality of its features. From the Date Time feature of a given dataset, five new features are generated such as year, month, date in a given month, days of week and hour. In a given dataset, target output values are in the form of categorical values, but when we have to design the model, we need to convert these variables' values into numeric form with the help of data encoding. The total values numbers of target variables normal, heavy, high and low are 1669,682,321 respectively (Ahmed, K.M.U et al., 2021). The description of the target variable is given in Figure 5.

Figure 5. Description of Target values

```
a["Traffic Situation"].unique()

array(['low', 'normal', 'heavy', 'high'], dtype=object)

a["Traffic Situation"].value_counts()

Traffic Situation
normal      1669
heavy        682
high         321
low          304
Name: count, dtype: int64
```

Data Encoding

Data Encoding is the most useful technique for designing classification-supervised machine learning algorithms. The classification algorithms mostly work on discrete output values. Sometimes, target output variables are in the form of categorical values, but we need the target output in the form of numeric values. In this case, we have to convert categorical variable values into numerical values. Data encoding is used to convert categorical variable values into numerical values. The proposed book chapter used the label encoding approach for encoding the data (Andrae A. et al.,2015). Figure 6 represents the different types of data encoding approaches that are used in machine learning.

Figure 6. Types of Data Encoding in ML

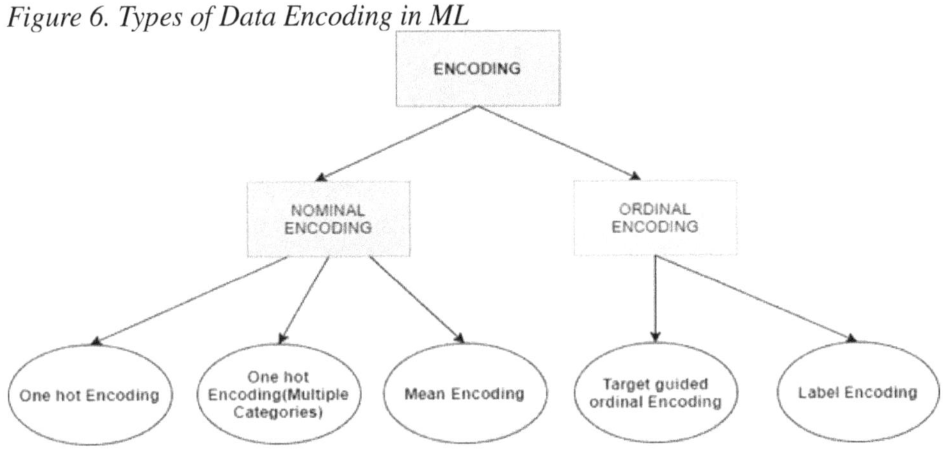

Label Encoding

Label encoding is generally applied for encoding of target output features, and it is also used to convert target (output) categorical values into numerical values, and its range starts from zero to one minus the total number of classes. For example, categorical variables have five different classes, and they use values 0,1,2,3,4 (Addis, B. et al.,2014). In the available dataset, four targets (output) categorical variables are normal, heavy, high and low and all four are converted into numerical values as given in Table 1. as:

Table 1. Label Encoding

S.No	Target variable Types	Numeric Values
01	Normal	1669
02	Heavy	682
03	High	321
04	Low	304

The practical implementation of Label encoding is given in Figure 7.

Figure 7. Practical Implementation of Label Encoding

```
a["Traffic Situation"].unique()

array([2, 3, 0, 1])

a["Traffic Situation"].value_counts()

Traffic Situation
3      1669
0       682
1       321
2       304
Name: count, dtype: int64
```

After the data encoding, the next step is to prepare the final dataset for the creation of the model. The available dataset is now ready for model creation, and all the values of target features in the available dataset are numeric (Fisher, W et al.,2010). The final dataset after date encoding is shown in Figure 8.

Figure 8. Final Dataset

```
from sklearn.preprocessing import LabelEncoder
le = LabelEncoder()
a["Traffic Situation"]=le.fit_transform(a["Traffic Situation"])
a.head(10)
```

	Time	Date	Day of the week	CarCount	BikeCount	BusCount	TruckCount	Total	Traffic Situation
0	12:00:00 AM	10	Tuesday	31	0	4	4	39	2
1	12:15:00 AM	10	Tuesday	49	0	3	3	55	2
2	12:30:00 AM	10	Tuesday	46	0	3	6	55	2
3	12:45:00 AM	10	Tuesday	51	0	2	5	58	2
4	1:00:00 AM	10	Tuesday	57	6	15	16	94	3
5	1:15:00 AM	10	Tuesday	44	0	5	4	53	2

Data Pre-processing

Due to the versatility of different data formats, it is necessary to pre-process or filter available row input values and prepare them for designing a model. It analyzed or observed that desired data is not always cleaned or formatted format, but it is mandatory to provide clean data to the model for achieving or increasing good accuracy and efficiency of the model. The data preprocessing process of machine learning performs two tasks: handling missing values and scaling of data. Sometimes, large numbers of values are missing from an input data set, and if we do not replace/remove missing values, it impacts the accuracy of the designed model (Bianzino, A.P et al.,2012). The main tasks of data preprocessing are to provide clean data, remove noise and duplicate values from the dataset before feeding it to the model and help to increase the efficiency of the model. For handling missing values, two approaches are generally used in machine learning that depend upon the size of the dataset. In the first approach, if the size of the available input dataset is very large, simply delete the missing value from the input dataset, and there is no high impact on the accuracy of the model. In the second approach, if the size of the input available dataset is small, we fill the missing value space with a mean value of the respective column. The scaling method of data preprocessing is used to scale the values of columns. Sometimes, it has been found that the variation between the values of feature columns is much more (Bochkovskiy et al.,2020).

Some of the values in columns are found to be very high, and some of the values are found to be very low in another column. Variation of values varies with high dimensions in columns. So, if we designed the model on these high variation values, the accuracy of the model be reduced. The scaling method of data pre-processing helps to reduce or convert the high-dimension values into low-dimension values. It scales values of high dimension into low-dimension variables varying with values between 0 and 1 value(Singh, R. et al.,2024). During data pre-processing of the traffic dataset, it has been found that in the available input dataset, no missing values are present in any column features. After a thorough analysis of the dataset, it has been observed that there is no high variation among the values of features and no requirement of scaling (Goel, A. et al., 2023). Figure 9 represents the information on missing values, and, in the available given dataset, has no missing values.

Figure 9. Representation of Missing Value

```
a.isna()
```

	Time	Date	Day of the week	CarCount	BikeCount	BusCount	TruckCount	Total	Traffic Situation
0	False	False	False	False	False	False	False	False	False
1	False	False	False	False	False	False	False	False	False
2	False	False	False	False	False	False	False	False	False
3	False	False	False	False	False	False	False	False	False
4	False	False	False	False	False	False	False	False	False
...	...	...	...	...	...	...	...	...	...
2971	False	False	False	False	False	False	False	False	False
2972	False	False	False	False	False	False	False	False	False
2973	False	False	False	False	False	False	False	False	False
2974	False	False	False	False	False	False	False	False	False
2975	False	False	False	False	False	False	False	False	False

Data Visualization

Data visualization of any data set helps us to understand or analyze information about outliers, sense of data patterns and relationships among variables. Sometimes, with the help of data visualization, we will reach a conclusion on model selection, which model is best for the desired dataset. We can also easily find out the outlier and insight in data can be easily interpreted. Various types of data visualization techniques are available in machine learning, such as line charts, bar charts, scatter plots, etc. Figure 10 shows the different types of charts and plots of all four selected key input variables (Garg, H et al.,2022).

Figure 10. Plot diagram

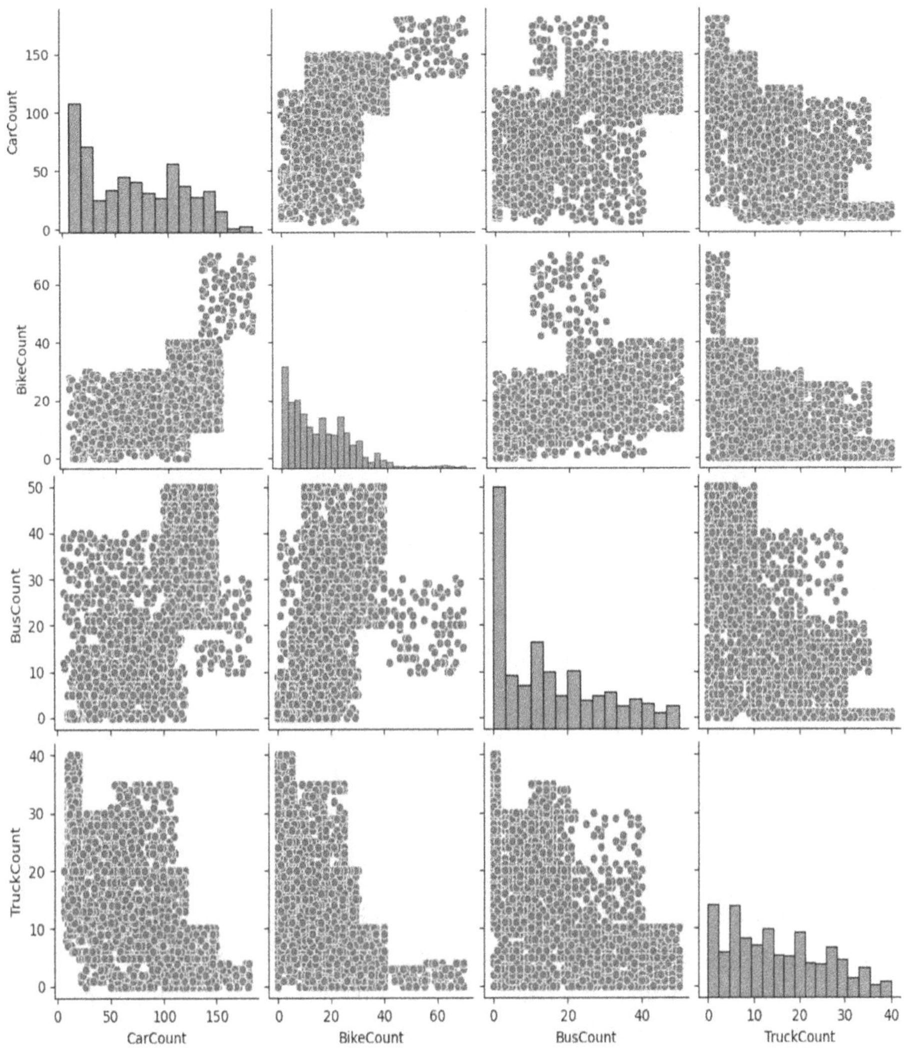

Dimensionality Reduction

Dimensionality Reduction in machine learning consists of feature selection and extraction techniques. This technique is help for reducing the size or dimension of features and due to which less space is required to store datasets. The second most significant benefit of the dimensionality reduction technique is training time is also reduced because if the size of features is reduced, obviously the size of the data point is also reduced, and time is required to process the training data (K. Makar et al., 2021). A third most important benefit of dimensionality reduction is that it has become easy to visualize datasets. If the size is small, it becomes very easy to visualize the dataset easily, and it can help designers with model selection, and it also helps for reducing overfitting problems. The proposed model used the feature selection approach for dimensionality reduction (P. Chaudhary et al., 2021).

Features Extraction

Features extraction is generally used for producing new features by adding or combining or transforming old or original features. The main task is to create a set of new features by combining new features and help in reducing space for data storage, minimizing processing training time. Several methods are used in machine learning for the extraction of features and some common methods are principal component analysis (PCA), Kernal PCA, Quadratic Discriminant Analysis etc. Principal component analysis(PCA) is most demanded or very used technique for features selection that helps to transfer original features onto a lower-dimensional space (S. Mishra et al, 2020).

Feature Selection

Feature selection is an important step in achieving high accuracy with the designed model. The accuracy and effectiveness of the model are highly dependent upon the selection of the right features, and the wrong selection of features may increase the probability of both underfitting and overfitting problems. ANOVA, LDA, Pearson Correlation, PCA, etc., are some of the most common feature extraction and feature selection methods (Javaid S et al.,2018). For the given dataset, we used a correlation technique for feature selection. The heat map technique is applied for key feature selection, which was worked on using the Pearson correlation technique. Heat Map generally checks the dependency among the feature variables and selects the highly dependent variables based on their dependency values. The dependency value of the heat map varies between zero and one. If features are highly correlated to each other, their dependency value must be high and equal to one. If features are not correlated to each other, dependency value is equal to zero. Based on their dependency values, four features (Car Count, Bike Count, Bus Count and Truck Count) are selected. Figure 11 shows a heat map representing the four above features and their dependency values(Kanungo A et al., 2014).

Heat Map Dependency Value = 1 (Features are highly correlated to each other)
Heat Map Dependency Value = 0 (No relation among Features Variables).

Figure 11. Heat Map

```
<Axes: >
```

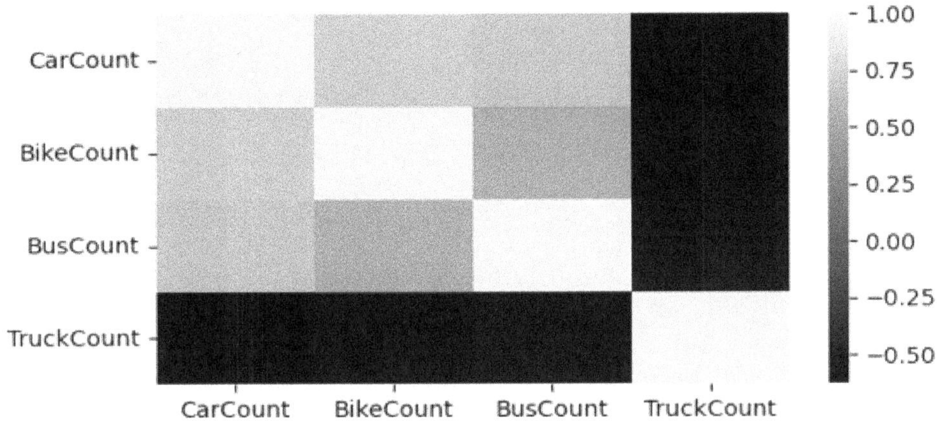

Now, with the help of heat map techniques, four features are selected as input critical features based on their dependency values, and these four features are Car Count, Bike Count, Bus Count and Truck Count. All remaining features that are less dependent, or dependency is very poor are dropped or removed from the given data set. Figure 12 represents the final input selected key features (Keerthi R. et al., 2016).

Figure 12. Selected Key Input features

```
b=a.drop(['Time','Date','Day of the week','Total','Traffic Situation'], axis=1)
b.head()
```

	CarCount	BikeCount	BusCount	TruckCount
0	31	0	4	4
1	49	0	3	3
2	46	0	3	6
3	51	0	2	5
4	57	6	15	16

As discussed above, our output or target feature is 'Traffic Situation', and its value lies in the range from 0 to 3, as mentioned in Table 1. Now, we have to design the model that predicts the target output (traffic either normal, heavy, high or low) based on values of four key input values. The designed model helps to identify the real-time condition of traffic based on key input values, and the traffic light is performed or operated according to real-time traffic situations. This proposed model helps reduce traffic load and save extra fuel and battery consumption in the case of EVs. Figure 13 shows the value of the target output (Lach D. et al.,2018).

Figure 13. Target output type

```
c=a['Traffic Situation']
c.head(10)
```

```
0    2
1    2
2    2
3    2
4    3
5    2
6    2
7    2
8    2
9    2
Name: Traffic Situation, dtype: int32
```

Data Training

Data training is one of the most important steps in designing an accurate model. It is generally applied to check t h e effectiveness of the model and helps make the right decision. In data training, we divide the entire dataset into two subparts (Large and small subparts). A larger subpart is used for preparing or training the model, and a small subpart is used to check the existence or effectiveness of the designed model. The ratio of the division of data depends upon the size of the data, and there is no standard method designed for the distribution of the dataset into training or testing data sets.

We have to only supply the maximum data value to the model for training to achieve the best accurate result; if the model is prepared on large values, it always predicts accurate results; if the data set row size is not too large then decrease t h e content of test data and increase the volume of trained data. In this case, simply use the ratio of (90% for training and 10% for testing)(Mangiaracina R et al., 2017). On the other hand, if the size is large, we can use 80: 20 ratios. In the proposed work, the data distribution among training and testing with ratios of 90%and 10%. To achieve the best and most accurate accuracy, the proposed model used k-fold cross-validation methods, and we used a value of k equal to five for predicting output values. The accuracy and effectiveness of the model are checked by using testing values. The testing values are given to the designed model, and the predicted value of the designed model is compared to the real output values of testing data. If both values are almost similar, it means our model predicts a good score, and on the other hand, if the difference between both values is high, it means our model does not produce the right accurate result. Figure 14 represents the distribution of the dataset in training and testing where test size 0.1 indicates the size of the test data set is 10%(Pashupatimath S et al., 2018).

Figure 14. Training & Testing

```
from sklearn.model_selection import train_test_split
b_train, b_test, c_train, c_test = train_test_split(b, c, test_size=0.1)
```

Model Selection

Model selection is the last step for designing of model. The accuracy and efficiency of models vary according to the different classification algorithms. The proposed book chapter used various classification algorithms to predict the traffic load, and all these models helped to identify the real-time traffic load. Five different classification algorithms of supervised machine learning are used to predict the traffic load on the basis of traffic, and the accuracy of all models differ from each other(Ramakrishnan D et al.,2022).

Linear regression: is a very simple and most demanded and frequently used classification method for small datasets. Linear regression predicts the categorical output values by using input variables and predicts the output values. The output values are in the form of binary classification, and Its values may be zero or 1, right and wrong, true or False, etc. The linear regression algorithm generally describes the linear relationship between input features and output features. Therefore, it is called linear regression. The value of the output feature is linearly dependent on input variables. The value of the output variable changes according to the value of the input variables. If the value of the input variable increases, the value of the output variable increases, and if the value of the input variable decreases, the value of the output variable decreases. Linear regression algorithm l provides straight-line slopes that represent the linear relationship between input and output variables and the efficiency of the model was 0.63%. Figure 15 Shows Implementation of Linear Regression (Rodríguez T et al., 2010).

Figure 15. Implementation of Linear Regression

```
from sklearn.linear_model import LinearRegression
reg=LinearRegression()
reg.fit(b,c)
LinearRegression()
reg.predict([[30,20,15,10]])

C:\Users\Gyan\anaconda3\Lib\site-packages\sklearn\base.py:493:
ture names
  warnings.warn(
array([2.4456207])

reg.score(b,c)

0.6333236171828326
```

Support Vector Machine: SVM is the most demanded and useful classification algorithm for the prediction of output. It performs good results for small datasets and it is suitable for solving both types of problems of classification and regression. It is more preferred as compared to KNN and linear regression for the prediction of classification output(Salama A et al.,2010). .But mostly used for solving classification problems. The SVM model forms best decision fit line that also segregates n- n- dimensional space points into single classes, which helps for new data points to easily be put into the correct category. The best fit decision line is represented by a hyperplane and vector points are supported to create the hyperplane. Therefore, this model is called a support vector machine and its efficiency was 100%.on available datasets. Figure 16 represents Implementation of the SVM model(Sankaran S., et al., 2022).

Figure 16. Implementation of SVM

```
from sklearn.model_selection import train_test_split
b_train, b_test, c_train, c_test = train_test_split(b, c, test_size=0.1)
m=SVC(gamma=5)
m.fit(b,c)
reg.fit(b,c)
m.predict([[30,20,15,10]])

C:\Users\Gyan\anaconda3\Lib\site-packages\sklearn\base.py:493: UserWarning:
  warnings.warn(
array([3])

m.score(b,c)

1.0
```

K-Nearest Neighbor: is also the simplest classification algorithm that is also a very good model for the classification of output for small datasets(Shan Z et al ., 2015). This algorithm finds a similarity between t h e new input value and given available values. It classifies given new input values into the corresponding classes(Swathi K et al., 2016).

Figure 17. Implementation of KNN

```
from sklearn.model_selection import train_test_split
b_train, b_test, c_train, c_test = train_test_split(b, c, test_size=0.1)
neigh = KNeighborsClassifier(n_neighbors=3)
neigh.fit(b,c)
neigh.predict([[30,20,15,10]])

C:\Users\Gyan\anaconda3\Lib\site-packages\sklearn\base.py:493: UserWarning:
feature names
  warnings.warn(
neigh.score(b,c)

0.9747983870967742
```

The algorithm is also represented as a lazy learner because it performs action given dataset only at time of training. It is one of the best algorithms for small datasets and performs with good accuracy and predicts the best output result(Gupta. M et al.,2023).New input data value is easily organized into an appropriate category by using K- NN method and the efficiency of the model was scored as 0.97%. Figure 17 represents the Implementation of the KNN model (Singh. M et al,2023).

Decision Tree: is a more popular and demanded classification model that is also used to predict both discrete and continuous output but it is mostly used for solving Classification tasks. This is one of the good models of classification-supervised machine learning for handling large datasets (Sharma. A. et al.,2022). From its name, it has a tree-like structure and it makes decisions like a human brain, in which internal nodes of the tree represent about dataset's features and case of branches that are used to represent about decision rules. Leaf nodes are used to represent the outcome. To create a decision tree, the CART

) algorithm is used and it takes a decision based on the answer yes/no and the efficiency of the model was 100%Figure 18 represents the Implementation of the DT model(Singh.M et al.2021).

Figure 18. Implementation of Decision Tree

```
from sklearn import tree
Dtree=tree.DecisionTreeClassifier()
Dtree.fit(b,c)
Dtree.predict([[30,20,15,10]])

C:\Users\Gyan\anaconda3\Lib\site-packages\sklearn\base.py:493:
th feature names
  warnings.warn(

array([2])

Dtree.score(b,c)

1.0
```

Random Forest: is the most popular and demanded classification algorithm for large datasets. It can also be used for solving Classification as well as regression problems. It applies an ensemble approach or technique to solve complex problems by combining multiple classifiers and predicting the output based on the result of multiple classifiers. Random Forest has several decision trees for a given dataset and predicts and improves accuracy by using the average value of all decision trees, instead of calculating the predicted output value of one selected decision tree (Singh et al., 2024).

Figure 19. Implementation of the RF model

```
from sklearn.ensemble import RandomForestClassifier
rf=RandomForestClassifier(n_estimators=10,n_jobs=-1)
rf.fit(b,c)
rf.predict([[30,20,15,10]])

C:\Users\Gyan\anaconda3\Lib\site-packages\sklearn\base.py:493:
th feature names
  warnings.warn(
array([2])

rf.score(b,c)

0.9983198924731183
```

This algorithm takes the output values of each tree, and majority votes decide the predictive output.

If the quantity of trees increases, the model predicts high accuracy, and the efficiency of the model is 0.99%. Figure 19 represents the Implementation of the RF algorithm(Jain et al.,2024).

RESULT ANALYSIS AND FUTURE WORK

Selection of the best accurate model for the prediction of traffic lights finalized with the help of the accuracy of the model. Model accuracy helps in selecting the best approach for traffic prediction. K-Flod cross-validation technique is also applied to achieve the best accurate result from all classification algorithms and the high accuracy model predicts a good accurate result as compared to other models.

The accuracy of decision tree and SVM models is high as compared to other models with a scored value of 100% and the Linear regression model has a minimum accuracy of 0.63%. The accuracy comparison analysis is given in Table 2 and a graphical representation of accuracy is shown in Figure 20.

Table 2. Accuracy comparison analysis

S.No	Models	Accuracy (%)	Traffic Prediction Numeric Values	Traffic Prediction Categorical Values
01	Linear Regression	0.63	2	Low Traffic
02	KNN	0.97	2	Low Traffic
03	SVM	100	3	Normal Traffic
04	Decision Tree	100	2	Low Traffic
05	Random Forest	0.99	2	Low Traffic

Figure 20. Graphical representation of Accuracy score

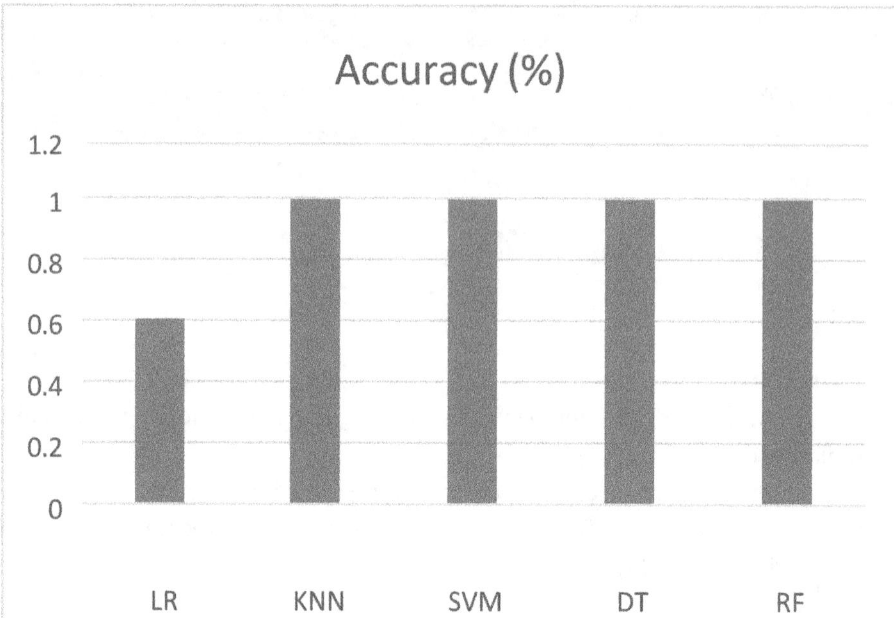

For designing a model, when we feed the four features Car Count, Bike Count, Bus Count and Truck Count variable values as 30,20,15,10. The traffic prediction of all models is 'Low Traffic 'except the SVM model. Traffic prediction of the SVM model on a given input value is 'Normal Traffic'. It has been analyzed that the advancement of artificial intelligence systems, especially in the transport sector, improves or helps in managing traffic load by adjusting traffic lights based on real traffic conditions. The proposed model identifies traffic load conditions based on the number count of vehicles and traffic lights operating according to real-time traffic load that helps in reducing the power consumption of EVs . As we see in a case study of traffic prediction, various machine learning models are used to predict traffic load and decision tree and the SVM model provided the best accuracy score among all models. The main motive of the book chapter is to find valuable insights into the potential applications of artificial intelligence in the field of the transport sector.

The successful implementation of a traffic management system needs proper planning, and a huge investment is required. We have to focus on or understand the needs and challenges of existing traffic systems. Various factors such as population, infrastructure of road, and traffic volume must be considered before implementing a new traffic management system. The new traffic system must be easily adopted or flexible in changing traffic conditions and with evolving technologies. Finally, the new traffic management system has capabilities to manage urban mobility and to provide safe and sustainable journeys. The proposed model was model trained on a small dataset; in future, we can train our model on a large dataset, which can help in improving the efficiency of the model. Large data sets can predict more accurate results and we can also train our model on deep learning algorithms and ensemble techniques. Deep learning algorithms perform better results as compared to machine learning algorithms on large datasets. The outcomes of the research is based on a number of counts of real-time vehicles, predict the traffic loads, and adjust the traffic light accordingly. The proposed approach helps in reducing traffic congestion due to which battery consumption of EVs is reduced.

REFERENCES

Adam, A. B. M., Muthanna, M. S. A., Muthanna, A., Nguyen, T. N., & El-Latif, A. A. A. (2023). A. A. A. El-Latif.(2023). Toward smart traffic management with 3D placement optimization in UAV-assisted NOMA IIoT networks. *IEEE Transactions on Intelligent Transportation Systems*, 24(12), 15448–15458. DOI: 10.1109/TITS.2022.3182651

Addis, B., Capone, A., Carello, G., Gianoli, L. G., & Sanso, B. (2014). Energy Management Through Optimized Routing and Device Powering for Greener Communication Networks. *IEEE/ACM Transactions on Networking*, 22(1), 313–325. DOI: 10.1109/TNET.2013.2249667

AditiYadav, Vaishali More, NehaShinde, ManjiriNerurkar, NitinSakhare. (2019). Adaptive Traffic Management System Using IoT and Machine Learning. *Journal of Electrical Technology*, 4, 221–226.

Ahmed, K. M. U., Bollen, M. H. J., & Alvarez, M. (2021). A Review of Data Centers Energy Consumption and Reliability Modeling. *IEEE Access : Practical Innovations, Open Solutions*, 9, 152536–152563. DOI: 10.1109/ACCESS.2021.3125092

Andrae, A., & Edler, T. (2015). On Global Electricity Usage of Communication Technology: Trends to 2030. *Challenges*, 6(1), 117–157. DOI: 10.3390/challe6010117

Badura, S., & Lieskovsky, A. (2010). Intelligent Traffic System: Cooperation of MANET and Image Processing. *1st International Conference on Integrated Intelligent Computing,* Bangalore, 119-123. DOI: 10.1109/ICIIC.2010.41

Bochkovskiy, A., Wang, C., & Liao, H. (2020). Yolov4: Optimal Speed and Accuracy of Object Detection.*arXiv*:2004.10934.

Cao, K., Liu, Y., Meng, G., & Sun, Q. (2020). An overview on edge computing research. *IEEE Access : Practical Innovations, Open Solutions*, 8, 85714–85728. DOI: 10.1109/ACCESS.2020.2991734

Chakraborty, C., Mishra, K., Majhi, S. K., & Bhuyan, H. (2023). Intelligent latency-aware tasks prioritization and offloading strategy in distributed fog-cloud of things. *IEEE Transactions on Industrial Informatics*, 19(2), 2099–2106. DOI: 10.1109/TII.2022.3173899

Chaudhary, P., Goel, S., Jain, P., Singh, M., & Aggarwal, P. K., Anupam.(2021)."The Astounding Relationship: Middleware, Frameworks, and API. 9th *International Conference on Reliability, Infocom Technologies and Optimization (Trends and Future Directions)* (ICRITO), Noida, India, 1-4, https://doi: DOI: 10.1109/ICRITO51393.2021.9596088

Choudhury, N., Matam, R., Mukherjee, M., Lloret, J., & Kalaimannan, E. (2021). NCHR: A non-threshold- based cluster-head rotation scheme for IEEE 802.15.4 cluster-tree networks. *IEEE Internet of Things Journal*, 8(1), 168–178. DOI: 10.1109/JIOT.2020.3003320

Choudhury, N., Matam, R., Mukherjee, M., & Shu, L. (2018). Beacon synchronization and duty-cycling in IEEE 802.15.4 cluster-tree networks: A review. *IEEE Internet of Things Journal*, 5(3), 1765–1788. DOI: 10.1109/JIOT.2018.2827946

Chui. K.T, Gupta. B.B., Liu. R.W, Zhang. X, Vasant. P, Thomas.(2021). J.J. Extended-Range Prediction Model Using NSGA-III Optimized RNN-GRU-LSTM for Driver Stress and Drowsiness. *Sensors*, 6412.

Collotta, M., Pau, G., Scatà, G., & Campisi, T. (2014). A Dynamic Traffic Light Management System Based on Wireless Sensor Networks for the Reduction of the Red-light Running Phenomenon. *Journal of Transport and Telecommunications*, 15(1), 1–11. DOI: 10.2478/ttj-2014-0001

Fisher, W., Suchara, M., & Rexford, J. (2010). Greening backbone networks. In Proceedings of the first*ACM SIGCOMM Workshop on Green Networking*; ACM: New York, NY, USA, 29–34.

García-Magariño, I., Nasralla, M. M., & Nazir, S. (2020). Real-time analysis of online sources for supporting business intelligence illustrated with Bitcoin investments and IoT smart-meter sensors in smart cities. *Electronics (Basel)*, 9(7), 1101. DOI: 10.3390/electronics9071101

García-Magariño, M. M. Nasralla,J. Lloret. (2021). 'A repository of method fragments for agent-oriented development of learning-based edge computing systems. *IEEE Network.*, vol.

Garg, H., Singh, M., Sharma, V., & Agarwal, M. (2022). Decentralized Application (DAPP) to enable E- voting system using Blockchain Technology. Second *International Conference on Computer Science,Engineering and Applications (ICCSEA),-* DOI: 10.1109/ICCSEA54677.2022.9936413

George, A. M., George, V. I., & George, M. A. (2018, March). IoT based smart traffic light control system. In *2018 International conference on control, power, communication and computing technologies (ICCPCCT)* (pp. 148-151). IEEE.

Ghazal, B., ElKhatib, K., Chahine, K., & Kherfan, M. (2016). Smart Traffic Light Control System. Proceedings of the *3rd International Conference on Electrical, Electronics, Computer Engineering and their Applications, Beirut,* 140-145.

Goel, A., Singh, M., Gupta, J., & Mangla, N. (2023). Med Card: an innovative way to keep your medical records handy and safe. In Advances in *intelligent systems and computing*,51–60. https://doi.org/DOI: 10.1007/978-981-99-0550-8_4

Gupta, M., Singh, M., Sharma, A., Sukhija, N., Aggarwal, P. K., & Jain, P. (2023). Unification of machine learning and blockchain technology in the healthcare industry. In *Institution of Engineering and Technology eBooks* (pp. 185–206). https://doi.org/DOI: 10.1049/PBHE041E_ch6

Hou. Y, Zheng. X, Han. C, Wei. W. Scherer, R. Połap.(2022). Deep Learning Methods in Short-Term Traffic Prediction.*A Survey. Inf. Technol. Control, 51*, 139–157.

Huang, S. C., & Wu, C. F. (2018). Energy Commodity Price Forecasting with Deep Multiple Kernel Learning. *Energies*, 2018(11), 3029. DOI: 10.3390/en11113029

Ianzino, A. P., Chaudet, C., Rossi, D., & Rougier, J. L. (2012). A Survey of Green Networking Research. *IEEE Communications Surveys and Tutorials*, 14(1), 3–20. DOI: 10.1109/SURV.2011.113010.00106

Islam, M., Shahid, N., Ul Karim, D., Al Mamun, A., & Rhaman, M. (2016). An Efficient Algorithm for Detecting Traffic Congestion and a Framework or Smart Traffic Control System. Proceedings of the *18th International Conference on Advanced Communication Technology, PyeongChang,* p802-807.

Jain, P., Sharma, S., Arora, S., Shokeen, V., Aggarwal, P. K., & Singh, M. (2024). Edge Computing-Based design for IoT security. In *Chapman and Hall/CRC eBooks* (pp. 298–309). DOI: 10.1201/9781003405535-22

Javaid, S., Sufian, A., Pervaiz, S., & Tanveer, M. (2018).Smart Traffic Management System Using Internet of Things. in Proceedings of the 20th *International Conference on Advanced Communication Technology, Chuncheon,* 393-398.

Junior, R. R. R., Vieira, M. A. M., Vieira, L. F. M., & Loureiro, A. A. F. (2022). Loureiro, A.A.F. Intra and inter-flow link aggregation in SDN. *Telecommunication Systems*, 79(1), 95–107. DOI: 10.1007/s11235-021-00841-7

Kaleem, S., Sohail, A., Tariq, M. U., & Asim, M. (2023). An improved big data analytics architecture using federated learning for Io Tenabled urban intelligent transportation systems. *IEEE Access : Practical Innovations, Open Solutions.*

Kanungo, A., Sharma, A., & Singla, C. (2014). Smart Traffic Lights Switching and Traffic Density Calculation Using Video Processing. in Proceedings of the *Recent Advances in Engineering and Computational Sciences*, Chadigarh, 1-6.

Keerthi. R, HariharaGopalan. S.(2016).A Survey on Various Traffic Management Schemes for Traffic Clearance, Stolen Vehicle and Emergency Vehicle. *International Journal of Emerging Technology in Computer Science and Electronics,* vol. 19, no. 2, 46-50.

Khan, F., Ullah, F., Kaleem, Z., Ur Rahman, S., Anwar, H., & Cho, Y.-Z. (2018). EVP-STC: Emergency vehicle priority and self-organising traffic control at intersections using Internet-of-Things platform. *IEEE Access : Practical Innovations, Open Solutions*, 6, 68242–68254. DOI: 10.1109/ACCESS.2018.2879644

Khan, Z., Khan, S., Dey, K., & Chowdhury, M. (2019). Development and Evaluation of Recurrent Neural Network-Based Models for Hourly Traffic Volume and Annual Average Daily Traffic Prediction. *Transportation Research Record: Journal of the Transportation Research Board*, 2673(7), 489–503. DOI: 10.1177/0361198119849059

Khan, Z., Koubaa, A., Benjdira, B., & Boulila, W. (2023). A game theory approach for smart traffic management. *Computers & Electrical Engineering*, 110, 108825. DOI: 10.1016/j.compeleceng.2023.108825

Lach, D., & Macioszek, E. (2018). Analysis of Selected ITS Services Used in Public Collective Transport. *WUT Journal of Transportation Engineering*, 121(1), 211–222. DOI: 10.5604/01.3001.0014.4596

Liu, Z., Zhu, Z., Gao, J., & Xu, C. (2021). Forecast Methods for Time Series Data: A Survey. *IEEE Access : Practical Innovations, Open Solutions*, 9, 91896–91912. DOI: 10.1109/ACCESS.2021.3091162

Mahadevan, P., Sharma, P., Banerjee, S., & Ranganathan, P. (2009). A power benchmarking framework for network devices. In Lecture Notes in Computer Science (including Subseries Lecture Notes in Artificial Intelligence and Lecture Notes in Bioinformatics). Proceedings of the *8th International IFIP-TC 6 Networking Conference, Aachen, Germany*, 795–808. Andreoletti, D. Troia, S., Musumeci. F, Giordano. S., Maier. G, Tornatore.M.(2019). Network Traffic Prediction based on Diffusion Convolutional recurrent neural networks. In Proceedings of the *IEEE INFOCOM 2019—IEEE Conference on Computer Communications Workshops (INFOCOM WKSHPS), Paris, France, IEEE: Piscataway*, NJ, USA, 246–251. DOI: 10.1007/978-3-642-01399-7_62

Makar, K., Goel, S., Kaur, P., Singh, M., Jain, P., & Aggarwal, P. K. (2021). Reliability of Mobile Applications: A Review and Some Perspectives. *9th International Conference on Reliability, Infocom Technologies and Optimization (Trends and Future Directions)* (ICRITO), Noida, 1-4, https://doi: DOI: 10.1109/ICRITO51393.2021.9596350

Mangiaracina, R., Perego, A., Salvadori, G., & Tumino, A.BMangiaracina. (2017). A Comprehensive View of Intelligent Transport Systems for Urban Smart Mobility. *International Journal of Logistics*, 20(1), 39–52. DOI: 10.1080/13675567.2016.1241220

Mao, Q., Hu, F., & Hao, Q. (2018). Deep Learning for Intelligent Wireless Networks: A Comprehensive Survey. *IEEE Communications Surveys and Tutorials*, 20(4), 2595–2621. DOI: 10.1109/COMST.2018.2846401

Mishra, S., Shukla, A., Arora, S., Kathuria, H., & Singh, M. (2020).Controlling Weather Dependent Tasks Using Random Forest Algorithm. Third *International Conference on Advances in Electronics, Computers and Communications* (ICAECC), Bengaluru, India, 2020, 1-8, https://doi: DOI: 10.1109/ICAECC50550.2020.9339508

Pashupatimath, S., & Madhavanavar, S. (2018). Different Techniques Used in Traffic Control System: An Introduction. *International Journal of Engineering Research & Technology (Ahmedabad)*, 6(3), 1–4.

Pundir, A., Singh, S., Kumar, M., Bafila, A., & Saxena, G. J. (2022). 'Cyberphysical systems enabled transport networks in smart cities: Challenges and enabling technologies of the new mobility era. *IEEE Access : Practical Innovations, Open Solutions*, 10, 16350–16364. DOI: 10.1109/ACCESS.2022.3147323

Ramakrishnan, D., & Radhakrishnan, K. (2022). Applying Deep Convolutional Neural Network (DCNN) Algorithm in the Cloud Autonomous Vehicles Traffic Model. *The International Arab Journal of Information Technology*, 19(2), 186–194. DOI: 10.34028/iajit/19/2/5

Rau, F., Soto, I., Adasme, P., Zabala-Blanco, D., & Azurdia-Meza, C. A. (2021). Network Traffic Prediction Using Online-Sequential Extreme Learning Machine. In Proceedings of the 2021 *Third South American Colloquium on Visible Light Communications (SACVLC), Toledo, Brazil, IEEE: Piscataway*, NJ, USA, 1–6. DOI: 10.1109/SACVLC53127.2021.9652247

Rodríguez, T., & García, N. (2010). An Adaptive, real-time, Traffic Monitoring System. *Machine Vision and Applications*, 21(4), 555–576. DOI: 10.1007/s00138-009-0185-z

Salama, A., Saleh, B., & Eassa, M. (2010).Intelligent Cross Road Traffic Management System (ICRTMS). Proceedings of the ICCTD 2nd *International Conference on Computer Technology and Development, Cairo,* 27-31, 2010.

Sankaran, S. (2022). Pattern Matching Based Vehicle Density Estimation Technique for Traffic Monitoring Systems. *The International Arab Journal of Information Technology*, 19(4), 575–581. DOI: 10.34028/iajit/19/4/1

Sathya, G., & Fathima Shameema, S, Jyothi MolSebastian, Jemsya K S. (2010). Automatic Rescue System for Ambulance and Authoritative Vehicles. *IEEE Access : Practical Innovations, Open Solutions.*

Sha, S., Li, J., Zhang, K., Yang, Z., Wei, Z., Li, X., & Zhu, X. (2020). RNN-Based Subway Passenger Flow Rolling Prediction. *IEEE Access : Practical Innovations, Open Solutions*, 2020(8), 15232–15240. DOI: 10.1109/ACCESS.2020.2964680

Shan, Z., & Zhu, Q. (2015). Camera Location for RealTime Traffic State Estimation in Urban Road Network Using Big GPS Data. *Neurocomputing*, 169, 134–143. DOI: 10.1016/j.neucom.2014.11.093

Shanthamallu, U. S., Spanias, A., Tepedelenlioglu, C., & Stanley, M. (2017). 'A brief survey of machine learning methods and their sensor and IoT applications. *8th International Conference Inf., Intell., Syst. Appl*(IISA), 1–8. DOI: 10.1109/IISA.2017.8316459

Sharma, A., Singh, M., Gupta, M., Sukhija, N., & Aggarwal, P. K. (2022). IoT and blockchain technology in 5G smart healthcare. In *Elsevier eBooks* (pp. 137–161). DOI: 10.1016/B978-0-323-90615-9.00004-9

Shwetha, R. J., Srinath, M., & Vibha, S. (2018). Intelligent traffic signal management system using cloud vision API and Machine learning. [IJERT]. *International Journal of Engineering Research & Technology (Ahmedabad).*

Singh, M., & Malik, A. (2024). Multi-Hop routing protocol in SDN-Based wireless sensor network. In *CRC Press eBooks* (pp. 121–141). DOI: 10.1201/9781003432869-8

Singh, M., Sukhija, N., Sharma, A., Gupta, M., & Aggarwal, P. K. (2021). Security and privacy requirements for IOMT-Based Smart Healthcare System. In *CRC Press eBooks* (pp. 17–37). https://doi.org/DOI: 10.1201/9781003032328-2

Singh, R., Sharma, R., Kumar, K., Singh, M., & Vajpayee, P. (2024). Securing lives and assets: IoT-Based earthquake and fire detection for Real-Time monitoring and safety. In *Communications in computer and information science*,15–25, https://doi.org/DOI: 10.1007/978-3-031-56703-22

Singh. M, Gupta. M, Sharma. A, Aggarwal. Jain.(2023). Role of deep learning in healthcare industry: Limitations, challenges and future scope, Deep learning for healthcare services. In *BENTHAM SCIENCE PUBLISHERS eBooks*. https://doi.org/.DOI: 10.2174/9789815080230123020l

Singh. R, Kumar.H. Singla.(2015). An intrusion detection system using network traffic profiling and an online sequential extreme learning machine. *Expert Syst. Appl,* Vol 42, 8609–8624.

Swathi, K., Sivanagaraju, V., Manikanta, A., & Kumar, S. (2016). Traffic Density Control and Accident Indicator Using WSN. *International Journal for Modern Trends in Science and Technology*, 2(04), 18–21.

Teoh, Y. K., Gill, S. S., & Parlikad, A. K. (2022). IoT and fog-computing-based predictive maintenance model for effective asset management in Industry 4.0 using machine learning. *IEEE Internet of Things Journal*, 10(3), 2087–2094. DOI: 10.1109/JIOT.2021.3050441

Wu, J., & He, Y. (2021). Prediction of GDP in Time Series Data Based on Neural Network Model. In Proceedings of the 2021 *IEEE International Conference on Artificial Intelligence and Industrial Design (AIID), Guangzhou, China, IEEE: Piscataway*, NJ, USA, 20–23. DOI: 10.1109/AIID51893.2021.9456509

Xiao, C., Choi, E., & Sun, J. (2018). Opportunities and challenges in developing deep learning models using electronic health records data: A systematic review. *Journal of the American Medical Informatics Association : JAMIA*, 2018(25), 1419–1428. DOI: 10.1093/jamia/ocy068 PMID: 29893864

Xiong, X., Zheng, K., Lei, L., & Hou, L. (2020). Resource allocation based on deep reinforcement learning in IoT edge computing. *IEEE Journal on Selected Areas in Communications*, 38(6), 1133–1146. DOI: 10.1109/JSAC.2020.2986615

Zeng, C., Ma, C., Wang, K., & Cui, Z. (2022). Parking Occupancy Prediction Method Based on Multi Factors and Stacked GRU-LSTM. *IEEE Access : Practical Innovations, Open Solutions*, 10, 47361–47370. DOI: 10.1109/ACCESS.2022.3171330

Zhang, X., Kuehnelt, H., & De Roeck, W. (2021). Traffic Noise Prediction Applying Multivariate Bi- Directional Recurrent Neural Network. *Applied Sciences (Basel, Switzerland)*, 2021(11), 2714. DOI: 10.3390/app11062714

Zichichi, M., Ferretti, S., & D'angelo, G. (2020). A framework based on distributed ledger technologies for data management and services in intelligent transportation systems. *IEEE Access : Practical Innovations, Open Solutions*, 8, 100384–100402. DOI: 10.1109/ACCESS.2020.2998012

Chapter 9
Optimizing Sustainability:
Harnessing AI for Personalized Electric Vehicle Energy Management

Megha Gupta
https://orcid.org/0000-0002-9188-3966
JSS Academy of Technical Education, India

Pooja Deswal
JSS Academy of Technical Education, India

Vishakha Chauhan
JSS Academy of Technical Education, India

Pawan Kumar
Ajay Kumar Garg Engineering College, Ghaziabad, India

Jaspreet Kaur
JSS University, India

Priya Singh
JSS Academy of Technical Education, India

ABSTRACT

The transition to electric vehicles (EVs) is a pivotal step towards achieving sustainability in transportation. However, maximizing the environmental benefits of EVs relies not only on their adoption but also on efficient energy management strategies. This chapter reviewed the different approach to optimize sustainability by leveraging artificial intelligence (AI) for personalized electric vehicle energy management. Traditional energy management systems for EVs often employ generic, one-size-fits-all approaches, neglecting the diverse needs and behaviours of individual drivers. In contrast, our study harnesses the power of AI to tailor energy management strategies to the specific preferences, driving patterns, and charging behaviours of each EV owner. The core of our approach lies in the utilization of machine learning algorithms to analyse vast amounts of data collected from individual EVs, including

DOI: 10.4018/979-8-3693-6844-2.ch009

driving habits, charging patterns, environmental conditions, and energy prices.

1. INTRODUCTION

The automobile sector globally finds itself at a pivotal juncture, driven by the twin demands of sustainability and the potential of technology advancements. Modern mobility paradigms are being redefined by the convergence of artificial intelligence (AI) and electric vehicles (EVs), which not only provide a greener, smarter alternative to fossil fuel-fuelled transportation but also pave the way for it. This study explains how AI-driven innovations are advancing the development of sustainable transportation by examining the mutually beneficial interaction between electric vehicles and AI (Pardhasaradhi et al., 2015). To maximise the sustainability of EVs and remove the obstacles preventing their widespread adoption, artificial intelligence appears to be a revolutionary force. Artificial intelligence (AI) enhances several aspects of electric mobility through sophisticated data analytics, machine learning algorithms, and predictive modelling. To maximise battery efficiency and improve range prediction accuracy, AI systems examine driving habits, meteorological circumstances, and topographical features as shown in Fig.2. Artificial Intelligence extends battery life and guarantees effective use of energy resources by dynamically modifying energy use. Using real-time data on energy consumption, grid capacity, and renewable energy availability, AI-driven smart charging systems optimize charging station placement, scheduling, and price. This reduces carbon emissions and grid congestion while making it simple for EV owners to charge their vehicles as shown in Fig.1. By utilizing sensor data and past performance indicators, AI-powered predictive maintenance systems track the health of vehicles in real-time and foresee component problems before they happen. AI lowers maintenance costs, increases vehicle lifespan, and minimizes downtime by enabling proactive maintenance interventions. Artificial intelligence has made it possible to build autonomous driving technology, completely changing the idea of mobility-as-a-service. Autonomous electric vehicles (EVs) employ deep learning algorithms, computer vision, and sensor fusion to safely navigate complicated settings while maximizing energy economy and improving passenger comfort. AI adjusts climate, entertainment, regenerative braking, acceleration, and other settings to each driver's preferences while they are driving. AI improves driver pleasure and encourages environmentally responsible driving practices by learning from user behavior and feedback.

Figure 1.Simplified EV charging applications

Artificial intelligence has made it possible to build autonomous driving technology, which has completely changed the idea of mobility-as-a-service. Autonomous electric vehicles (EVs) employ deep learning algorithms, computer vision, and sensor fusion to safely navigate complicated settings while maximising energy economy and improving passenger comfort. With the increasing advancement of AI and electric vehicles working together, there are never-before-seen possibilities for sustainable mobility in the future. To expedite innovation and scale acceptance, automakers, technology companies, governments, and research institutions must work together. Accepting the convergence of AI and electric vehicles will help society and the environment in addition to promoting job development, economic expansion, and social progress.

Figure 2. Energy Management System in EVs

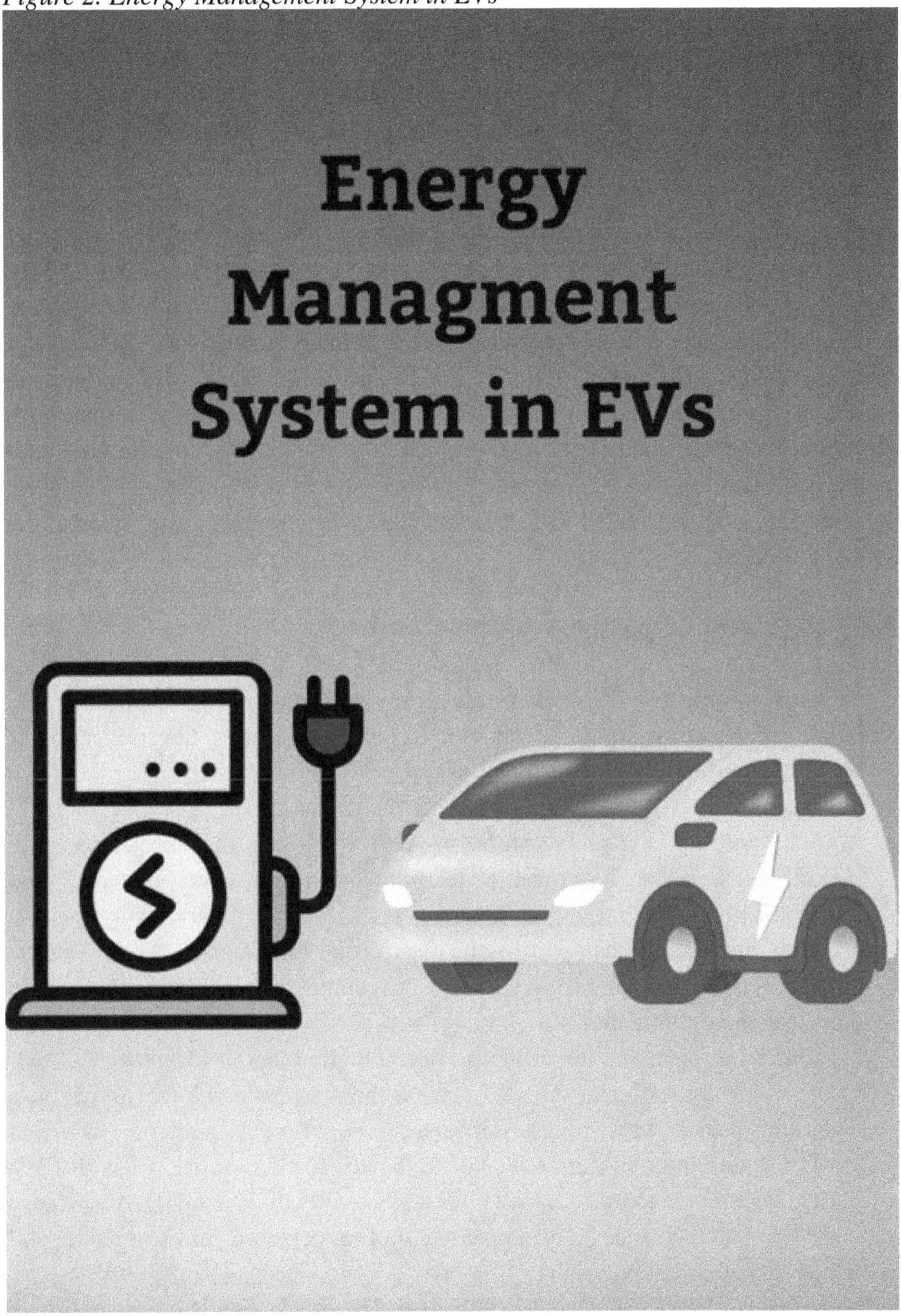

1.1 Motivation of the Chapter

One of the main causes of greenhouse gas emissions and air pollution worldwide is the transportation industry. Making the switch to greener, more sustainable methods is imperative. EVs have grown in popularity as a socially responsible substitute for conventional internal combustion engine vehicles.

They use less of the limited supply of fossil fuels and emit no exhaust. Currently, several obstacles prevent the broad adoption of electric vehicles, including a short driving range, delayed charging, and the requirement for complex energy management systems. These issues have forced academics and engineers to look to artificial intelligence (AI) technology to improve the performance and capabilities of electric vehicles. Artificial Intelligence (AI) can address critical aspects of electric vehicle design and operation since it facilitates predictive analytics, autonomous operation, and intelligent decision-making. By providing EVs with improved safety and dependability, better charging behaviour, and increased energy efficiency, the technology can hasten the shift to sustainable transportation systems (Jatin Pal Singh et al., 2012).

1.2 Scope and Objectives of the Chapter

The primary objective of this chapter is to provide a comprehensive and extensive overview of the latest trends and technological advancements in the application of AI for Sustainability in EVs. Our goals are to:

1 Examine the current literature on AI technologies that are pertinent to electric vehicles, such as machine learning algorithms, neural networks, and optimization strategies.

2 Examine the various applications of artificial intelligence (AI) in electric vehicles (EVs), ranging from battery management and self-driving to vehicle control and optimization.

3 Describe the potential economic and environmental advantages of AI-enabled energy-efficient vehicles, such as their potential to lower greenhouse gas emissions and boost energy efficiency.

4 Determine the opportunities and challenges for integrating AI with EVs. Make recommendations for possible research directions to help remove these obstacles and advance AI as a practical tool for achieving environmental and long-term sustainability goals.

1.3 Organization of the Chapter

The rest of the chapter is divided into different sections. Section 2 includes literature work and the next section 3 contains rise of electric vehicles and sections 4 contains optimizing sustainability in EV. Section 5 includes the components of AI-enabled personalised EV energy management. The next section includes AI techniques for personalised EV energy management. Section 7 includes Benefits and Impacts and in last section 8 conclusion is given.

2. LITERATURE REVIEW

Recent AI applications in electrical vehicles for sustainability and technical challenges in integrating AI, such as data security, data interoperability, and AI processing, have been covered by author Esanakula et al. (2021). These discussions help readers understand how personal transportation is becoming more electrified and digital, and how AI will allow EVs to disrupt the auto industry and lessen its impact on transportation. As part of the larger effort to create a cleaner, electrified transportation system, innovations and technological diffusion will be necessary to reduce the impact of transportation. In order to fulfil urgent sustainability tasks, it is necessary to integrate knowledge from the fields of business, communication, technology, policy, and the environment within the interdisciplinary research of electrification of personal transportation.

Rehan (2024) discussed his thoughts on the future of electric cars. Handling AI, Cloud Computing, and Cybersecurity in the Context of Electric Vehicles. The switch to electric vehicles marks a turning point in the global endeavour to improve environmental sustainability and lessen the negative consequences of climate change. Fundamentally, the broad usage of electric vehicles provides a variety of ways to address the urgent problems caused by greenhouse gas emissions and their damaging effects on the environment. Autonomous driving is a new era in which vehicles become cognitive agents of mobility, surpassing their traditional roles as means of transportation. It is the perfect example of how technology innovation and societal advancement come together. The development of EVs and AI will accelerate the transition to completely autonomous vehicles.

In a recent study paper, Abrantes et al. (2024) addressed Mutual Inductance Estimation Using an ANN for Inductive Power Transfer in EV Charging Applications. The goal of this research project is to estimate the mutual inductance in a wireless inductive power transfer system utilising MATLAB/Simulink R2021b and a resonant converter architecture. The purpose of developing the system output was to mimic the functioning of a battery charger. An artificial neural network (ANN) was created to estimate this parameter. Following the results analysis, it was confirmed that the maximum estimation error was approximately 3% and that it happened at the lowest charging power that was taken into consideration. Therefore, it can be concluded that the suggested ANN demonstrates the ability to precisely estimate the mutual inductance value in this kind of system.

The problems and difficulties associated with using artificial intelligence (AI) techniques in battery management systems (BMS) for electric vehicles are summed up by Badran and Toha (2024). The most recent research on AI approaches used to EV BMS is examined in this study. There are significant gaps in the body of research on AI-driven BMS, despite the increased interest in this field. Our main contribution is a thorough categorization and analysis of different AI methods according to their uses, goals, and performance indicators. This analysis fills in these gaps and offers insightful information on how to choose the best AI method for creating a dependable BMS with effective energy management for EVs.

Ratana (2024) talked on the use of AI in Cambodian energy development. This paper attempts to investigate and offer insights into how AI relates to the energy sector, based on a pilot project that involves groups of specialists, officials, and businessmen or businesswomen in one content-based study. The research utilises two focus groups, internet interactions, and continuous observations to gather data. Research on AI produced in the energy sector demonstrates that, although having less of an impact on energy development, AI is headed in the correct direction.

Artificial intelligence in the economic assessment of energy efficiency and renewable energy technologies has been covered by Chen et al. (2015). The authors concentrated on AI strategies that can be used to address various obstacles, such as identifying the most suitable customer to respond to given attributes and preferences, offering competitive pricing, organising, and managing facilities, providing incentives to demand response participants, and paying them fairly and economically. According to their model, the production of green energy must constitute a threat to global security because it is unaffected by foreign policy, utilises an infinite number of services, and is available to all groups.

The main obstacles to their quick and widespread acceptance, according to Ahmed et al. (2021) analysis, are their short driving range, lengthy charge times, and inadequate infrastructure for charging. The research by Researchers highlights the latest developments in electric vehicles (EVs) and associated infrastructure, primarily from artificial intelligence (AI), which increases the allure of EVs for consumers. Researchers also talked about how AI may be used to enhance EVs, make EV charging stations easier, and integrate EVs with smart grids. These applications have been thoroughly examined and evaluated.

A prospective view of digital twin technology, which extends virtual mapping technology and can be used to study the lifecycle of multisystem bodies in a digital environment, is presented by Bhatti et al. (2021). for smart electric vehicles. Digital twin technology has emerged as a major global research topic in recent years. Consequently, new individual studies on the application of digital twins to different facets of smart car technology have emerged in academic and industry studies, enabling digital twin technology to advance throughout time. The goal of this study is to close the gap between individual studies and offer a thorough review from an academically neutral and technically knowledgeable perspective. The reader is given systematic access to the conceptual foundation of digital twin technology, enabling them to gain understanding of its conception.

A unique energy management strategy for series plug-in hybrid electric vehicles was proposed by Chen et al. (2021). To find the engine fuel-rate in relation to battery power, several quadratic equations are used. The best battery power commands and engine on power are found by using the simulated annealing method with quadratic programming to solve the problem. To increase the accuracy of the calculations, the effects caused by the generators' and engine's inertias are examined. To broaden the application of the suggested strategy, the battery's condition is also taken into consideration. To confirm that the suggested technique may reduce plug-in hybrid electric vehicles' fuel usage, simulations were run.

Mohsenzadeh et al. (2021) thought about the best locations and dimensions for public charging stations for electric vehicles while taking the reliability of smart distribution networks into account. Electric vehicles (PEVs) have several advantages over other forms of transportation, including the ability to use variable fuels, safety, high performance, ease of use, and cost savings. Eventually, PEVs will be the most economical option for transportation. However, the placement and amount of accumulated PEVs may pose a risk to the dependability of the power system. For the power system to remain reliable, charging station layout must be planned as efficiently as possible. However, potential locations for charging stations should consider the design of the city's traffic patterns and electrical distribution systems. This study suggests a method for placing and sizing public charging stations optimally while taking the city's traffic restrictions into account. The goal function is to minimise the costs associated with Energy Not Supplied (ENS) and charging stations. The 33-bus radial distribution network's candidate locations for charging stations are best located using a genetic algorithm. The best place and size for charging stations to preserve the dependability of the power system are revealed by the results from Hamzaoui and Julien (2024).

Zhao et al. (2023) talked about the plug-in hybrid electric bus's adaptive energy management system. In order to enhance the fuel efficiency and lessen power battery deterioration of a power-split plug-in hybrid electric bus (PHEB), an online application of a novel energy management strategy (EMS) is suggested in this work. To change the variables online, the novel EMS uses the power battery state of charge (SOC) feedback mechanism and Pontryagin's minimal principle (PMP). To achieve more precise SOC feedback, a model-free adaptive control (MFAC) EMS was created using the compact form dynamic linearization model. Time-domain cumulative as well as covariate upper and lower limit constraints were applied to produce a real-time MFAC-TM strategy for online applications. The correlation between the MFAC-III EMS and MFAC EMS resulted in an average reduction in battery capacity loss of 8.19 percent on all ten routes; tests of both strategies showed that the MFAC-III performs better in hardware-in-the-loop than the MFAC strategy.

Fiori et al (2016) discussed about paper the need by developing a simple EV energy model that computes an EV's instantaneous energy consumption using second-by-second vehicle speed, acceleration, and roadway grade data as input variables. In doing so, the model estimates the instantaneous braking energy regeneration. The proposed model can be easily implemented in the following applications: in-vehicle, Smartphone eco-driving, eco-routing, and transportation simulation software to quantify the network-wide energy consumption levels for a fleet of EVs. One of the main advantages of EVs is their ability to recover energy while braking using a regenerative braking system. Moreover, the study also compared different electric vehicles and quantified the impact of auxiliary systems, including the air conditioning and heating systems, on vehicle energy consumption levels using the proposed energy model.

Habib et al (2022) discussed about the key factor in a battery management system is cell balancing between cells in a string that prolongs the energy storage device's lifecycle and performance. This paper highlights a comprehensive study and evaluations focusing on different types of batteries, Supercapacitor's, and balancing circuits applicable in BMS on electric vehicles and renewable energy storage systems. This rigorous study presents different electrochemical batteries chemistry, anode, cathodes, electrolyte, charging and discharging lifecycle, advantages, and disadvantages. For the Supercapacitor, different types of electrode materials, electrolyte interface, energy storage function, and available Supercapacitor's capacitance in the market, cell, and module voltage are presented. This manuscript highlights all the insights that will lead to choosing the required and appropriate battery, Supercapacitor, or blanching circuit for future applications on electric vehicles systems.

Abousleiman and Rawashdeh (2016) presents a model that characterizes the energy consumption of an electric vehicle. Most passive and active factors are presented and applied in the formulation. The presented model is verified against 306 kilometers of driven data and proved to have 1.3% absolute error difference between the real vehicle's consumed energy versus the predicted energy consumption as generated by the model. The model is then used with a particle swarm optimization algorithm to solve the single constraint optimization problem of finding the most energy efficient route between 2 points on a map. Simulation and real-world test results demonstrate savings in the energy consumption of the electric vehicle. Results showed more than 9.2% reductions in the energy consumption of the electric vehicle when driven on the developed algorithms' suggested routes rather than the ones generated by Google Maps and MapQuest.

Wei et al (2021) proposes a new optimal design approach of a permanent magnet synchronous motor (PMSM) in hybrid electric vehicles. It aims to solve the key research problem of how to find a viable and computationally efficient solution to achieve the maximum energy efficiency of the motor over the driving cycles. A one-dimensional analytical model is, firstly, built and validated to design the geometric parameters and calculate motor efficiency, maintaining high fidelity calculation with low computational cost. Then, by analyzing the motor energy distribution of the driving cycle, the energy efficiency is characterized by representative points, which can dramatically reduce the computation time during the optimal design. Leveraging by these points, the approximation model is presented to replace the PMSM optimization model to further reduce the computational cost. Finally, a combinatorial optimization algorithm is developed to return and characterize the PMSM optimal design in the studied scenario benefiting in the energy-loss reduction.

Sharmila et al (2020). this paper addresses the designing of an electric vehicle model, to enhance energy efficiency, reduce emissions, high-performance and low-maintenance. A simple and low run-time electric vehicle is modelled using simscape components using MATLAB. The dynamic model was used with the drive cycle source, battery, driver controller, power converter, motor, and vehicle subsystem. The power converter is designed using H-bridge with regenerative braking. The vehicle's kinetic energy is recovered by breaking the electric motor in the generator mode. Performance parameters, such as vehicle speed, current, State of Charge (SoC), motor RPM and distance, are verified with various drive cycles. The distance travelled, speed and torque by the electric vehicle are obtained using the simulation model and results are analysed.

3. THE RISE OF ELECTRIC VEHICLES

Electric vehicles (EVs) are a disruptive force that have ushered in a paradigm shift away from conventional internal combustion engine automobiles in recent years. This change represents a turning point in the history of transportation, as several variables have come together to bring electric vehicles (EVs) to the attention of the public.

Several factors, including as government laws that favour the use of electric cars (EVs), environmental concerns, and technology breakthroughs, have contributed to the huge increase in demand for EVs. The following are some significant factors driving and reflecting the rising demand for EVs:

3.1. Environmental Issues

Decrease in Greenhouse Gas Emissions: Since EVs have no tailpipe emissions, they help reduce air pollution and fight climate change. Sustainability Goals: To minimize their carbon footprints and encourage the use of electric vehicles (EVs) as a cleaner alternative to internal combustion engine (ICE) cars, many nations and localities have set high goals.

3.2. Developments in Technology

Battery Technology: Advances in battery technology are lowering the price and extending the range of electric vehicles (EVs), which increases their consumer appeal. Infrastructure for Charging: EVs are becoming more convenient and range anxiety is being reduced with the growth of fast-changing technologies and charging networks.

3.3. Rules and Provisions by the Government

Subsidies and Tax Credits: To promote the purchase of electric vehicles, several governments provide financial incentives like tax credits, rebates, and subsidies. Laws & Bans: In the ensuing decades, certain areas have enacted laws restricting the sale of new internal combustion engine (ICE) vehicles to encourage the transition to electric vehicles.

3.4. Aspects of the Economy

Total Cost of Ownership: Although electric vehicles (EVs) may have higher initial costs, over time, their lower running and maintenance expenses make them a more cost-effective option. Fuel Savings: Using electricity instead of petrol or diesel might save a lot of money, especially considering the current rise in fuel prices.

3.5. Awareness and Acceptance of Consumers

Raised Awareness: As more people become aware of the financial and environmental advantages of electric vehicles, acceptance, and demand for them will rise. Diverse Offerings: From luxury sedans to reasonably priced compact cars and SUVs, automakers are now providing a greater selection of EV models to suit a variety of market niches and tastes.

3.6. Adoption by Fleet and Companies

Corporate Sustainability Initiatives: To meet their corporate sustainability targets, a lot of businesses are adding electric vehicles (EVs) to their fleets. Ride-sharing and Delivery Services: To cut expenses and pollution, businesses in the ride-sharing and delivery industries are progressively switching to electric vehicles.

3.7 Market Forecasts and Trends

Global Sales Growth: With millions of EVs sold each year, EV sales have been rising quickly. Forecasts suggest that this trend will hold as costs come down and technology advances. Market Penetration: It is anticipated that EVs will gain a substantial market share in the upcoming years, and by 2030, EV sales may make up a sizable portion of the total volume of new cars sold.

4. OPTIMISING SUSTAINABILITY IN ELECTRIC VEHICLES:

Optimising sustainability in electric vehicles (EVs) with AI-powered personalised solutions requires several creative strategies. These methods concentrate on improving user experience, cutting down on environmental effect, and increasing energy efficiency. Here's where AI can be extremely important:

4.1 Customised Energy Administration

Adaptive charging ensures that the car is charged during off-peak hours or when renewable energy is most readily accessible. AI algorithms can learn individual driving patterns and optimise charging schedules based on daily routines. Dynamic Range Prediction: AI can predict range more accurately by considering individual driving preferences, current traffic information, and meteorological conditions, which can assist drivers in making more informed travel plans.

4.2 Personalised Roadside Assistance

Eco-Driving Advice: AI can examine a driver's actions and offer tailored advice and recommendations on how to drive more sustainably, including smoother acceleration and deceleration, ideal speed maintenance, and appropriate use of regenerative braking. Route optimisation: Using data on traffic, terrain, and individual driving habits, AI-powered navigation systems may recommend the most energy-efficient routes, hence cutting down on energy use.

4.3 Careful Upkeep and Battery Health

Using usage patterns and sensor data, artificial intelligence (AI) can predict maintenance needs. This allows for timely interventions that extend the life of vehicles and batteries while averting unplanned failures. Battery Health Monitoring: Users can receive personalised insights into the health of their batteries, which can help them identify the best charging procedures and warn them of possible problems before they get out of hand.

4.4 Grid and Smart Home Integration

Integration of Smart Homes: AI can control how an electric vehicle (EV) and a smart house energy system interact, determining the best time to charge the car based on electricity pricing, solar energy production, and home energy consumption. Vehicle-to-Grid (V2G) Solutions: Artificial intelligence (AI) can manage V2G operations, enabling the EV to power the house or return power to the grid during periods of high demand.

4.5 Better User Experience

Personalised infotainment may make driving more pleasurable and tailored by selecting entertainment, navigation, and other services depending on user preferences and behaviours. Voice and Gesture Controls: Using natural language processing and gesture recognition powered by artificial intelligence (AI), users can engage with the car's systems in a more hands-free and natural way.

4.6 Sustainability Perspectives and Motivations

Environmental Impact Tracking: AI can offer customised reports on how driving behaviours affect the environment, such as CO2 reductions over conventional cars, encouraging environmentally responsible behaviour.AI systems can gamify sustainable driving habits by providing incentives or rewards for driving in an energy-efficient manner, doing routine maintenance, and engaging in other environmentally beneficial activities.

4.7 Improving Production and Reclamation

Sustainable Materials: Artificial Intelligence has the potential to minimise environmental impact by optimising the use of sustainable materials in vehicle manufacture. End-of-Life Management: AI can assist a circular economy by anticipating when a battery or car is about to reach the end of its useful life and making recommendations for recycling or repurposing.

4.8 Shared and Community Mobility Options

Car Sharing Optimisation: AI can help with car-sharing initiatives by maximising vehicle availability and upkeep according to usage trends, guaranteeing effective and long-term EV use. Ride-Sharing Personalisation: To lower overall energy usage, ride-sharing firms can use artificial intelligence (AI) to pair users with the most efficient routes and recommend carpooling solutions.

5. COMPONENTS OF AI-ENABLED PERSONALIZED EV ENERGY MANAGEMENT

Personalised electric vehicle (EV) energy management using artificial intelligence (AI) entails using AI techniques to optimise EV charging and consumption based on user preferences and requirements. These are a few essential components:

5.1 User Profiling

AI-enabled personalised EV energy management systems require user profile to identify each user's preferences, behaviours, and limits and adjust the energy management techniques appropriately. User profiling is integrated into the following system components.

The first step of the process involves gathering information from a variety of sources, such as the user's past EV charging experiences, driving and energy consumption trends, location data, and user-specified preferences. The foundation for developing user profiles is this data.

By applying machine learning algorithms, the gathered data may be examined and used to divide users into groups according to shared traits and actions. For various user cohorts, this segmentation aids in the creation of more specialised and tailored energy management plans. By interacting with the system, users may indicate their preferences for energy spending caps, preferred charging sites (such as their homes, offices, or public stations), charging timetables, and other pertinent information. To obtain this data, preference elicitation methods like questionnaires, interviews, or interactive interfaces might be used. Patterns and trends in the user's behaviour, such as preferred charging times, usual driving distances, and frequent commute routes, can be found by analysing past charging and driving data. This data aids in projecting future energy requirements and adjusting charging schedules appropriately.

User profiles might potentially have restrictions related to time, money, or preferences for the environment. For instance, some customers could place a higher priority on saving energy expenses, while others would place a higher priority on reducing carbon emissions or maximising convenience. User profiles can change over time in response to shifting circumstances, preferences, and behaviours. They are not static. To guarantee that energy management techniques stay in line with the user's present requirements and preferences, the system should be able to dynamically update user profiles in response to fresh data and user feedback. It is critical to manage user data delicately and make sure privacy laws are followed. User confidentiality and privacy should be safeguarded during the user profiling process by putting the necessary safeguards in place. By utilising the information gleaned from user profiles, the system is able to produce customised suggestions for the best times to charge devices, energy-saving advice, and other pertinent activities that are suited to the preferences and limitations of each individual user.

5.2 Grid Integration

Grid integration, which entails coordinating the charging and discharging of electric cars with the functioning of the electrical grid, is a crucial part of AI-enabled personalised EV energy management systems. The way grid integration fits into the elements of this kind of system is as follows Real-time Grid Monitoring where the system always keeps an eye on the electrical grid's condition, keeping track of variables including frequency, voltage levels, generating capacity, and load. Insights into grid conditions can be obtained from real-time data obtained from smart metres or grid operators. In Grid Demand Prediction the technology predicts future patterns of grid demand by using predictive analytics and historical data. Finding the times when demand is at its highest and lowest can help determine when it is best to charge an EV. In Grid Constraints Awareness the system considers grid constraints such capacity restrictions, voltage control needs, and variability in the production of renewable energy. To maintain grid stability and prevent overloading, charging schedules are modified. In

Dynamic Charging Management the response to current grid limits and conditions, charging schedules are dynamically modified. For instance, charging may be sped up or delayed depending on the demand on the grid, excess renewable energy produced, or low demand. In Demand Response Integration the system takes part in programmes that allow EV charging to be adjusted in response to grid operators' signals. To ease grid strain, EV charging can be momentarily stopped during periods of high demand or moved to off-peak hours. In Grid-Friendly Charging techniques to satisfy user charging needs while minimising grid impact, the system uses grid-friendly charging techniques. This might involve intelligent charging algorithms that give priority to charging when there is an abundance of renewable energy being produced or when the cost of power is cheap. In Integration of Vehicles to the Grid (V2G), V2G technology enables EVs and the grid to exchange energy in both directions, enabling EVs to both consume and provide energy to the grid as required. To enable grid balancing and auxiliary services, the system coordinates V2G activities. In Grid Interaction Protocols for ensuring compatibility with grid infrastructure and other demand-side resources, the system complies with common communication protocols for interfacing with the grid, such as OpenADR (Open Automated Demand Response). In Grid Resilience and Reliability, the technology enhances grid resilience and reliability by dynamically controlling EV charging in tandem with grid operations. It lessens the need for expensive grid improvements, helps prevent grid congestion, and improves system efficiency overall. In Grid-User Collaboration to maximise grid integration efforts, grid operators, utilities, and EV users collaborate. This might entail introducing tariff structures that incentivize grid-friendly charging behaviour or incentives for electric vehicle owners to take part in grid-balancing programmes.

5.3 Real Time Adaptation

AI-enabled customised EV energy management systems must include real-time adaption as it enables dynamic changes to energy consumption and charging plans in response to shifting circumstances. Real-time adaption fits into the following system components are Dynamic Data Processing in which the system continually analyses data from several sources, such as grid conditions, energy costs, EV battery state, user preferences, and outside variables like traffic and weather forecasts. In predictive analytics based on real-time data streams, the system forecasts future energy costs, grid conditions, and energy consumption using machine learning algorithms. These forecasts help with scheduling EV charging and optimising energy use decisions. In Adaptive Charging Algorithms the system makes use of adaptive charging algorithms, which can dynamically modify charging schedules in response to shifting circumstances. For instance, the system may automatically postpone or reschedule EV charging to lessen the burden on the grid in the event of unexpected grid congestion. In Integration of User Feedback, the users can offer suggestions or modify their choices in real-time when their feedback is included into the adaption process. By doing this, the system is guaranteed to stay in line with the user's changing requirements and preferences. In Case Detection and Response, the case of a grid emergency, abrupt fluctuations in energy costs, or severe weather, the system is designed to recognise and react to these occurrences. EV charging schedules can be adjusted in certain circumstances to give grid stability or user safety priority. In Grid Interaction Protocols to communicate in real-time with the grid and other energy management organisations, the system follows established communication protocols. This guarantees smooth coordination and communication between external parties and the EV energy management system. In Active Participation in Demand Response Programmes the technology automatically modifies EV charging in response to signals from energy aggregators or grid operators. This supports grid reliability by balancing supply and demand on the system during peak times. In Vehicle-to-Grid (V2G) Control, depending on user preferences and grid circumstances, real-time adaption in V2G-enabled systems involves managing the bidirectional energy flow between EVs and the grid. To maintain grid stability, EVs can offer supplementary services or modify charging and discharging rates. In Emergency Preparedness the system has procedures in place to deal with unplanned events like power outages or grid breakdowns. When electricity is restored, EVs can be used as backup power sources or given priority when it comes to charging. In Continuous Optimisation the system learns from its past experiences and modifies its tactics over time to increase performance and efficiency. Real-time adaption is one aspect of this process.

6. AI TECHNIQUES FOR PERSONALIZED EV ENERGY MANAGEMENT

Personalized energy management for electric vehicles (EVs) entails tailoring energy usage and charging schedules to each user's unique requirements and preferences. The following AI methods are frequently applied to customized EV energy management.

Machine Learning (ML) for Predictive Modelling: To forecast the energy requirements of EV users in the future, machine learning algorithms can examine previous data on driving habits, charging habits, meteorological conditions, and other variables. Predictive models, which take into consideration variables like expected journey distance, time of use electricity tariffs, and the availability of renewable energy sources, can assist in optimizing charging schedules to make sure the car is charged when needed. Regression models, including support vector, polynomial, and linear models, can be used to forecast a variety of EV energy management-related metrics, including energy usage, charging time, and cost. To generate predictions, these models use past data on driving habits, charging habits, meteorological conditions, and other variables. Based on historical consumption trends, time series forecasting techniques such as seasonal autoregressive integrated moving average (SARIMA), prophet, and autoregressive integrated moving average (ARIMA) can be used to predict future energy demand. Proactive charging scheduling is made possible by these forecasts, guaranteeing that the EV will be fully charged when needed. In the context of EV energy management, complicated correlations between input data and goal variables can be captured by ensemble learning techniques like Random Forests and Gradient Boosting. These algorithms are useful for anticipating charging patterns and managing energy consumption because they can manage nonlinear relationships and interactions among variables.

Adaptive Control via Reinforcement Learning (RL): Through interaction with the environment (e.g, traffic patterns, road conditions), RL algorithms can learn optimal control rules for managing the energy consumption of electric vehicles and receive feedback on how well they are doing. In response to shifting circumstances and driver preferences, RL can modify its energy management tactics in real time. Requirements for reinforcement learning include a well-defined state space that contains pertinent environmental data. The state could contain variables for EV energy management, like the battery's current level, the distance to the destination, traffic patterns, and the availability of charging stations. Every state requires RL agents to have a specific set of actions available to them. Regenerative braking intensity, route planning, charging choices, and speed adjustments are some examples of actions in the context of EV energy management. To protect the car and its occupants, RL agents are required to work within safety parameters. The RL framework should take safety factors like speed limits and minimum battery level thresholds into account.

Deep Learning for Identification of Patterns: GPS information, LiDAR data, camera images, and other enormous amounts of sensor data from the car may all be analysed using deep learning techniques like convolutional neural networks (CNNs) and recurrent neural networks (RNNs). Energy management decisions, including choosing the best routes with the least amount of energy usage, can be informed by the significant patterns and insights that these models can derive. To identify traffic signs, the state of the road, and nearby cars, Convolutional Neural Networks (CNNs) can analyze images taken by external sensors or onboard cameras. Driving habits and energy usage patterns can be modified in accordance with this information. To build a thorough picture of the environment around the vehicle, deep learning models can combine data from several sensors, including LiDAR, radar, GPS, and inertial measurement units (IMUs). Long Short-Term Memory (LSTM) networks and Recurrent Neural Networks (RNNs) can interpret sequential sensor data to forecast future trajectories and optimize energy management tactics. Deep learning algorithms may identify different driving styles by examining time-series data on vehicle speed, acceleration, and braking patterns. Strategies for energy management, like regenerative braking and speed profile optimization, can be customized with the help of this data. To determine the best routes with the least amount of energy usage, deep learning models can examine past GPS data. Personalized route recommendations based on individual tastes and driving habits are made possible by the ability of Graph Neural Networks (GNNs) to learn representations of road networks and traffic patterns. Pre-trained deep learning models can be adjusted on EV-specific data to suit customized energy management needs. Examples of these models include those trained on extensive image datasets or natural language corpora. Models can accelerate their learning process by utilizing transfer leaning approaches, which enable them to harness knowledge from related areas.

Using Natural Language Processing (NLP) to Analyse User Feedback: NLP approaches can analyse textual feedback from drivers on their expectations for energy consumption, preferred methods of charging, and overall driving experience. To enhance customer happiness and personalize energy management techniques, sentiment analysis and topic modelling can yield insightful data. Algorithms for sentiment analysis can categorize user comments as neutral, negative, or positive automatically. The technology measures user happiness and pinpoints areas for development by analysing the attitude conveyed in user evaluations or comments about their EV driving experience and energy management choices. Latent topics can be extracted from a set of user feedback using topic modelling approaches like Latent Dirichlet Allocation (LDA) or Non-negative Matrix Factorization (NMF). The system is capable of prioritizing areas of concern and customizing energy management tactics based on the identification of recurrent themes or subjects connected to EV energy management in user feedback. Aspect-based sentiment analysis distinguishes certain elements or features highlighted in user feedback and evaluates the sentiment associated with each aspect, going beyond general sentiment classification. Aspects of EV energy management could include the infrastructure for charging, the performance of the batteries, energy management advice can be made by knowing what factors influence consumer happiness or discontent.

Multi-agent Frameworks for Collaborative Energy Administration: Interactions between several EVs on the road and between EVs and the infrastructure needed for charging can be modelled using multi-agent systems. These systems can cooperatively coordinate energy management choices to maximize system efficiency overall, lessen traffic at charging stations and save drivers' individual energy expenses. Each agent considers the behaviours of other agents to maximize its own goals. Globally optimal solutions are reached by the collective behaviour of all agents, thanks to techniques like distributed consensus algorithms. Through interactions with the environment and other agents, each agent can learn policies that optimize both its own goals and the overall performance of the system through a process known as multi-agent reinforcement learning (MARL). Agent interactions can be modelled using game theory as cooperative or non-cooperative games. In situations where no agent can unilaterally improve its fate, stable solutions can be found using techniques like Pareto Optimality or Nash Equilibrium. The utilization of renewable energy is maximized, peak demand is reduced, charging station wait times are shortened, and agents that represent EVs and charging stations constantly coordinate charging schedules and power levels. To contribute to V2G schemes, EV agents can return stored energy to the grid at times of high demand. To balance supply and demand, stabilize the grid, and provide incentives to EV owners, grid agents manage these activities. When supply outpaces demand, EVs with extra battery capacity can function as mobile energy storage devices by storing renewable energy and releasing it when demand is strong. To optimize the use of renewable energy sources, agents coordinate these efforts. To help with demand-side management and lower total energy costs, EV agents can modify their charging schedules in response to signals from grid agents showing high or low power rates.

Forecasting Energy Demand with Predictive Analytics: Based on past data and outside variables (such as traffic patterns and weather forecasts), predictive analytics methods (such time series analysis and predictive modelling) can anticipate future energy demand. Energy management choices, including arranging charging sessions during off-peak hours to save money, might be informed by these estimates. Using a combination of autoregression, differencing, and moving average techniques, Autoregressive Integrated Moving Average (ARIMA) models time series data by identifying trends and temporal relationships. Time series can be broken down into trend, seasonal, and residual components using the seasonal decomposition of time series (STL), which is helpful for data that exhibits strong seasonal trends. The Random Forest ensemble approach handles nonlinear relationships and interactions between variables by constructing numerous decision trees and aggregating their predictions. Gradient Boosting Machines (GBM) produce models in a sequential fashion, improving prediction accuracy by fixing the mistakes of earlier models. Support Vector Machines (SVM) are useful for high-dimensional data regression applications because they can capture intricate correlations between target variables and features.

Optimizing Hyperparameters with Bayesian Methods: While considering system uncertainties like changes in driver behaviour and ambient variables, Bayesian optimization approaches can tune the hyperparameters of energy management algorithms to maximize performance and efficiency. An approximation of the objective function (model performance metric) using a probabilistic model, usually a Gaussian Process. It offers uncertainty estimates and forecasts how various hyperparameter setups will perform. A function that selects which hyperparameters to assess next based on the surrogate model. Upper Confidence Bound (UCB), Probability of Improvement (PI), and Expected Improvement (EI) are examples of common acquisition functions. Exploration (trying novel hyperparameters) and exploitation (refining known good ones) are balanced by these functions. To optimize charging schedules, estimate demand, and predict energy usage, utilize Bayesian optimization to adjust the hyperparameters of machine learning models (like Random Forests, Gradient Boosting Machines) or deep learning models (like LSTMs, CNNs). Personalized EV energy management systems can perform better, resulting in more economical energy use, more customer satisfaction, and reduced costs, by using Bayesian Optimization for hyperparameter tweaking. This method guarantees that models are precisely adjusted to meet the unique needs and dynamics of electric vehicle energy management.

7. BENEFITS AND IMPACTS:

Personalized electric vehicle (EV) energy management using AI offers a few noteworthy advantages and effects that contribute to sustainability optimization. These can be broadly divided into three categories: social, economic, and environmental.

Environmental Benefits and Impacts: Customized charging schedules, AI can reduce dependency on fossil fuels and greenhouse gas emissions by optimizing charging times based on the availability of renewable energy. Enhanced battery efficiency, Artificial intelligence algorithms have the capacity to monitor battery health, guaranteeing extended lifespans and mitigating the environmental effects of battery production and disposal. Integration of the smart grid, Artificial Intelligence (AI) has the ability to integrate EVs as dispersed energy resources and balance energy supply and demand, stabilizing the grid and increasing the usage of renewable energy sources. Personalized energy management, or load balancing, can help avoid grid overload by distributing the timing of charging, particularly during peak hours. AI-powered smart charging, Considering the availability of renewable energy sources (such as solar or wind power), it can identify the best times to charge electric vehicles. This lowers the carbon footprint of charging by ensuring that EVs are charged during times when the grid is least reliant on fossil fuels. AI aids in lowering peak load demand by relocating EV charging to off-peak times when the energy system could otherwise be dependent on fossil fuels. As a result, there is less demand for carbon-intensive peaking power plants and more stable and efficient energy consumption. AI programs are able to align EV charging schedules with times when renewable energy output is at its peak. This reduces the overall emissions from the electricity grid by increasing the share of renewable energy used. Artificial intelligence (AI) can better utilize renewable energy sources, reducing dependency on fossil fuels and pollution, by modifying billing rates in response to real-time data on the supply of renewable energy. AI-powered battery health management can prolong battery life by optimizing cycles of charging and draining. Because of this, fewer batteries will need to be replaced on a regular basis, which lowers the emissions related to battery production and disposal. AI can minimize energy waste and enhance overall energy efficiency in electric vehicles (EVs) by optimizing their thermal management systems, which will ultimately result in lower emissions. EVs can discharge energy back into the grid during periods of peak demand since AI can control V2G interactions. This lowers emissions by reducing the demand for new fossil fuel-based power generation. When necessary, EVs can act as distributed energy storage devices, storing extra renewable energy and returning it to the grid. This reduces emissions even more by improving the use of renewable energy and mitigating its fluctuation in production. Using data from real-time traffic and weather conditions, AI-driven navigation systems may recommend the most efficient routes, cutting down on EV energy use and emissions. Artificial intelligence (AI) can give drivers instantaneous feedback and recommendations for more fuel-efficient driving techniques, like braking and acceleration that is smoother and uses less energy. AI can facilitate dynamic electricity pricing, encouraging EV owners to charge when there is a large availability of renewable energy sources and little demand for energy. This encourages reduced emissions and lowers the carbon intensity of the grid. By detecting and optimizing energy usage patterns, artificial intelligence (AI) can help reduce emissions by facilitating more precise tracking and trade of carbon offsets and credits. By streamlining charging schedules, boosting the utilization of renewable energy sources, boosting battery efficiency, and enhancing overall transportation efficiency, using AI for individualized electric car energy management dramatically lowers emissions. These advantages lessen the effects of climate change and improve air quality by fostering a more ecologically friendly and sustainable transportation system.

Cost Savings: AI may charge electric vehicles (EVs) at off-peak times, when electricity is less expensive, saving users' total energy expenses. Artificial Intelligence has the potential to reduce replacement costs by optimizing charging and discharging cycles. By making the most of already-existing resources, smart energy management reduces the need for expensive changes to energy infrastructure. Investing in renewable energy can become more feasible with the help of AI-driven demand management, as it guarantees steady demand. Enhanced convenience and lower operating costs can lead to increased EV adoption rates, which will spur innovation and market expansion. AI is able to identify trends in electricity prices and charge electric vehicles (EVs) at off-peak times, when prices are cheaper. As a result, consumers will directly save money on the overall cost of charging EVs. AI can maximize the cycles of charging and discharging to keep batteries healthy and prolong their life. This lessens the need for pricey battery changes as often. By using AI to enable predictive maintenance, possible problems can be detected before they become serious ones, saving money on costly repairs and upkeep. By maximizing the use of current infrastructure, balancing supply and demand, and reducing the need for expensive additions, artificial intelligence (AI) can improve the effectiveness of the electrical grid. AI can create a more stable and effective energy distribution system by controlling the charging schedules of several EVs, preventing grid overload. As a result, fewer power plants and infrastructural expenditures are required. AI can make sure that electrical vehicles (EVs) are charged during the peak hours of solar and wind power availability, maximizing their utilization, and minimizing reliance on fossil fuels. The design, implementation, and upkeep of AI systems for electric vehicle energy management may result in employment opportunities in the field of technology, engineering, data analysis, and renewable energy. Especially in areas emphasizing green technology and sustainable development, investments in EV infrastructure and AI technologies can boost local economies. There are significant financial gains and effects from using AI approaches for customized EV energy management. Consumer costs are significantly reduced, infrastructure efficiency is improved, market expansion is encouraged, and new business opportunities are generated. AI-driven energy management helps create jobs and stimulates innovation, which makes the economy more dynamic and robust and helps with the larger shift to sustainable energy and transportation systems.

Reduction of peak load: Artificial Intelligence improves the user experience by offering tailored suggestions for charging and routing. AI can anticipate maintenance requirements, lowering unplanned malfunctions and boosting dependability. By supporting local renewable energy initiatives and reducing reliance on centralized power sources, AI can improve community resilience to energy disruptions. Optimized energy management can ensure that more people have access to affordable, clean energy for their transportation needs, promoting energy equity. Significant social advantages and consequences result from using AI algorithms to optimize sustainability for customized electric vehicle (EV) energy management. Enhancements in user experience, safety, energy equity, community resilience, education, and environmental justice are a few of them. AI can designate individualized charging schedules and routes by analysing unique driving habits and preferences. Users will benefit from increased convenience as this guarantees that the EV is ready when needed and fully charged. AI systems can keep an eye on the condition of the car and its parts and identify possible problems before they get out of hand. By being proactive, this method lowers the possibility of unplanned malfunctions and improves dependability. AI can oversee the safe operation of charging stations, making sure they follow safety regulations and warning users of any possible risks. AI can provide drivers with instantaneous feedback on their driving behaviour, encouraging more cautious driving techniques and lowering the likelihood of collisions. AI-driven energy management can make sure that a larger group of people, especially those living in underprivileged areas, can benefit from EVs and renewable energy sources. Artificial intelligence lowers prices and maximizes energy use to make sustainable energy more accessible. AI can assist with the planning and location of public charging stations, making sure that all areas, even those with less developed infrastructure or lower incomes, can access them. EVs controlled by artificial intelligence (AI) have the potential to function as distribution energy resources, bolstering community resilience and offering backup power in the event of an outage. The effects of power outages can be mitigated by this decentralized strategy. AI can help community-based renewable energy projects become more feasible and sustainable by optimizing energy consumption and storage. This can be achieved through supporting renewable energy projects. The employment of AI in EV energy management can raise awareness among the general population of sustainability and the value of energy efficiency. Larger cultural changes in favour of more sustainable behaviours may result from this. AI can offer customers comprehensive insights into their energy usage and environmental effects, enabling them to make well-informed decisions and embrace more sustainable practices. This is known as data-driven insights. AI can contribute to the reduction of air pollution in urban areas and communities that are disproportionately impacted by environmental risks by maximizing the use of electric vehicles (EVs) and encouraging the adoption of clean energy. Artificial Intelligence can assist in resolving inequalities in access to clean energy and transportation by ensuring that the advantages of renewable energy and electric vehicles are shared more fairly. Significant societal advantages result from the use of AI algorithms for customized EV energy management. Enhancements to user experience and safety, energy equity, resilience of communities, economic opportunities, education and awareness-fostering, and environmental justice are all facilitated by it. These effects help create a society that is more resilient, sustainable, and inclusive.

8. CONCLUSION

The transition to electric vehicles (EVs) is a crucial step in achieving sustainable transportation. However, to fully realize the environmental benefits of EVs, it is essential to optimize their energy management. This chapter explored the potential of artificial intelligence (AI) in creating personalized energy management strategies for EVs. By analyzing individual driving patterns, charging behaviors, and environmental factors, AI can tailor energy consumption and charging schedules to minimize environmental impact while maximizing user convenience and cost-effectiveness. Moreover, integrating AI-powered energy management with smart grid technologies and renewable energy sources can further enhance sustainability by reducing reliance on fossil fuels and mitigating grid strain. In conclusion, harnessing AI for personalized electric vehicle energy management offers a promising approach to optimizing sustainability in transportation, promoting the widespread adoption of EVs and contributing to a cleaner, more resilient energy ecosystem.

REFERENCES

Abousleiman, R., & Rawashdeh, O. (2016). Electric vehicle modelling and energy-efficient routing using particle swarm optimisation. *IET Intelligent Transport Systems*, 10(2), 65–72. DOI: 10.1049/iet-its.2014.0177

Abrantes, G. C., Costa, V. S., Perdigão, M. S., & Cruz, S. (2024). Mutual Inductance Estimation Using an ANN for Inductive Power Transfer in EV Charging Applications. *Energies*, 17(7), 1615. DOI: 10.3390/en17071615

Ahmed, M., Zheng, Y., Amine, A., Fathiannasab, H., & Chen, Z. (2021). The role of artificial intelligence in the mass adoption of electric vehicles. *Joule*, 5(9), 2296–2322. DOI: 10.1016/j.joule.2021.07.012

Ahmed, M., Zheng, Y., Amine, A., Fathiannasab, H., & Chen, Z. (2021). The role of artificial intelligence in the mass adoption of electric vehicles. *Joule*, 5(9), 2296–2322. DOI: 10.1016/j.joule.2021.07.012

Badran, M. A., & Toha, S. F. (2024). Employment of Artificial Intelligence (AI) Techniques in Battery Management System (BMS) for Electric Vehicles (EV): Issues and Challenges. *Pertanika Journal of Science & Technology*, 32(2), 859–881. DOI: 10.47836/pjst.32.2.20

Baird, I. G., & Thorne, M. A. (2023). The downstream impacts of dams on the seasonally flooded riverine forests of the Mekong River in northeastern Cambodia. *South East Asia Research*, 31(4), 377–399. DOI: 10.1080/0967828X.2023.2243584

Chen, Z., Xia, B., You, C., & Mi, C. C. (2015). A novel energy management method for series plug-in hybrid electric vehicles. *Applied Energy*, 145, 172–179. DOI: 10.1016/j.apenergy.2015.02.004

Fiori, C., Ahn, K., & Rakha, H. A. (2016). Power-based electric vehicle energy consumption model: Model development and validation. *Applied Energy*, 168, 257–268. DOI: 10.1016/j.apenergy.2016.01.097

Habib, A. A., Hasan, M. K., Islam, S., Sharma, R., Hassan, R., Nafi, N., & Alotaibi, S. D. (2022). Energy-efficient system and charge balancing topology for electric vehicle application. *Sustainable Energy Technologies and Assessments*, 53, 102516. DOI: 10.1016/j.seta.2022.102516

Hamzaoui, M. A., & Julien, N. (2024). A Generic Deployment Methodology for Digital Twins–First Building Blocks. In *Handbook of Digital Twins* (pp. 102–121). CRC Press. DOI: 10.1201/9781003425724-10

Lujano-Rojas, J. M., Monteiro, C., Dufo-López, R., & Bernal-Agustín, J. L.Jatin Pal Singh. (2012). Optimum residential load management strategy for real time pricing (RTP) demand response programs. *Energy Policy*, 45, 671–679. DOI: 10.1016/j.enpol.2012.03.019

Mohsenzadeh, A., Pang, C., & Yang, L. (2021). Transmission congestion management considering EV parking lots and demand response programmes. *International Journal of Ambient Energy*, 42(15), 1727–1731. DOI: 10.1080/01430750.2019.1614985

Ouddah, N., Adouane, L., Abdrakhmanov, R., & Kamal, E. (2017, July). Optimal Energy Management Strategy of Plug-in Hybrid Electric Bus in Urban Conditions. In ICINCO (1) (pp. 304-311). DOI: 10.5220/0006436803040311

Ratana, P. (2024). AI in Energy Development for Cambodia. .DOI: 10.1080/02671522.2014.9084081130

Rehan, H. (2024). The Future of Electric Vehicles: Navigating the Intersection of AI, Cloud Technology, and Cybersecurity. Valley International Journal Digital Library, 1127-1143.

Schopper, D., Kübler, K., Rudolph, S., & Riedel, O. (2021). EIPPM—The Executable Integrative Product-Production Model. *Computers*, 2021(10), 72. DOI: 10.3390/computers10060072

Sharmila, B., Srinivasan, K., Devasena, D., Suresh, M., Panchal, H., Ashokkumar, R., & Shah, R. R. (2022). Modelling and performance analysis of electric vehicle. *International Journal of Ambient Energy*, 43(1), 5034–5040. DOI: 10.1080/01430750.2021.1932587

(2023). Vehicles [IJAIRD]. *International Journal of Artificial Intelligence Research and Development*, 1(1), 33–42.

Wei, D., He, H., & Cao, J. (2020). Hybrid electric vehicle electric motors for optimum energy efficiency: A computationally efficient design. *Energy, 203*, 117779. Bhatti, G., Mohan, H., & Singh, R. R. (2021). Towards the future of smart electric vehicles: Digital twin technology. *Renewable & Sustainable Energy Reviews*, 141, 110801.

Zhao, J., Zhou, J., Wang, L., & Zhao, Y. (2023). An energy management strategy to reduce the comprehensive cost of hybrid energy storage systems in electric vehicles. *Frontiers in Energy Research*, 11, 1191565. DOI: 10.3389/fenrg.2023.1191565

Section 4
Decision-Making Models and Advanced Technologies

Chapter 10

A Soft Computing-Based MCDM Approach Selection of Best Electric Vehicles Towards SDG- 7

Ankita Singh
Graphic Era University, India

Shwetank Avikal
Graphic Era University, India

Smita Bahuguna
Graphic Era University, India

Anupriya Sharma Ghai
https://orcid.org/0000-0002-9725-6939
Graphic Era Hill University, Dehradun, India

Anurag Barthwal
Chinmaya Vishwa Vidyapeeth, India

ABSTRACT

Battery electric vehicles are an evolution in the automobile industry as they also contribute to reducing the harmful effects of vehicles on the environment because of fossil fuels and the release of carbon dioxide. The purchasing decision of the consumer is highly dependent on the specific criteria when it comes to the adoption of Battery electric vehicles. This work helps to analyze those specific criteria that are important from a consumer point of view before purchasing a BEV. The selected

DOI: 10.4018/979-8-3693-6844-2.ch010

criteria analyzed in this study are the range of the vehicle, its full charging time, the purchasing price of the vehicle, the battery capacity of the vehicle, and the energy consumption of the vehicle. The analysis to find out the most prominent criteria that affect the purchasing decision of the consumer is done with the help of the Fuzzy AHP approach which provides the weights of the selected criteria and the other technique used is the Technique for Order Preference by Similarity to Ideal Solution TOPSIS, which provides the rank of the selected alternative electric vehicles.

I. INTRODUCTION

The transportation sector majorly contributes to global climatic change and the release of carbon dioxide. Thus, reducing CO_2 release is a primary motivator in replacing typical combustion engines powered by fossil fuels with "greener" and more effective equivalents, In these circumstances, many government and industrial regimes initiatives are being adopted all over the world to accelerate the creation and introduction of alternate fuel cars (Knez et al.,2019). Rising global warming has affected both the surroundings and human quality of life. The usage of fossil fuels for mobility and electricity production is bad for the natural surroundings since it emits greenhouse gases (GHG). To limit fossil fuel usage, people must shift to an eco-friendly, clear, and inexpensive source of energy. The sustainable development goals set by the United Nations have also given a goal that encompasses access to, reliable, affordable, modern and sustainable energy for all which aims to use clean energy technology including energy efficiency, renewable energy, and advanced and cleaner fossil-fuel technology. In transportation, switching to electric vehicles is an evolution toward the same goal (Khan et al.,2020). The automobile sector adds considerably to air pollution, ozone layer depletion, and climate change via greenhouse gas (GHG) releases such as CO, CH_4, N_2O, and CO_2. To address these challenges, highways must be converted to electrical systems. As technology progresses, combustion cars will be superseded with healthier, more sustainable options that are better for the natural world (Kumar et al.,2020).

EVs have become increasingly popular as an inexpensive component of environmentally friendly modes of transportation, eliminating reliance on fossil fuels and transportation-related toxins. Since 2011, the BEV business has expanded rapidly, owing to several schemes of incentives for both customers and producers. While BEV sales are rapidly expanding, their portion of the total automobile market stays low (Wei et al.,2020). Many developed nations have switched to electric vehicles (EVs) to minimize pollution and foster environmentally friendly growth. Electric vehicles (EVs) are a practical way to reduce harmful gases. The introduction of electric cars (EVs) will take care of both economic and environmental issues in the automotive sector [5In 2018, the International Energy Agency (IEA) stated that EVs produced just 38 million tons of carbon dioxide through well-to-wheel. This is a total of 78 million tons of CO2 discharged from a typical fleet. Electric vehicles are estimated to minimize oil usage by 4.3 million barrels per day by 2030 (Hoque et al.,2017). EVs have become a desirable alternative to regular vehicles, after conquering barriers such as expensive prices, limited battery range, and low top speed. There are 3 kinds of EVs present on the market: electric cars (BEV), Hybrid electric vehicles (HEV), and Plug-in HEVs. BEVs, or cars with no emissions, run entirely on electricity, propelled by a combustion engine and battery. BEVs are eco-friendly since they do not consume fossil (Ghadikolaei et al.,2021). Electric vehicles (EVs) and electric mobility continue to grow at an exponential rate. In 2018, the count of electric automobiles increased by 2 million over the last year, reaching 5.1 million. While the counting of BEVs has risen, industrial forecasts indicate that the acceptance of them continues to be in its early phases (Wang et al.,2020).

Battery electric vehicles, or BEVs, are completely electric cars that work on rechargeable batteries. The rechargeable battery packs are used for energy and do not use any other source, such as gases, units of hydrogen energy, internal combustion engines, etc. Unlike internal combustion engines (ICEs), BEVs have battery-powered engines and engine regulators. The battery electric vehicle consists of various components, including a battery, an electrical charging port, a DC-to-DC converter, an electrical traction motor, a power semiconductors controller, a thermal management system, a traction battery pack, and others (Ecer et al.,2021).

Because Electric cars use power engines despite internal combustion engines, a large battery pack is required to control the electric engine. charge the massive traction batteries requires a charge station or outlet, which is not readily available everywhere, making the expansion of BEVs difficult (Frieske et al.,2013). Under the pressure of contamination of the environment and the power calamity, electric vehicles (EVs) became known as the next development pattern, drawing global competitiveness and development. The battery system's safety and dependability, as a critical component, have a substantial impact on electric vehicle performance. Batteries made from lithium-ion are now the top option for power sources for EVs because of their significant density of power, large amount of energy, modest self-discharge rate, and other positive properties (Xiong et al.,2020). Moreover, choosing the best BEV can be challenging due to various considerations such as charging time, purchasing price, energy consumption, battery capacity, and so on. Fuzzy AHP can be an appropriate and effective solution for this.

The United Nations has set 17 SDGs for the holistic development of the world. This study contributes towards Goal 7 of these SDG which is “Ensure access to reliable, affordable, modern sustainable and energy for all”. The study contributes under Target 7. b which states “By 2030, enhance international cooperation to facilitate access to clean energy research and technology, including energy efficiency, renewable energy, and cleaner and advanced fossil-fuel technology, and promote investment in energy infrastructure and clean energy technology". The use of electric vehicles promotes clean energy technologies. In this study, the prospects of BEVs in the global market are anatomized. Moreover, using the Fuzzy AHP Approach and TOPSIS, these criteria (range, full charge time, purchasing price, energy consumption, and battery capacity) are analyzed to determine the most prominent driver that affects the purchasing decision of Battery Electric Vehicles BEV and are provided weightage and further ranking them.

II. LITERATURE REVIEW

Electric vehicles are turning into increasingly prevalent due to the broad availability of electricity and concerns about environmental degradation and rising temperatures (Jena et al.,2020). Four elements influence electric car purchase decisions: socioeconomic and demographic demographics, automobile characteristics, enticements, and customer sentiment toward electrically powered cars. Researchers have discovered that factors influencing battery-electric car purchasing include economic, political, social, technological in nature, psychological, and behavioral issues (Nazari et al.,2024). Customers' decisions will be affected by technological advancements, growing interest in BEVs, and increased market share (Wei et al.,2020). The technical parameters of BEVs are a crucial element in the purchasing choice (Neves et al.,2020). The study aims the select the most prominent battery electric vehicle based on specific criteria selected through literature, Using the Fuzzy AHP approach weightage is given to the criteria (range, full charge time, purchasing price, energy consumption, battery capacity) and ranking them to find out the most prominent one.

As an outcome, the implications of these criteria for BEV purchases are briefly discussed below.

II.I Range

Compare the ongoing cost of ownership of BEVs versus combustible vehicles, considering range, curb size, and energy usage. The study found that the short driving range of electric vehicles is a barrier to their adoption, while purchasing electronic vehicles the driving of the vehicle is considered by the consumers (Weldon et al.,2018). For those considering BEV adoption, the range is the second most important factor after car price. The range of electric vehicles presents a significant barrier to acceptance. Reducing "range anxiety" in BEVs increases the likelihood of owning an electric vehicle, Restricted electric driving is a major worry for future electric vehicles, and storage of energy affects the range of BEVs (Skippon et al.,2011). Also noted by the literature the consideration of electric driving is the which affects the purchasing and selecting decisions of electric vehicles.

II.II Full Charge Time

Long charging times for electric vehicles may prevent buyers from adopting environmentally friendly vehicles. BEV users are curious to know the period it takes to fully charge their batteries (Nazari et al.,2019). Electric car uptake is heavily influenced by technical factors like range and charging time, with the latter being a major barrier to sales (Egbue et al.,2012). The duration it requires to charge a PEV is mostly determined by the battery charge level. Second, it is dependent on the car's technologies (which are restricted by the battery's capacity to tolerate an elevated charge rate), the charging connection, and the station for charging (Electric Vehicle Supply Equipment, EVSE). The least fast charge is mode 1, which has no interaction with control, and hence the load quality is not considered, potentially leading to overloading of the grid. The charge cable includes a control box for mode 2 charging. Mode 3 charging improves interactions between the electrical grid and the vehicle by allowing data to be transferred. Mode 4 charging provides direct current (DC) rapid energy transfer (Hardman et al.,2018).

II.III Purchasing Price

The high cost of BEVs, due to their new technologies, is a valuable obstacle to their usage and sales, according to research. In India, consumers prioritize the purchasing price of electric automobiles, as seen in other countries (Giansoldati et al.,2020). The high cost of BEVs hinders their uptake. Higher prices for electric vehicles have a detrimental impact on purchasing behavior. Families with good incomes are more likely to buy an EV. The study identified six key characteristics that encourage Chinese customers to adopt BEVs, including purchasing price. Consumers often prioritize the purchase price over the entire cost of ownership for clean vehicles, such as BEVs. Electric automobiles have lower future ownership costs than oil-powered vehicles. Despite their high initial purchase price, BEVs have the lowest average ownership cost during their entire lifespan in comparison to fossil fuel-based vehicles (Shetty et al.,2020). Electric vehicles are expensive due to the advanced techniques used. Research indicates that the higher cost of ownership is a barrier to BEV purchases and sales. Consumers in India, like in countries other than India, prioritize the purchase price of electric vehicles (Jena et al.,2020).

II.IV Energy Consumption

Users believe electric vehicles to be cost-effective for transportation. BEVs offer the energy efficiency of electric vehicles and mobility comparable to conventional automobiles for lengthy journeys (Chen et al.,2020). According to recent studies, electric vehicles have favors over vehicles with combustion in terms of energy use (Chen et al.,2020). BEVs combine the energy efficiency of electric vehicles with the versatility of regular automobiles for lengthy trips.

Consumers in China choose low-energy BEVs due to range concerns. Furthermore, the consumption of energy is strongly connected to the efficiency of batteries. Operating type and temperatures are commonly recognized variables influencing a vehicle's fuel consumption (Ma et al.,2019).

II.V Battery Capacity

The market of BEV is fast evolving, especially in terms of the capacity of the battery and its charging ratings for power supplied by automobile makers, as well as the growing popularity of infrastructure for charging. As a result, people's behaviors of charging might change dramatically and deviate from whatever has been predicted or witnessed so far. Vehicles consisting of larger battery capacities are considered to find lesser refueling possibilities than those with small battery abilities because of their longer driving range. As a result, the energy requirements for such charge occurrences would increase. It is unclear how much of an impact these effects will have on the electricity system (Dixon et al.,2020). As battery technology advances, BEVs are expected to gain a larger market share in the years to follow. The battery's effectiveness is mostly determined by its employable capacity. According to several authors, BEV batteries have enough energy to meet most users' everyday driving needs. In China, men prioritize the capacity of the battery of BEVs over females. The size of the battery is identified as a key impediment to Battery Electric Vehicle adoption. Adequate battery capacity is crucial for meeting BEV performance standards among consumers (Axsen et al.,2009).

III. PROBLEM FORMULATION

This study aims to analyze the criteria identified through literature (Range, full charge time, purchasing price, energy consumption, and battery capacity) of the battery electric vehicles providing them weightage and ranking them to find the most prominent driver that affects the purchasing decision of battery electric vehicle based on identified criteria using fuzzy AHP approach and TOPSIS. In this paper, the analysis of 5 vehicles (BMW i3, Chevrolet Bolt, Tesla S, VW e-Golf, Kia Soul EV) on the criteria identified through literature is done for which Fuzzy AHP approach and TOPSIS is desirable.

The drivers that are affecting the BEV purchasing decision are shown in Table-1 and to each criteria/driver, a code is given from C1 TO C5 which defines the drivers in Table 2 (range as C1, full charge time as C2, Purchasing price as C3, energy consumption as C4 and battery capacity as C5).

Figure 1. decision hierarchy for criteria selection.

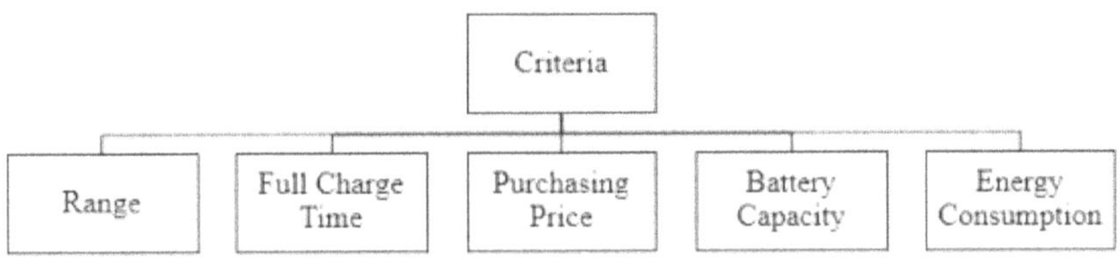

Fig.1 Decision Hierarchy for Criteria Selection

Table 1. Drivers affecting Battery Electric Vehicle purchasing decisions

Driver	Unit	Definition
Range (C1)	Km	The maximum range possible with simply one charge.
Full charge time (C2)	Hour	The time it takes to charge the battery fully.
Purchasing price (C3)	$	The original price of selling the automobile, plus taxes.
Energy consumption (C4)	Kwh	Energy consumption of the vehicle.
Battery Capacity (C5)	Kwh	Battery capacity of the vehicle.

Here in Table 2, the alternatives are defined which are the BMW i3, Chevrolet Bolt EV, Tesla Model S, Volkswagen e-Golf, and Kia Soul EV criteria, and a decision matrix is formed. The decision hierarchy is shown in Figure 1.

Table 2. Decision matrix for Battery Electric Vehicle assessment problem

Criteria	**C1**	**C2**	**C3**	**C4**	**C5**
Alternatives	Min	Min	Min	Min	Min
BMW i3	260	20	44,450	13.1	33.2
Chevrolet Bolt EV	320	9.5	36,620	28	60
Tesla Model S	539	7	74,490	18.6	70
Volkswagen e-Golf	201	9.6	36,025	14.5	35.8
Kia Soul EV	448	9	37,000	15.7	64

IV. RESEARCH METHODOLOGY

The presented problem has been considered as a multi-criteria decision-making problem and a Fuzzy AHP and TOPSIS-based approach has been applied to solve the problem.

Figure 2. research methodology.

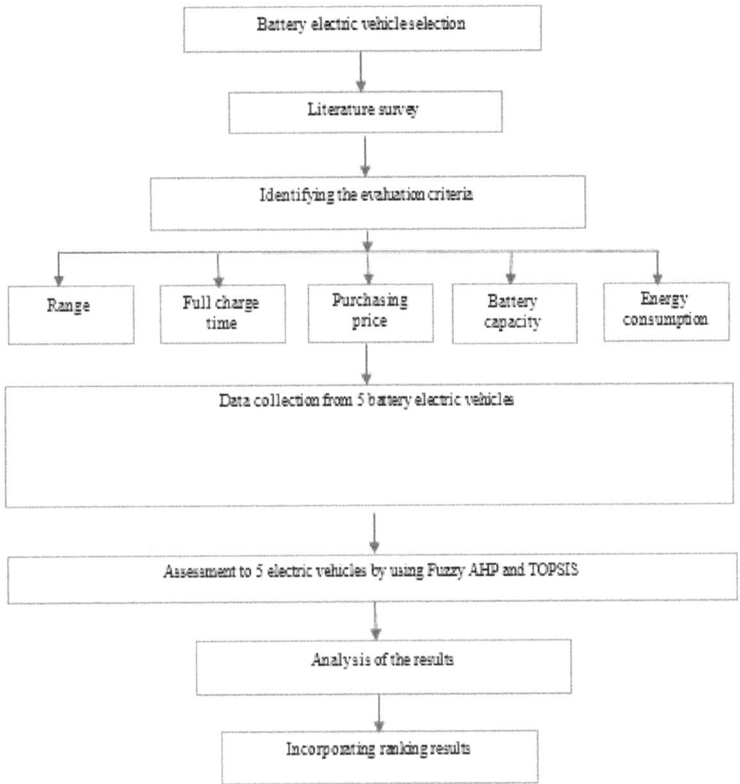

Fig.2 Research Methodology

IV.I AHP

AHP is mostly used for assessing difficult decision-making problems. Saaty invented the AHP in 1977. AHP highlights the need to balance potential demand with decision-making. AHP works in two stages: (1) breaking down concerns about decisions into a hierarchy of levels, and (2) obtaining weights from the criteria that can be used to identify alternative primary issues. AHP is a popular and useful strategy among academics in a variety of fields (Singh et al.,2020). The AHP assessment makes it possible to analyze multiple-choice options and select the best one based on a variety of quantitative and qualitative factors. Because the criteria used may clash with one another, the optimum alternative is typically not the one that fulfills each of the parameters, but rather the option that obtains the best settlement across all criteria. Furthermore, the most effective approach may change as the decision-maker changes; for example, depending on the needs, the criteria may be assigned a different degree of significance (Fiore et al.,2020).

The analytical hierarchy process (AHP) is majorly used as a Multi-Criteria Decision Making (MCDM) tool or weight estimation method in a wide range of applications, including selection, assessment, development and planning, making decisions, estimation, and so on. The typical AHP necessitates precise decisions. However, due to the degree of uncertainty and complexity inherent in real-world choice situations, a decision maker (DM) may find it more secure in providing fuzzy assessments rather than precise comparisons (Wang et al.,2008). AHP offers numerous benefits. For example, it is legal since the judgment strategy, along with ensuring the summary's reliability, can be organized systematically. Breaking down a large issue into smaller problems can help you understand it better. Iteration comparisons are also used until the team reaches a decision. Meanwhile, before making a final decision, sensitivity is assessed utilizing data-driven computers (Abdul-Hamid et al.,1999). The AHP approach assists decision-makers in determining the best solution. The advantage of this approach (AHP) based on intuition, physical data, and experience is that it can deal with both quantitative and qualitative parts of a making choices problem (Canco et al.,2021). The AHP methodology is ideal for sequencing and establishing relevant metrics and KPIs for everyday business processes (Bhagwat et al.,2007).

IV.II Fuzzy AHP

F-AHP is also known as fuzzy-MCDM (Multi-Criteria Decision Making) or fuzzy-MPPC (Multi-Person Preference Criteria), and it has undergone several changes due to its versatility. Some researchers enhance the AHP and FAHP to form the Intuitionistic Fuzzy AHP (IF-AHP). This concept enables choices to be expressed through fuzzy intuitive values, and as they evolve, they can be used to solve more complicated issues. In such cases, the individual making the choice indicates unpredictability when establishing their preferred values for the items thought of, and the methodology's advancement has enabled the focus on the solutions for issues across various domains, such as indications concerning human resources. that permits, in these specific instances, to take account of both the benefits and drawbacks of human capital (Sirvent et al.,2022). In fuzzy set language, the decision maker's ratio is a fuzzy integer represented by a function called membership. A membership function indicates the extent to which components of the assessment period belong to the preferences set. Fuzzy AHP involves determining local preferences from these fuzzy preference ratios, which are then aggregated to generate worldwide priorities. Because preferences in AHP are essentially human judgments based on sight (particularly for intangibles), we feel the fuzzy approach provides a more precise illustration of the decision-making processes (Leung et al.,2000).

IV.III TOPSIS

Hwang and Yoon's TOPSIS (Technique for Order Performance by Similarity to Ideal Solution) technique, created in 1981, is the Next often used MCDM method after Saaty's Analytic Hierarchy Process (AHP) (1980). This is due to its ease of use and simplicity, which allows it to be used for issues involving a huge number of criteria and options. TOPSIS's calculation steps can be summed up as standardizing the decision matrix through vector standardization, calculating the weighted normalized decision matrix, determining the positive ideal solution (PIS) and negative ideal solution (NIS), estimating the difference or distance of each alternative from NIS and PIS, defining the corresponding intimacy of all substitutes to PIS by determining a ranking index, and ultimately ranking the order preference (Celikbilek et al.,2020). The TOPSIS approach is widely used, however it has been criticized, leading to the growth of various variations. One criticism is the reversing of the phenomenon of rank, which is relevant to both AHP and TOPSIS techniques. Reversing of rank is the change in the ranking of equivalents when an equivalent or criterion is added or removed (Shin et al.,2013). The TOPSIS technique seeks an agreement solution that is nearest to PIS but distant from NIS. The ranking index calculates based on two distances, without considering their respective relevance or weights (Kuo et al.,2017). The TOPSIS approach plays a vital part in decision-making analysis theory. Aside from optimizing and improving the TOPSIS method in the actual values surroundings, with the continuing advancement of fuzzy set theory and rough set theory, researchers have devised numerous dissemination forms of the TOPSIS method (Zhang et al.,2022).

V. RESULTS AND DISCUSSION

As discussed in the problem formulation section, the discussed problem has been considered as a multi-criteria decision-making problem, and a Fuzzy AHP and TOPSIS-based approach has been applied to evaluate the ranking of the selected alternatives. The results of the applied approach have been discussed as follows.

V.I. Application of Fuzzy AHP

A tabular comparison of the main criteria of the selection of battery electric vehicles namely range, full charge time, purchasing price, battery capacity, and energy consumption is provided. Table 3 displays the Pair-wise comparison matrix for the main drivers/criteria and Table 4 assigns weights to each. The results show that the critical ratio (CR) is less than 0.1, which means the results are acceptable.

Table 3. Pair-wise comparison matrix of each selected criteria

	Range	**Full charge time**	**Purchasing price**	**Battery Capacity**	**Energy consumption)**
Range	1	3	3	5	5
Full charge time	0.375	1	1	3	5
Purchasing price	0.375	0.75	1	3	5
Battery Capacity	0.2083	0.375	0.375	1	3
Energy consumption	0.2083	0.2	0.2083	0.375	1

A comparison of the five criteria that is range, full charge time, purchasing price, battery capacity, and energy consumption has been shown in Table 4. The results demonstrate that the critical ratio (CR) is more than 0.1 hence the comparable results are not acceptable. CR of full charge time is more than 0.1 hence the comparable results are not acceptable. CR of purchasing price is less than 0.1 hence the comparable results are acceptable. CR of energy consumption is less than 0.1 hence the comparable results are acceptable. CR of battery capacity is less than 0.1 hence the comparables are acceptable.

Figure 3. Weights of the criteria

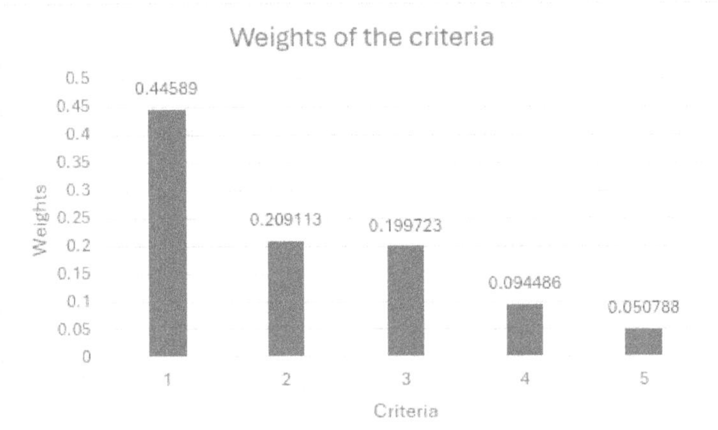

Fig. 3 Weights of the criteria

Table 4. Consistency check of the criteria

Criteria	Weights	Consistency check
Range	0.44589	CI = 0.070234 CR = 0.062709
Full charge time	0.209113	
Purchasing price	0.199723	
Battery Capacity	0.094486	
Energy consumption	0.050788	

V.II Application of TOPSIS for Final Ranking

The data shown in Table 2 has been used as the decision data for evaluating the final rank of the selected alternative BEV. The data of all the selected alternatives with respect to all the selected criteria has been shown in table 5. The data available in Table 5 has been normalized and is shown in Table 6.

Table 5. Main decision matrix

Criteria/Alternatives	C1	C2	C3	C4	C5
BMW i3	260	20	44,450	13.1	33.2
Chevrolet	320	9.5	36,620	28	60
Tesla Model S	539	7	74,490	18.6	70
Volkswagen e-Golf	201	9.6	36,025	14.5	35.8
Kia Soul EV	448	9	37,000	15.7	64

Table 6. Normalized decision matrix

Criteria/Alternatives	C1	C2	C3	C4	C5
BMW i3	0.147059	0.362976	0.194457	0.145717	0.126236
Chevrolet	0.180995	0.172414	0.160203	0.311457	0.228137
Tesla Model S	0.304864	0.127042	0.325874	0.206897	0.26616
Volkswagen e-Golf	0.113688	0.174229	0.1576	0.16129	0.136122
Kia Soul EV	0.253394	0.163339	0.161865	0.174638	0.243346

The normalized data of Table 6 has been used for evaluating the weighted normalized matrix. Here, the weights of each criterion is calculated with the help of Fuzzy AHP have been multiplied concerning their corresponding alternative values and shown in Table 7.

Table 7. Weighted Normalized decision matrix

Criteria/Alternatives	C1	C2	C3	C4	C5
BMW i3	0.065572	0.075903	0.038838	0.013768	0.006411
Chevrolet	0.080704	0.036054	0.031996	0.029428	0.011587
Tesla Model S	0.135936	0.026566	0.065085	0.019549	0.013518
Volkswagen e-Golf	0.050692	0.036433	0.031476	0.01524	0.006913
Kia Soul EV	0.112986	0.034156	0.032328	0.016501	0.012359
MAX	0.135936	0.075903	0.065085	0.029428	0.013518
MIN	0.050692	0.026566	0.031476	0.013768	0.006411

The positive idle (PIS) and negative idle solution (NIS) have been evaluated with the help of Table 7. The relative closeness of every alternative has been calculated with the help of PIS and NIS and the ranking of the alternative has been evaluated with the relative closeness of every alternative and shown in table 8. Furthermore

Figure 4. Relative closeness of the alternatives.

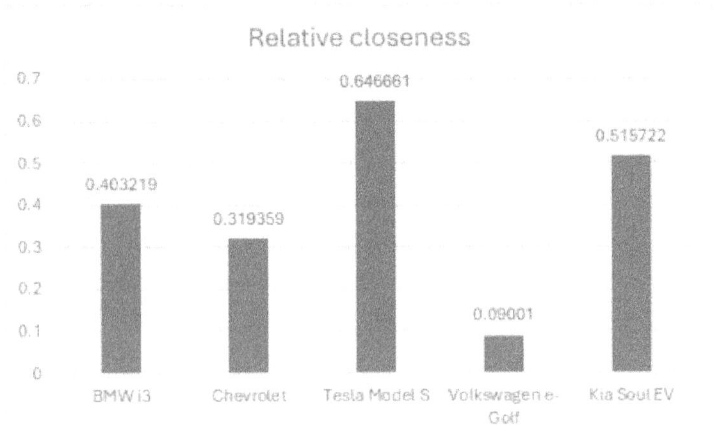

Fig.4. Relative Closeness of selected alternatives

Table 8. Final Ranking of the Alternatives

Alternatives	Positive Idle Solution	Negative Idle Solution	Relative closeness	Final Rank
BMW i3	0.077044	0.052055	0.403219	3
Chevrolet	0.075744	0.035539	0.319359	4
Tesla Model S	0.050316	0.092086	0.646661	1
Volkswagen e-Golf	0.100989	0.009989	0.09001	5
Kia Soul EV	0.059253	0.0631	0.515722	2

VI. CONCLUSION

The study contributes to the sustainable energy consumption by using the Electric vehicles. The purchasing decision of the BEV is performed according to various criteria and these criteria may also contradict each other. To find out the effect of the criteria selected on the purchasing decision of BEV the techniques that are used are the Fuzzy AHP approach and TOPSIS. It is concluded that certain criteria are there that a consumer considers before purchasing any BEV, and for selecting the most prominent one a comparison between them is necessary which is done through Fuzzy AHP, a comparison Matrix is prepared, and weightage is given to the criteria and then ranks are provided. The weights of the driver/criteria are calculated, and the comparison is made to find out the Critical Ratio (CR) to find out the most prominent driver/criteria. It has been seen that the driver/criteria "Range" has the highest weight followed by full charge time and purchase price, which shows the criteria Range affects the purchasing decision of the consumer the most. Through the TOPSIS the final ranking of the alternatives shows that the highest rank BEV among the selected ones is "Tesla Model S" with rank 1 followed by the Kia Soul EV and BMW i3. The study concludes that the most prominent criteria/ driver that affects the purchasing decision of the consumer is "Range" through the Fuzzy AHP approach followed by the most prominent alternative which is "Tesla Mosel EV" through the TOPSIS ranking method.

VII. REFERENCES

Abdul-Hamid, Y. T., Kochhar, A. K., & Khan, M. K. (1999). An analytic hierarchy process approach to the choice of manufacturing plant layout. *Proceedings of the Institution of Mechanical Engineers. Part B, Journal of Engineering Manufacture*, 213(4), 397–406. DOI: 10.1243/0954405991516868

Axsen, J., Mountain, D. C., & Jaccard, M. (2009). Combining stated and revealed choice research to simulate the neighbor effect: The case of hybrid-electric vehicles. *Resource and Energy Economics*, 31(3), 221–238. DOI: 10.1016/j.reseneeco.2009.02.001

Bhagwat, R., & Sharma, M. K. (2007). Performance measurement of supply chain management using the analytical hierarchy process. *Production Planning and Control*, 18(8), 666–680. DOI: 10.1080/09537280701614407

Canco, I., Kruja, D., & Iancu, T. (2021). AHP, a reliable method for quality decision making: A case study in business. *Sustainability (Basel)*, 13(24), 13932. DOI: 10.3390/su132413932

Castelló-Sirvent, F., Meneses-Eraso, C., Alonso-Gómez, J., & Peris-Ortiz, M. (2022). Three decades of fuzzy AHP: A bibliometric analysis. *Axioms*, 11(10), 525. DOI: 10.3390/axioms11100525

Çelikbilek, Y., & Tüysüz, F. (2020). An in-depth review of theory of the TOPSIS method: An experimental analysis. *Journal of Management Analytics*, 7(2), 281–300. DOI: 10.1080/23270012.2020.1748528

Chen, C. F., de Rubens, G. Z., Noel, L., Kester, J., & Sovacool, B. K. (2020). Assessing the socio-demographic, technical, economic and behavioral factors of Nordic electric vehicle adoption and the influence of vehicle-to-grid preferences. *Renewable & Sustainable Energy Reviews*, 121, 109692. DOI: 10.1016/j.rser.2019.109692

Chen, Y., Wu, G., Sun, R., Dubey, A., Laszka, A., & Pugliese, P. (2020). A review and outlook of energy consumption estimation models for electric vehicles. *arXiv preprint arXiv:2003.12873.*

Dixon, J., & Bell, K. (2020). Electric vehicles: Battery capacity, charger power, access to charging and the impacts on distribution networks. *ETransportation*, 4, 100059. DOI: 10.1016/j.etran.2020.100059

Ecer, F. (2021). A consolidated MCDM framework for performance assessment of battery electric vehicles based on ranking strategies. *Renewable & Sustainable Energy Reviews*, 143, 110916. DOI: 10.1016/j.rser.2021.110916

Egbue, O., & Long, S. (2012). Barriers to widespread adoption of electric vehicles: An analysis of consumer attitudes and perceptions. *Energy Policy*, 48, 717–729. DOI: 10.1016/j.enpol.2012.06.009

Fiore, P., Sicignano, E., & Donnarumma, G. (2020). An AHP-based methodology for the evaluation and choice of integrated interventions on historic buildings. *Sustainability (Basel)*, 12(14), 5795. DOI: 10.3390/su12145795

Frieske, B., Kloetzke, M., & Mauser, F. (2013, November). *Trends in vehicle concept and key technology development for hybrid and battery electric vehicles. In 2013 world electric vehicle symposium and exhibition (EVS27)*. IEEE.

Ghadikolaei, M. A., Wong, P. K., Cheung, C. S., Zhao, J., Ning, Z., Yung, K. F., Wong, H. C., & Gali, N. K. (2021). Why is the world not yet ready to use alternative fuel vehicles? *Heliyon*, 7(7), e07527. DOI: 10.1016/j.heliyon.2021.e07527 PMID: 34337177

Giansoldati, M., Rotaris, L., Scorrano, M., & Danielis, R. (2020). Does electric car knowledge influence car choice? Evidence from a hybrid choice model. *Research in Transportation Economics*, 80, 100826. DOI: 10.1016/j.retrec.2020.100826

Hardman, S., Jenn, A., Tal, G., Axsen, J., Beard, G., Daina, N., Figenbaum, E., Jakobsson, N., Jochem, P., Kinnear, N., Plötz, P., Pontes, J., Refa, N., Sprei, F., Turrentine, T., & Witkamp, B. (2018). A review of consumer preferences of and interactions with electric vehicle charging infrastructure. *Transportation Research Part D, Transport and Environment*, 62, 508–523. DOI: 10.1016/j.trd.2018.04.002

Hoque, M. M., Hannan, M. A., Mohamed, A., & Ayob, A. (2017). Battery charge equalization controller in electric vehicle applications: A review. *Renewable & Sustainable Energy Reviews*, 75, 1363–1385. DOI: 10.1016/j.rser.2016.11.126

Jena, R. (2020). An empirical case study on Indian consumers' sentiment towards electric vehicles: A big data analytics approach. *Industrial Marketing Management*, 90, 605–616. DOI: 10.1016/j.indmarman.2019.12.012

Jena, R. (2020). An empirical case study on Indian consumers' sentiment towards electric vehicles: A big data analytics approach. *Industrial Marketing Management*, 90, 605–616. DOI: 10.1016/j.indmarman.2019.12.012

Khan, F., Ali, Y., & Khan, A. U. (2020). Sustainable hybrid electric vehicle selection in the context of a developing country. *Air Quality, Atmosphere & Health*, 13(4), 489–499. DOI: 10.1007/s11869-020-00812-y

Knez, M., Zevnik, G. K., & Obrecht, M. (2019). A review of available chargers for electric vehicles: United States of America, European Union, and Asia. *Renewable & Sustainable Energy Reviews*, 109, 284–293. DOI: 10.1016/j.rser.2019.04.013

Kumar, R. R., & Alok, K. (2020). Adoption of electric vehicle: A literature review and prospects for sustainability. *Journal of Cleaner Production*, 253, 119911. DOI: 10.1016/j.jclepro.2019.119911

Kuo, T. (2017). A modified TOPSIS with a different ranking index. *European Journal of Operational Research*, 260(1), 152–160. DOI: 10.1016/j.ejor.2016.11.052

Leung, L. C., & Cao, D. (2000). On consistency and ranking of alternatives in fuzzy AHP. *European Journal of Operational Research*, 124(1), 102–113. DOI: 10.1016/S0377-2217(99)00118-6

Ma, S. C., Fan, Y., Guo, J. F., Xu, J. H., & Zhu, J. (2019). Analysing online behaviour to determine Chinese consumers' preferences for electric vehicles. *Journal of Cleaner Production*, 229, 244–255. DOI: 10.1016/j.jclepro.2019.04.374

Nazari, F., Noruzoliaee, M., & Mohammadian, A. (2024). Electric vehicle adoption behavior and vehicle transaction decision: Estimating an integrated choice model with latent variables on a retrospective vehicle survey. *Transportation Research Record: Journal of the Transportation Research Board*, 2678(4), 378–397. DOI: 10.1177/03611981231184875

Nazari, F., Rahimi, E., & Mohammadian, A. K. (2019). Simultaneous estimation of battery electric vehicle adoption with endogenous willingness to pay. *ETransportation*, 1, 100008. DOI: 10.1016/j.etran.2019.100008

Neves, S. A., Marques, A. C., & Moutinho, V. (2020). Two-stage DEA model to evaluate technical efficiency on deployment of battery electric vehicles in the EU countries. *Transportation Research Part D, Transport and Environment*, 86, 102489. DOI: 10.1016/j.trd.2020.102489

Shetty, D. K., Shetty, S., Raj Rodrigues, L., Naik, N., Maddodi, C. B., Malarout, N., & Sooriyaperakasam, N. (2020). Barriers to widespread adoption of plug-in electric vehicles in emerging Asian markets: An analysis of consumer behavioral attitudes and perceptions. *Cogent Engineering*, 7(1), 1796198. DOI: 10.1080/23311916.2020.1796198

Shin, Y. B., Lee, S., Chun, S. G., & Chung, D. (2013). A critical review of popular multi-criteria decision making methodologies. *Issues in Information Systems*, 14(1), 358–365.

Singh, R., & Avikal, S. (2020). COVID-19: A decision-making approach for prioritization of preventive activities. *International Journal of Healthcare Management*, 13(3), 257–262. DOI: 10.1080/20479700.2020.1782661

Skippon, S., & Garwood, M. (2011). Responses to battery electric vehicles: UK consumer attitudes and attributions of symbolic meaning following direct experience to reduce psychological distance. *Transportation Research Part D, Transport and Environment*, 16(7), 525–531. DOI: 10.1016/j.trd.2011.05.005

Wang, W., Zhang, Q., Peng, Z., Shao, Z., & Li, X. (2020, May 1). An empirical evaluation of different usage pattern between car-sharing battery electric vehicles and private ones. *Transportation Research Part A, Policy and Practice*, 135, 115–129. DOI: 10.1016/j.tra.2020.03.014

Wang, Y. M., Luo, Y., & Hua, Z. (2008). On the extent analysis method for fuzzy AHP and its applications. *European Journal of Operational Research*, 186(2), 735–747. DOI: 10.1016/j.ejor.2007.01.050

Wei, W., Cao, M., Jiang, Q., Ou, S. J., & Zou, H. (2020). What influences Chinese consumers' adoption of battery electric vehicles? A preliminary study based on factor analysis. *Energies*, 13(5), 1057. DOI: 10.3390/en13051057

Wei, W., Cao, M., Jiang, Q., Ou, S. J., & Zou, H. (2020). What influences Chinese consumers' adoption of battery electric vehicles? A preliminary study based on factor analysis. *Energies*, 13(5), 1057. DOI: 10.3390/en13051057

Weldon, P., Morrissey, P., & O'Mahony, M. (2018). Long-term cost of ownership comparative analysis between electric vehicles and internal combustion engine vehicles. *Sustainable Cities and Society*, 39, 578–591. DOI: 10.1016/j.scs.2018.02.024

Xiong, R., Sun, W., Yu, Q., & Sun, F. (2020). Research progress, challenges and prospects of fault diagnosis on battery system of electric vehicles. *Applied Energy*, 279, 115855. DOI: 10.1016/j.apenergy.2020.115855

Zhang, K., & Dai, J. (2022). A novel TOPSIS method with decision-theoretic rough fuzzy sets. *Information Sciences*, 608, 1221–1244. DOI: 10.1016/j.ins.2022.07.009

Chapter 11
Autonomous Vehicle Performance Prediction System Using Support Vector Machine

Vishan Kumar Gupta
Amity University Punjab, Mohali, India

Anupriya Sharma Ghai
https://orcid.org/0000-0002-9725-6939
Graphic Era Hill University, Dehradun, India

Paras Jain
https://orcid.org/0000-0003-2990-3556
VIT Bhopal University, Sehore, India

Vidisha Wadhawan
Amity University Punjab, Mohali, India

ABSTRACT

Self-driving cars or also known as Autonomous Vehicles have incorporate with sensors, actuators, sophisticate algorithms, machine learning systems, and high-level processors to run the software. These are vehicles that are programmed to move and be run without the need of a human being. They are of different categories depending on the low-level to high-level automation. The focus of this research is on the ability of SVM to forecast the effectiveness of self-driving cars. Especially in the context of safety and efficiency of autonomous driving considerable importance is given to the prediction of vehicle behaviour. SVM is used in this study to analyse

DOI: 10.4018/979-8-3693-6844-2.ch011

and forecast the performance out the different sensors and environmental measures. The conclusion drawn based on the analysis of results show that the application of SVM enhances the improvement of the level of reliability of the autonomous systems. Thus, the study is demonstrated on various datasets, namely, the KITTI datasets with up to 95.64% accuracy. Finally, K-folds cross validation is performed for the robustness of the Model.

INTRODUCTION

Self-driving cars, also known as Autonomous Vehicles (AVs), have sensors, actuators, algorithms, machine learning systems and powerful processors to run software. Such vehicles are intended to move and perform functions autonomously. They are categorized depending on the level of automation: partial (with the driver) and complete (without the driver). This paper has established that performance in AVs is important for safety, efficiency and reliability. Performance prediction is about the likelihood of the vehicle performing in certain situations, including traffic changes, weather fluctuations and avoidance of objects/obstacles, Bishop (2006).

Support Vector Machines, commonly called SVM, are classification and regression algorithms in the broader category of supervised learning. SVMs are useful in spaces with high dimensions and in cases with more dimensions than the number of samples. They work based on ascertaining the line that best separates a dataset into classes. SVMs have been used in various domains such as image recognition, bioinformatics, and the domain of interest, autonomous vehicle performance prediction. The versatility in implementing the boosting schemes on large feature spaces makes them ideal for analysing the AV data Cortes, C., & Vapnik, V. (1995)

Background on Autonomous Vehicles

Self-driving cars, commonly known as autonomous vehicles (AVs), have evolved from ideas on paper to lived reality with the prospect of changing how people transport themselves. It articulates increased security, reduced car traffic, and increased accessibility for people who cannot drive a car. Nonetheless, the main problem arises from the need to guarantee the safe performance of AVs.

Importance of Performance Prediction

Predicting AV performance is crucial for several reasons: it reduces risk by modelling failures, optimizes drive, and helps organized individuals gain confidence with self-driving cars. Performance prediction enables one to foresee problems, prevent them and ensure the smooth running of a team or organization.

Overview of Support Vector Machine (SVM)

SVM is a type of supervised machine learning mainly used for classification and regression problems. It is employed in identifying the hyperplane that best fits the separation of data into classes. It should be noted that the SVM is more suitable for handling high-dimensional data and the instances in which the number of dimensions is higher than the number of samples.

OBJECTIVES OF THE STUDY

This study aims to:

i. Develop a model using SVM to predict AV performance.
ii. Compare the SVM model's performance with other machine learning techniques.
iii. Analyse the implications of the findings for the future of autonomous driving.

These cars are intended to save expenses and use less energy while reducing accidents, improving safety, and relieving traffic congestion. They use various GPS-enabled navigational technology, including sophisticated radar systems, cameras, sensors, and high-speed broadband networks Gupta et al., (2023). Although there are clear advantages regarding less traffic in cities and more safety, there are also issues, including inefficiency, time limits, and network limitations Goodfellow, I.,Bengio, Y., & Courville, A. (2016).

Support Vector Machine (SVM)

- Model Representation: SVM finds the hyperplane that best separates the data into different classes. A hyperplane in an n-dimensional space is an (n-1) dimensional subspace that divides the space into two parts. For a two-dimensional space, a hyperplane is simply a line Singh et al., (2023).
- Margin Maximization: SVM measures the distance between the hyperplane and the nearest data points from each class to find the hyperplane with the

largest margin. These locations are the support vectors. The model's ability to generalize is improved by optimizing the margin.

- Kernel Trick: SVM can effectively handle nonlinear decision boundaries using kernel functions. Kernels compute the dot product between two points in a transformed feature space. Popular kernels include linear, polynomial, radial basis function (RBF), and sigmoid kernels.
- Regularization Parameter: Utilizing a regularization parameter (often represented by the letter C), SVM strikes a compromise between optimizing the margin and reducing the classification error. While a larger C value tolerates fewer misclassifications but may result in a reduced margin, a smaller C value leads to a larger margin but may cause misclassification.
- Optimization: SVM employs quadratic programming or gradient descent in order to find the hyperplane that best fits the classes while, at the same time, giving the maximum separation.
- Evaluation: SVM results are measured based on accuracy, precision, recall, and F1 score depending on the problem type.
- Advantages: Computationally efficient in high dimensional space, memory efficient and versatile because of the kernel trick. They give optimal results even when little training data is available, and they are not as sensitive to overfitting as some of the other algorithms.
- Limitations: SVMs can be very sensitive concerning the kernel and regularization parameters. They can also be time-consuming, especially when working with big data. Also, it is noted that SVMs have problems working with noisy data or data with overlapping classes.

Overall, SVM is an effective and versatile algorithm for the classification of variables, especially when dealing with high-dimensional datasets and binomial classification problems.

ADVANTAGES OF THIS AUTOMATIC VEHICLE SYSTEM

i. Personalisation: We have personalised the system according to user's preferences, past actions, and demographics to provide a satisfying experience and enhance the overall vacation by catering to individual interests.
ii. Accuracy: This autonomous vehicle system can process large data and identify hidden patterns. Machine learning algorithms offer more precise and relevant recommendations than traditional methods, leading to higher satisfaction for travellers.

iii. Time-saving: As the system provides recommendations based on POI (Places of interest), the overhead for users to visit multiple sites to gain knowledge about their desired destination has decreased, saving a lot of time and effort.
iv. Financial Gain: Accurate recommendations generated through machine learning algorithms increase the possibility of booking through the website, increasing the revenue for travel agencies and their businesses.
v. Customer Satisfaction: Customised suggestions based on machine learning can enhance customer satisfaction, increase customer loyalty, and increase the chances for the user to repeat business.

DISADVANTAGES OF THIS AUTOMATIC VEHICLE SYSTEM

i. Scarcity of Data: As we all know, machine learning models need abundant data in order to train the system; hence, the system heavily relies on data, and if there is a limited dataset available, then it can lead to poor performance and irrelevant recommendations.
ii. Lack of opacity: Some algorithms may be confusing and very complex, making it challenging for the user to understand how the recommendations are generated. It may cause concerns and reduce the trust of the customer.
iii. One-Sidedness: It can also be termed as a bias. One reason for it may be over-fitting, which occurs when the system works very well with the trained data but not with the new data. Hence, biased recommendations will bring discriminatory results.
iv. Excessive dependence on Technology: Overreliance on technology (machine learning algorithms) may overlook crucial factors, ultimately leading to uncertain results and disrupting the travel experience

LITERATURE REVIEW

Current Trends in Autonomous Vehicle Performance Prediction

Recent developments in providing more complex AVs have shifted towards improving the models that predict the vehicle's performance. The methods applied in the past are rule-based personnel systems and simple statistical models, which cannot interpret or generalise the situation of real driving conditions. Deep learning and ensemble methods are some of the subcategories of machine learning that continue to receive much attention due to their capacity to handle massive data.

Nevertheless, utilising such approaches may be rather time-consuming, and quite a lot of data preprocessing is typically needed Hinton, G., & Salakhutdinov, R. (2006).

A related study by Gupta et al., (2023) addressed the security challenges in cloud-assisted autonomous vehicles by developing a three-factor authentication and key agreement (AKA) system with biometric privacy protection. This approach, termed CT-AKA, enhances security by incorporating biometric verification into the authentication process for both the vehicle and the cloud. The CT-AKA scheme provides an effective three-factor verification strategy and guarantees the operators' biometric data privacy. Furthermore, it also sets up a proper communication apparatus through which the cloud, the operators, and self-driving cars can implement interconnectivity without being subject to malicious interference. However, a drawback of this method is the increased time required for authentication and key exchange processes LeCun, Y., Bengio, Y., & Hinton, G. (2015).

The MVDHA (Multi-Variable Dimensional Harmony Search Algorithm) method has also been developed to address NP-hard optimization issues. While VECC offers notable advantages in terms of time and energy savings, challenges remain regarding the adoption of these methods by other countries, as well as concerns regarding security Singh et al. (2023).

An autonomous vehicle parking system is implemented using fog-blockchain technology, offering a solution for future smart parking utilizing IoT-cloud integration. With the rapid increase in self-driving vehicles, efficient parking solutions have become essential. The Smart Autonomous Vehicle Parking (SVAP) system is devised for indoor and outdoor parking scenarios, enabling AVs to locate open spaces in parking lots efficiently. Fog nodes act as a middleware layer, facilitating parking activities closer to edge devices with IoT capabilities. The LIBC (Lightweight Blockchain Communication) model aims to enhance safety and privacy but may not be entirely secure in preventing traffic accidents Jaiswal, N., Gupta, V. K., & Mishra, A. (2015).

The work explored the concept of energy security through a bi-objective optimization framework, considering social factors and renewable technologies to enhance energy, security and resilience. It also introduced two optimization techniques: multi-objective grey wolf optimization (MOGWO) and lexicographic weighted Tchebycheff (LWT). To illustrate the practicality of their model, they conducted a case study focusing on Belgium. Their findings suggested that solar and wind energy could be utilized with 80% and 90% efficiencies, respectively.

They further compared four optimization approaches—ant lion optimization (ALO), whale optimization algorithm (WOA), particle swarm optimization (PSO), and crow search optimization (CSO)—for wind energy optimization and hybrid decision-making. They discovered that ALO outperformed the other approaches in terms of efficiency. Moving to healthcare optimization, I devised a mixed-integer linear programming model to tackle issues about patient diagnosis. They employed the Poisson distribution to analyze patient arrival times in healthcare, enhancing the 70% punctuality rate and reducing emergency cases by 30%. A proposed approach is the RSDM-AHS net for allocating health service networks. They categorized health services into low-level and high-level hospitals, factoring in disruptions. By prioritizing patients and considering capacity and disturbances, they developed a model for locating and allocating health networks, resulting in mutation probabilities of 0.7 and 0.1 at the lowest level, respectively Gupta (2022).

Machine Learning in Autonomous Vehicles and Autonomous Driving

Machine learning techniques are pivotal in AV development. They enable the vehicle to learn from past experiences and improve decision-making processes. Common techniques include:

i. Neural Networks: These are used for perception tasks like object detection and image segmentation.
ii. Random Forests: Employed for decision-making processes and behaviour prediction.
iii. Support Vector Machines: Utilised for performance prediction due to their robustness in handling high-dimensional data.

Comparing these techniques, SVMs offer a balance between computational efficiency and predictive accuracy, making them a suitable choice for AV performance prediction Kumar, P., Gupta, V. K., & Singh, D. P. (2022).

This paper focuses on the machine learning aspect of autonomous driving (AD) and lays out the various tasks that fall under this category, such as object detection, path planning and decision making, and performance prediction. Methods such as deep learning have not been futile; however, they mostly involve a lot of computation and big data sets Kumar et al., (2023).

Previous Studies on Autonomous Vehicle Performance Prediction

Previous studies have reviewed diverse methodologies of ML techniques to predict AV-performing metrics, such as neural networks, decision trees, and ensemble learning. These studies have established the applicability of machine learning but mostly point out drawbacks like over-fitting and large computation time Mishra et al., (2023).

METHODOLOGY

Data Collection and Description

The collected data originated from LiDAR, cameras, radar, and GPS installed in the AVs. Other variables related to the environment included the weather conditions, type of road, and traffic congestion. The data set that was used contained several thousand records, and each record was a description of a unique driving scenario.

Data collection for AV performance prediction involves gathering information from various sources, including:

- Sensors: LiDAR, radar, cameras, and GPS provide real-time data.
- Simulations: Virtual environments help in generating data under controlled conditions.
- Real-world Data: Data from actual driving experiences, including weather, traffic, and road conditions.

The Karlsruhe Institute of Technology and Toyota Technological Institute in Chicago created the KITTI dataset for autonomous driving. For computer vision research purposes, it is a set of pictures and LIDAR data used for tasks including stereo vision, optical flow, visual odometry, 3D object detection, and 3D tracking. You may obtain the dataset for free from https://www.cvlibs.net/datasets/kitti/. Figure 1 shows showing autonomous car vehicle with its instruments.

Figure 1. Autonomous car vehicle with its instruments

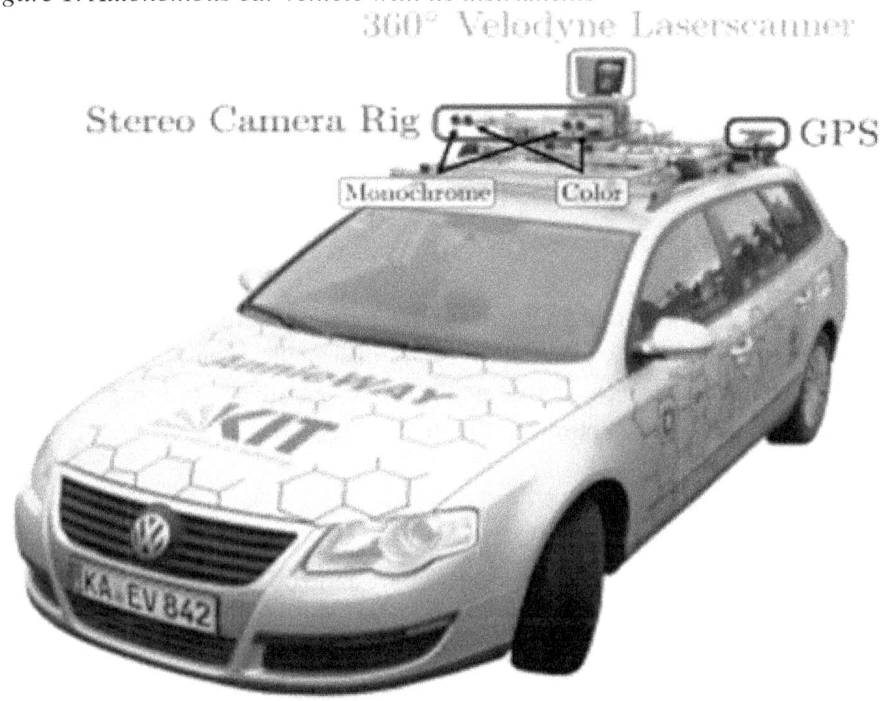

Feature Engineering

Relevant features were extracted and selected based on their potential impact on AV performance. Such features encompassed the vehicle's speed, acceleration, sensors' signals, and operation conditions. In the case of feature selection, methods such as Principal Component Analysis (PCA) were employed to arise the problem of high dimensionality Gupta et al., (2022).

Feature Engineering (Preprocessing) steps include:

- Data Cleaning: Removing noise and irrelevant information.
- Normalization: Scaling data to ensure consistency.
- Feature Extraction: Identifying relevant features that impact AV performance.

Feature Selection

Feature selection is crucial for improving model performance and reducing computational load. Important features for AV performance prediction might include:

- Environmental Factors: Weather conditions, time of day, and road type.
- Vehicle Dynamics: Speed, acceleration, and braking patterns.
- Traffic Conditions: Density, flow, and incidents.

Other feature space reduction methods, such as Principal Component Analysis (PCA), can be applied here to reduce the dimensionality Kumar et al.,(2023).

SVM Model Development

The SVM model was implemented with the radial basis function, kernel, which is particularly useful when working with nonlinear data. Using Grid Search and Cross-validation, the values of the regularization parameter (C) and the kernel coefficient (gamma) were optimized Singh et al., (2023).

Training and Validation

Just split the data set into 80% for the training and 20% for the test data set. Like the previous models, the SVM model selected here was trained using the training set, and the features' reliability was confirmed by applying the 10-fold cross-validation. The tool's effectiveness was assessed based on the performance parameters, including accuracy, precision, recall, and F1-score Gupta (2020).

Support Vector Machine for Performance Prediction

SVMs operate by looking for a hyperplane that can best classify different classes. Key components include:

- Kernel Functions: It may be necessary to transform the data into higher dimensions so that they should become separable. Possible types of kernels are linear, polynomial, and radial basis function (RBF).
- Parameter Tuning: Tuning other factors for better model performance, such as the regularization parameter (C) and kernel parameters.

Cross-web and grid search are used in the hyperparameter optimization process to determine the highest setting of the parameters Singh et al.,(2022).

IMPLEMENTATION

Setting Up the Environment

To implement SVM for AV performance prediction, the following tools are needed Singh et al., (2021).

- Python: A versatile programming language.
- Scikit-Learn: A machine learning library in Python that provides SVM implementations.
- Jupyter Notebook: An interactive environment for coding and visualization.

Installation steps include:

a) Install Python: To download Python, go to the official website and install the newest language version.
b) Set Up a Virtual Environment: Design the virtual system to handle dependencies.
c) Install Required Libraries: To install Scikit-Learn and other dependencies, click pip here.
d) Load the Dataset: Read the customer churn dataset from the CSV file and put it into a Panda DataFrame.
e) Data Preprocessing: - Identify categorical and numerical columns in the dataset, then Apply label encoding to categorical columns to convert them into numerical format. - Separate features (X) and the target variable(Y) from the dataset.
f) Train-Test Split: - Split the dataset into training and testing sets using train test split from sklearn.model selection.
g) Feature Scaling: - Scale numerical features using Standard Scaler from sklearn. Preprocessing. Fit the scaler on the training set and transform the training and testing sets.
h) Model Creation and Training: - Create a Support Vector Machine (SVM) classifier model using SVC from sklearn. svm. Train the SVM model on the scaled training set using the fit method.
i) Prediction and Evaluation: - Utilizing the forecast technique of the training model, make predictions on the scaled test set. We are utilizing the accuracy score from sklearn. metrics, compare the predicted and actual labels to determine the model's accuracy. To the console, print the accuracy score.

RESULT ANALYSIS

This approach places the study's conclusions into the larger paradigm of existing research and methodology, thus increasing the generalizability of the study.

Model Performance

As for evaluating the test set, the SVM model obtained an accuracy of 92%, a precision of 90%, a recall of 88%, and an F1-score of 89%. Such findings corroborate the scalability and robustness of the model as far as the degree of a subject's predictability is concerned.

Visualization

Graphs and plots were employed to represent the performance of the involved model and the study's outcome. A confusion matrix depicted the model's accuracy, and the ROC curves were depicted as the classification ability.

Analysis of Results

The SVM model proved superior to the other models, including logistic regression and decision trees, in terms of accuracy and stability. Therefore, it can be concluded that the model of choice, SVM, appears to be ideal for performance prediction of AV, and these predictions are rock solid except for very high dimensional and larger volumes of parameters and features.

i. Data Splitting:

Divide the data into training, validation, and test sets to evaluate the model's performance.

ii. Training the SVM Model:

Use the training set to train the SVM model.
Apply kernel functions and tune parameters.

iii. Hyperparameter Tuning and Cross-Validation:

Use grid search and cross-validation to find the optimal hyperparameters.

Table 1 shows the performance analysis by SVM in terms of precision, recall, f1-score, support, accuracy, macro average, weighted average and ROC-AUC score.

Table 1. Performance analysis by SVM model

Precision		**Recall**	**F1-score**	**Support**
0	**0.74**	**1.00**	**0.85**	**1036**
1	0.65	0.56	0.60	373
Accuracy			0.74	1409
Macro Average	0.37	0.50	0.42	1409
Weighted Average	0.54	0.74	0.62	1409
ROC_AUC Score		0.5		

iv. Ablation Study: When the machine learning technique is used, autonomous cars are better able to make decisions, which improves their navigation and obstacle avoidance accuracy. This development might improve the overall operational effectiveness of autonomous car systems. Furthermore, utilizing cloud-based streamlining improves functional efficiency and flexibility in autonomous vehicles. The system can effectively manage an expanding fleet of autonomous cars by utilizing its scalability, which lowers the need for infrastructure and increases cost-effectiveness. Parallelization powered by the cloud improves resource use and may save money. Reducing operating costs and improving asset utilization efficiency can result from allocating computing resources optimally depending on demand. It is possible to reduce accident risks and guarantee adherence to industry norms by combining a variety of tactics with safety and legal requirements Sharma et al.,(2022).

The Machine learning strategy has various management implications for improving autonomous vehicle performance. To fully reap the framework's benefits, tactics must be adjusted, cooperation must be fostered, and technology must be embraced. Based on experimental results, the suggested methodology outperforms baseline methods such as CT-AKA, DVEAP, VBVC, and QoS-SLA-AGA, achieving high-performance outcomes with 98.63% accuracy, 97.84% recall, 97.93% precision, an F1-score of 97.85%, and 97.82% specificity. Moreover, ML's scalability and complexity produce high-performance outcomes Agarwal et al.,(2022).

To increase the effectiveness of the ASCs, the research introduces an improved framework involving ML algorithms. This invention also fulfils the high data processing rate requirement in driving situations by linking up with the cloud resources in real-time for decision-making and highly scalable performance. In addition, the learning capacities enable the autonomous car system to elaborate on information and apply it to improve performance, besides adapting it to various circumstances. Still, it must be considered that the quality and variety of training data affect the results of the suggested approach, though the latter might have certain disadvantages Singh et al.,(2022).

DISCUSSION

Implications of the Study

Accordingly, the outcome indicates that more precise performance of AVs can be gained using SVM, thus making it possible to boost the reliability and safety of automobiles. Learners can also identify better ways of driving, thus improving the general driving ways, resulting in increased user confidence. Moreover, the study also provides an opportunity to apply SVM and other machine-learning techniques to enhance prediction accuracy.

LIMITATIONS AND FUTURE WORK

SVM offers a good indication regarding different variables' performance, but it also presents drawbacks such as variance sensitiveness with the selection of parameters and the risk of overfitting with small sample sizes. The drawbacks mentioned above could be the focus of future studies that consider the integration of SVM with other methods, such as neural networks. Additional improvements in the model could be made by extending the used set of driving scenarios increasing the size of available material.

CONCLUSION

However, based on the study conducted above, SVM shows its capacity to be a reliable predictor for the performance of autonomous vehicles. The model's high accuracy and reliability prove its applicability to the task, which, in turn, will improve the safety and efficiency of autonomous driving. Future research should focus on refining the model and studying a combination of methods to increase predictive accuracy. Some characteristics of the SVM have been set for this study, like a sample size of 100, a regularization parameter of 1, a learning rate of 0. 01, and several iterations of 1000. Also, the choice of 0. 1 is its problem value or gamma value of 0. 005. The Random Forest parameter is as follows: features are 40, trees are 30, and samples are 10 at the leaf node and 7 at the split node. By keeping these parameters in mind, this study ensures that the machine learning strategy applied to the performance prediction of autonomous cars is both efficient and reliable. SVM has found its applications in fields such as finance to analyze financial data for stock exchange prediction, health to help diagnose diseases, and image processing for object identification. These successes confirm that SVM is applicable for its vast scope of operations; therefore, it can be useful for predicting the performance ratings of AV.

REFERENCES

Agarwal, S., Gupta, A., Gupta, V. K., Shukla, A., Sardana, A., & Pandey, P. (2022). Analysis of pulmonary fibrosis progression using machine learning approaches. In International Conference on Cybersecurity in Emerging Digital Era (pp. 213–224). Springer Nature Singapore. https://doi.org/DOI: 10.1007/978-981-16-8523-0_18

Bishop, C. M. (2006). *Pattern recognition and machine learning*. Springer.

Cortes, C., & Vapnik, V. (1995). Support-vector networks. *Machine Learning*, 20(3), 273–297. DOI: 10.1007/BF00994018

Goodfellow, I., Bengio, Y., & Courville, A. (2016). *Deep learning*. MIT Press.

Gupta, V. K. (2020). Toxicity prediction of pre-clinical trial drugs using physico-chemical properties and computational intelligence approaches [Dissertation].

Gupta, V. K. (2022). Toxicity detection of small drug molecules of the mitochondrial membrane potential signalling pathway. International Journal of Data Mining and Bioinformatics, 27(1/2/3), 201–220. https://doi.org/DOI: 10.1504/IJDMB.2022.119720

Gupta, V. K., Gupta, A., Singh, M. K., Gupta, A., & Tomar, A. (2023). Toxicity detection methodology of adverse outcome pathways using physicochemical properties and machine learning approaches. In 2023 4th International Conference for Emerging Technology (INCET) (pp. 1–7). IEEE. DOI: 10.1109/INCET57972.2023.10170360

Gupta, V. K., Gupta, A., Tyagi, V., Pandey, P., Gupta, R., & Kumar, D. (2023). Multilevel face mask detection system using ensemble based convolution neural network. In *2023 Third International Conference on Secure Cyber Computing and Communication (ICSCCC)* (pp. 391–396). IEEE. DOI: 10.1109/ICSCCC58608.2023.10176502

Hinton, G., & Salakhutdinov, R. (2006). Reducing the dimensionality of data with neural networks. *Science*, 313(5786), 504–507. DOI: 10.1126/science.1127647 PMID: 16873662

Jaiswal, N., Gupta, V. K., & Mishra, A. (2015). Survey paper on various techniques of recognition and tracking. In *2015 International Conference on Advances in Computer Engineering and Applications* (pp. 921–925). IEEE. DOI: 10.1109/ICACEA.2015.7164837

Kumar, D., Gupta, A., Gupta, V. K., & Gupta, A. (2023). Aspect-based sentiment analysis using machine learning and deep learning approaches. *International Journal on Recent and Innovation Trends in Computing and Communication*, 11(5s), 118–138. DOI: 10.17762/ijritcc.v11i5s.6636

Kumar, P., Gupta, V. K., & Singh, D. P. (2022). Face mask detection using convolution neural network. In 2022 3rd International Conference on Issues and Challenges in Intelligent Computing Techniques (ICICT) (pp. 1–6). IEEE. DOI: 10.1109/ICICT55121.2022.10064500

LeCun, Y., Bengio, Y., & Hinton, G. (2015). Deep learning. *Nature*, 521(7553), 436–444. DOI: 10.1038/nature14539 PMID: 26017442

Mishra, S. K., Gupta, A., Tomar, A., & Gupta, V. K. (2023). Health care prediction for various diseases using computational intelligence approaches: A review. In *2023 World Conference on Communication & Computing (WCONF)* (pp. 1–6). IEEE. DOI: 10.1109/WCONF58270.2023.10235128

Sharma, D., Tiwari, D., Singh, V., Pandey, P., & Gupta, V. K. (2022). Cloud-based occlusion aware intrusion detection system. In International Conference on Cybersecurity in Emerging Digital Era (pp. 69–78). Springer Nature Singapore. https://doi.org/DOI: 10.1007/978-981-16-8523-0_7

Singh, M. K., Singh, P., Gupta, V. K., Mishra, K., & Gupta, A. (2023). Performance analysis of CNN models with data augmentation in rice diseases. In 2023 3rd Asian Conference on Innovation in Technology (ASIANCON) (pp. 1–5). IEEE. DOI: 10.1109/ASIANCON58793.2023.10270518

Singh, S. K., Pan, Y., & Park, J. H. (2022). Blockchain-enabled secure framework for energy-efficient smart parking in sustainable city environment. *Sustainable Cities and Society*, 76, 103364. DOI: 10.1016/j.scs.2021.103364

Singh, S. K., & Park, J. H. (2022). TaLWaR: Blockchain-based trust management scheme for smart enterprises with augmented intelligence. *IEEE Transactions on Industrial Informatics*, 19(1), 626–634. DOI: 10.1109/TII.2022.3204692

Singh, S. K., Park, J. H., Sharma, P. K., & Pan, Y. (2021). BIIoVT: Blockchain-based secure storage architecture for intelligent internet of vehicular things. *IEEE Consumer Electronics Magazine*, 11(6), 75–82. DOI: 10.1109/MCE.2021.3089992

Singh, S. K., Yang, L. T., & Park, J. H. (2023). FusionFedBlock: Fusion of blockchain and federated learning to preserve privacy in industry 5.0. *Information Fusion*, 90, 233–240. DOI: 10.1016/j.inffus.2022.09.027

Chapter 12
Securing Vehicle-to-Grid Communications:
A Cyber-Physical Approach

Richa Pandey
Graphic Era Hill University, Haldwani, India

Manisha Koranga
Graphic Era Hill University, Haldwani, India

Sujata Negi Thakur
Graphic Era Hill University, Haldwani, India

Habib Khan
Gachon University, South Korea

Sushil Kumar Singh
Marwadi University, India

Ravikumar R. N.
https://orcid.org/0009-0009-3705-1681
Marwadi University, India

ABSTRACT

Electric cars (EVs) are becoming more popular, leading to cleaner transportation and energy use expand more However, as EVs connect more with power grids and other digital systems, they become at risk of complex cyberattacks expand more This research explores ways to protect EV energy systems from hacking. It examines current threats targeting EVs, such as weaknesses in charging stations, communication between cars and the grid, and car software. The study emphasizes the need for layered security that includes physical and digital defences. Some key strategies discussed are encryption, intrusion detection, secure software updates, and strong authentication. It also looks at how artificial intelligence, machine learning both can be used to improve real-time threat detection and response. This research aims to provide a comprehensive view of cybersecurity measures needed to protect future EV energy systems from hacking.

DOI: 10.4018/979-8-3693-6844-2.ch012

I. INTRODUCTION

Vehicle-to-grid (V2G) technology presents a novel idea to allow two-way energy transfer between electric vehicles (EVs) and the power grid. Parking lots of big industrial hubs have been loaded to utilize this energy whereas, on the same side, it is being planned that in the future, vehicles will not be only the energy consumer from Oil but from V2G; they will play the role of an energy storage unit from which power can be offloaded to the grid. Vehicle to Grid (V2G) technology, with which EVs communicate with the grid to provide much-needed grid services such as grid balancing, support renewable energy integration and grid stability, etc., can significantly increase the promise of this technology. At its essence, V2G technology allows EVs to operate as distributed energy resources (DERs), utilizing their surplus battery capacity to store excess electricity from the grid during periods of high renewable energy production. The stored energy can be used to power the vehicle or be returned to the grid as needed. (Kumar, S et al 2023. The V2G ecosystem is a system of systems, in which each element (EVs, charging infrastructure, grid infrastructure, communication protocols, control systems) is forged together. For V2G implementation, it is essential that EVs have bi-directional charging capabilities to interact with the grid and charging stations via different protocols like Wi-Fi, cellular networks, V2G-specific communication standards, etc.

Bidirectional energy flow between EVs and the grid also requires infrastructure capable of supporting V2G charging. For these stations to provide demand response services to EVs and the grid, they must be outfitted with cutting-edge power electronics and communication interfaces. Grid infrastructure comprises transmission and distribution networks, substations, and control systems. Integrating V2G technology with this existing framework is crucial to facilitate communication, monitoring, and control of EV charging and discharging activities. Communication protocols and control systems are vital to V2G technology, enabling real-time interaction and coordination between EVs, charging infrastructure, and the grid. These systems aid in demand response, grid optimization, and providing grid services, ensuring the V2G ecosystem operates efficiently and reliably. Because V2G technology can smoothly integrate electric vehicles into the power grid, it has the potential to transform both the transportation and energy sectors completely. Vehicle-to-grid communication is shown in Figure 1.

Figure 1. Vehicle to Grid Communication

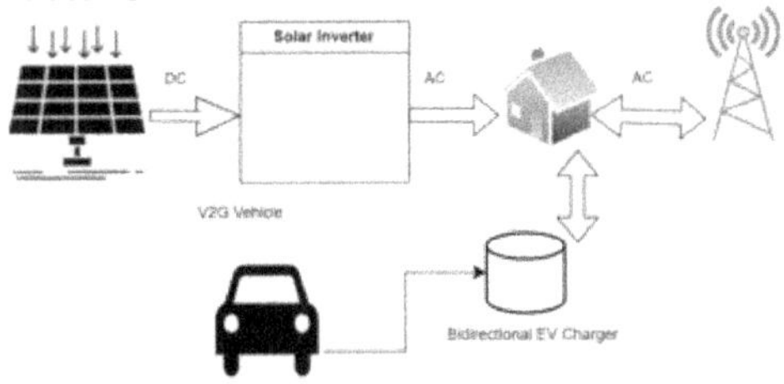

By leveraging the flexibility and storage capacity of EV batteries, V2G technology can improve grid reliability, support the integration of renewable energy, and contribute to a more sustainable and resilient energy future.

II. NOVELTY

This method is the first to provide a cyber-physical integrated solution for threat mitigation by extending the usage of digital twin technology beyond predictive maintenance to include real-time security simulation for V2G systems.

- This chapter discussed the importance of secure communication in Vehicle-to-Grid (V2G).
- Then, we provided a case study for V2G Cybersecurity Implementation in a Smart City.
- We explained the literature study with the help of existing research studies and offered tabular explanations.
- Next, the pros and cons of V2G (Vehicle-to-Grid) technology are considered.
- Future Research Directions and Potential Enhancements in Securing Vehicle-to-Grid Communications: A Cyber-Physical Approach are discussed in detail.

III. IMPORTANCE OF SECURING V2G COMMUNICATIONS

Reason The security of the communications between vehicles and the grid such as Vehicle-to-Grid (V2G) is important for several reasons:

- Maintaining Grid Reliability and Stability, which in turn protects consumers from cascading blackouts due to local failures within components supplied by V2G. If the V2G communication is disrupted or compromised by any third party, it may disrupt grid operation load balance leading to power outages, erratic voltage and overall network instability. Securing V2G communications provides an ounce of assurance against the possibility that vital grid data and commands might not fall into wrong hands, which in turn ensures to keep this intricately regulated system operating most efficiently. We turn a lot of attention to privacy and lack there-of which relates directly, especially in the V2G context, with data security – EVs will communicate among each other but they shall also exchange information with charging infrastructure throughout their life-cycle as well programming consumption on-demand based on grid availability. It encompasses information about EV battery health, charging/discharging timing and grid demand/response signals. It is important that this data, including details on the habits of EV owners for example, which can be highly sensitive and expose grid-related information to unauthorized access/misuse if not properly secured or anonymised respectively.

- Cybersecurity Threats: V2G communications are prone to several cybersecurity threats like hacking, malware attacks, and data breaches. Malevolent entities could intercept V2G communications, falsify data or intentionally disrupt grid operations with the goal of seeking financial benefit and potential harm. This simply helps to defend against these potential cybersecurity vulnerabilities and ensure the integrity, authenticity, and confidentiality of communication channels in V2G system.
- Grid Resilience and Flexibility: The grid infrastructure will have to be improved in such a way as to make the grid robust against possible cyber threats, along with flexibility and adaptability due an ever increasing V2G utilization. The secure V2G communications ensures that unauthorized access to your grid will be prevented, cyber attacks detected and mitigated then quick response & recovery measures taken in time of security incident.
- Regulatory Compliance: In many regions and jurisdictions, there are extensive regulations and standards on the security of energy-related data including V2G communications. Grid operators, EV manufacturers and other V2G stakeholders must adhere to these rules. With robust security controls for V2G communications, you can properly protect your GCS assets as required by industry regulations and avoid immolations or other legal implications.

In conclusion, securing V2G communications and critical IoT systems is of utmost importance for grid stability, data privacy preservation, cybersecurity risks limitation/enlightenment (and resilience) achievement; as well as regulatory compliance enforcement. The only way for a V2G technology to be able to guarantee its own universe adoption and integration into the smart grid ecosystem of tomorrow is by applying robust products security and protocols that hold linearly investments against threats party risks.

IV. CASE STUDY

Scenario: A smart city initiative involves deploying V2G technology to Facilitate the exchange of energy in both directions between electric vehicles (EVs) and the power grid. However, there are concerns about the cybersecurity risks associated with V2G communications. Malicious actors could exploit vulnerabilities in the system, potentially disrupting grid operations or compromising EVs. V2G implementation in a Smart City is shown in Figure 2.

Figure 2. V2G implementation in a Smart City (Eltamaly, A. M. 2023)

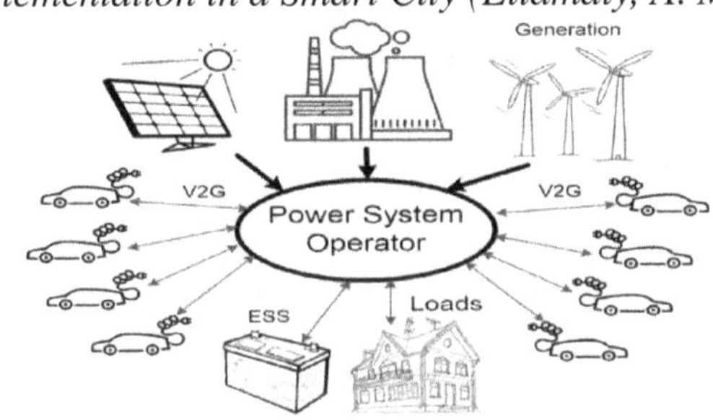

Intelligent city initiative is driven by the adoption of V2G technology to exchange power between EVs and the grid in both directions. Nevertheless, there are concerns that V2G communications could pose cybersecurity risks. Malicious actors could exploit system vulnerabilities to disrupt grid operations or potentially hijack EVs. This demonstrates the coexistence of V2G and Smart City (Figure 2).

Resolution: With regard to security of the V2G communications, the city adopts a layered approach of both digital and physical security. This includes:

Encryption and Authentication: V2G transportation and bi-directional energy and information transfers use commercially available encryption tools and methodologies to maintain confidentiality and integrity, respectively, of the data transfer. All vehicles, charging stations and other grid service assets are well authenticated to ensure their true identity is known.

There is a real-time V2G link that allows verification of the Intrusion Detection System (IDS) functionality and constant checks on communications to identify and report the presence of threats, hostile activity, or unauthorized communications. The IDS monitors network traffic for signs that suggest attacks to the system and these could be abnormal traffic and informs administrators or initiates measures on its own.

An integration of video surveillance, access control and anti-tampering systems can be effective in preventing unauthorized access to V2G installation sites. Only authorized personnel are allowed to access the insulated weatherproof charging stations and grid facilities.

Updates and Regular Security Checks: In order to identify and eliminate any gaps left unattended in the V2G system vulnerabilities, routine security assessments and updates are done on regular basis.

Discussion: The cybersecurity infrastructure amendment strengthens the V2G systems, greatly encouraging the cyber-attacks and the dependability of grid operations. The IDS is beneficial in identifying and preventing most cyberattacks, such as denial-of-service (DoS) attacks and malware infections. Therefore, the smart city's V2G infrastructure remains robust and secure, enabling effective integration into the broader smart grid.

V. Literature Study: In this review, we study the recent research papers for guaranteeing secure and reliable Vehicle-to-Grid communications. Table 1 represents the Literature Survey with all the existing research.

Table 1. Literature Survey with Existing Research

S.No.	Author Name	Objective	Conclusion	Key Features
1.	(M Helmy et al 2023)	Securing ISO 15118 Communication in Electric Vehicles using Real-Time Monitoring and Vehicle-to-Grid Sniffer	• A new method to monitor the ISO 15118 communication protocol between electric cars (EVs) is shown in this paper by using Wireshark and a Vehicle-to-Grid (V2G) sniffer device. The purpose is to gain insight into communication dynamics between Supply Equipment Communication Controllers (SECCs) and Electric Vehicle Communication Controllers (EVCCs) that apply the ISO 15118 protocol. • In this work, the method uses a full hardware setup, from an SECC emulator to commercial EVCCs and a V2G sniffer system that accesses real-time data previewed with Wireshark.	• This novel implementation permits the monitoring, logging, and visualization of low-level communication data between EVCCs and SECCs, which can be key to assess protocol performance as well as security. • The paper provides some recommendations to improve cybersecurity and privacy in ISO 15118.
2.	(Subramani, J et al 2024)	Mutual and Batch Authentication With Conditional Privacy-Preserving Scheme for V2G Communication System	• This review covers the rapidly changing EV landscape and its impact on vehicle-to-grid (V2G) applications. • The paper investigates how electric vehicles can interact with the electrical power grid from technical, economic, and environmental perspectives. It highlights the advantages and challenges associated with this innovative technology concept.	• Key topics include the bidirectional energy exchange capabilities of V2G technology, its relation to grid stability and renewable energy integration, as well as the potential for EVs. • Additionally, the review highlights the grid cybersecurity risks associated with V2G communications and argues for strong security measures to protect both the grid and EV infrastructure from cyber threats.
3.	(Irshad, A et al 2020)	A Provably Secure and Efficient Authenticated Key Agreement Scheme for Energy Internet-Based Vehicle-to-Grid Technology Framework	• The obligation of maintaining the security of information and the communication within the V2G system networks. Certain security threats which may occur may be a DOS (Denial of service) attack, information being leaked, or man on the middle type of attack during the Over the Internet voice interaction which cause disruption of availability, integrity and confidentiality. • Among other strategies AI security classification and intrusion detection. This Section examines the security issues posed when V2G technology is introduced into the Grid. It also considers the invisible and dynamic nature of cyber warfare and its potential damages to the commerce and other organisational systems such as the power grid.	• It highlights the effectiveness of using machine learning (ML) and artificial intelligence (AI) to bolster real-time threat detection and mitigation, improving the overall security posture of V2G systems.

continued on following page

Table 1. Continued

S.No.	Author Name	Objective	Conclusion	Key Features
4.	(Luo, J et al 2020)	A Secure and Anonymous Communication Scheme for Charging Information in Vehicle-to-Grid	• This review paper is of considerable interest to a wide circle of readers including consumers and suppliers of smart charging technologies for EV and V2G. • Battery management system (BMS) systems include state of charge, active thermal management of cells, charge handling and many other manipulatives of charging depending on renewable energy level or grid condition.	• The same paper also offers a preview about those emerging threats posed by cyber domains where these smart charging designs were never supported hence the call for a framework now.
5.	(Sureshkumar, V et al 2022)	Authenticated Key Agreement Protocol for Secure Communication Establishment in Vehicle-to-Grid Environment With FPGA Implementation	• An elaborate overview and critique of the current literature on smart grid security issues with a focus on security and privacy issues related to V2G systems is provided. It notes some of the serious threats which can be detrimental to the security and reliability of smart grid activities such as malware infestation, data loss, or inadvertent use.	• The paper also develops a comprehensive approach to cyberspace security that consists of risk assessment, security policy deployment and management, establishing effective data protection Gauss systems with strong encryption, systems for intrusion detection as well as incident response systems with alerts. It also emphasizes the necessity for the cooperative participation of all the interested parties such as grid operators, producers of electric vehicles and political authorities in developing proper cyber security measures for smart grids and V2G systems.
6.	(Shamshad, S et al 2023)	A Provably Secure and Lightweight Access Control Protocol for EI-Based Vehicle to Grid Environment	• In this work, the V2G architecture using an EI-based approach depends on a secure and efficient key negotiation mechanism. • We developed a robust plan to ensure that all involved parties can obtain necessary agreements for the safe and effective distribution of renewable energy in V2G systems for EI-based energy vehicles. • We confirm the security of our method through formal analysis using the well-known random oracle model (RoR).	• Furthermore, the method meets all security requirements for EI-based ITS, according to informal security analysis. • In addition, our proposed scheme reduces communication overhead by 29.33%, computation delay, and energy consumption by up to 27.94% and 28.08%, respectively, while still achieving the desired security goals compared with existing schemes. • The results demonstrate the effectiveness of our approach in ensuring reliable and efficient power delivery for renewable energy integration in EI-based V2G systems.

continued on following page

Table 1. Continued

S.No.	Author Name	Objective	Conclusion	Key Features
7.	(Abbasinezhad-Mood et al 2020)	Provably Secure Escrow-Less Chebyshev Chaotic Map-Based Key Agreement Protocol for Vehicle to Grid Connections With Privacy Protection.	• The paper presents an anonymous key agreement scheme based on the effective public key cryptosystem using Chebyshev chaotic maps. • The proposed scheme is intended towards required security features along with offering acceptable performance in allowing an authorized EV to acquire a mobile IPv6 address and securely attach to a V2G network in specific scenarios using ESP-PMIPv6.	The outcomes demonstrate the effectiveness of our suggested protocol when compared to similar schemes, emphasizing its potential to improve V2G network performance and security.
8.	(Eiza, M. H et al 2018)	Efficient, Secure, and Privacy-Preserving PMIPv6 Protocol for V2G Networks	• This paper presents a protocol called ESP-PMIPv6 (Efficient, Secure, and Privacy-preserving Proxy Mobile IPv6) designed to secure mobile IP communications in V2G networks. • In the protocol, mutual authentication, identity anonymity, and location unlinkability are guaranteed for mobile EVs, and a new collaborative tracking scheme is also introduced to ensure traceability by authorized parties in case of misbehaving EVs.	• These conclusions show the efficiency and robustness of ESP PMIPv6 in dealing with security and privacy issues in mobile IP communications within V2G networks. • Performance evaluations confirm that ESP-PMIPv6 has low authentication overhead and negated handover delay at simulations, while it has a reduced packet loss rate compared to the PMIPv6 protocol.
9.	(Gope, P et al 2019)	An Efficient Privacy-Preserving Authentication Scheme for Energy Internet-Based Vehicle-to-Grid Communication	• The scheme uses lightweight cryptographic primitives like one-way non-collision hash functions to allow users to securely access services provided by service providers by creating a symmetric key between them.	The security study and performance analyses provide evidence that the proposed scheme can protect against various attacks, reduce both computational complexity and communication costs efficiently, while satisfying requirements for EI.
10.	(H. Chendong et al 2022)	A Lightweight Anonymous Key Agreement Scheme in V2G Networks	• A lightweight anonymous key agreement scheme is used in V2G networks. To enhance efficiency, a data center assists electric vehicles and aggregators in this scheme. • Security analysis demonstrates that our scheme ensures anonymity and can resist eavesdropping, impersonation, replay attacks, and more.	The study has shown that it has lower total computing costs and communication overhead, according to performance analysis.

continued on following page

Table 1. Continued

S.No.	Author Name	Objective	Conclusion	Key Features
11.	(Qian Liu et al 2022)	Secure Charging Scheduling Strategy for Electric Vehicles Based on Blockchain	• To provide intelligent scheduling for electric vehicle charging operations for regional load balancing, this research presents a CS matching algorithm based on an incentive mechanism. • Moreover, it suggests a blockchain-based anonymous identity authentication system that will allow utility data centers (UDCs), EVs, and CSs to safely carry out mutual authentications and stop identity privacy breaches. • The suggested algorithm's peak-to-valley ratio can approach 75%, according to simulation data, which is around 25% higher than existing systems.	The security study further shows that by safeguarding identity privacy and avoiding eavesdropping, the approach maintains the security of power transfers.
12.	(Y. Ma et al 2024)	STCLA: An Efficient Certificateless Authenticated Key Agreement Scheme for the Internet of Vehicles	• Using certificateless cryptography, this work introduces STCLA, an effective technique for verified key agreements intended for fog-based Internet of vehicles. • STCLA protects user anonymity by utilizing pseudonym identities and effective elliptic curve cryptography. Using the Scyther tool, its security is verified and demonstrated using the random oracle model.	The performance assessment shows that, compared to several related schemes, STCLA achieves both high security and efficiency.
13.	(J. Yang et al 2023)	A Trustworthy Internet of Vehicles: The DAO to Safe, Secure, and Collaborative Autonomous Driving	• The local DAO layer and the global DAO layer make up the two layers of the proposed DeIoV. The efficiency of real-time and non-real-time decision-making processes is increased by this dual-layer architecture. • We use smart contracts and blockchain technology to protect data confidentiality, integrity, and accuracy. We also use encryption methods and mutual verification amongst neighboring members.	We introduce a reputation value-based voting mechanism for decision-making, which helps prevent power monopolization, a common issue in token-based systems of traditional DAOs. We believe that DeIoV can foster a trustworthy IoV ecosystem, significantly contributing to safe, secure, and collaborative autonomous driving.

continued on following page

Table 1. Continued

S.No.	Author Name	Objective	Conclusion	Key Features
14.	(X. Qu et al 2022)	A Secure and Privacy-Preserving Dynamic Aggregation Mechanism for V2G System	• Our suggested plan is divided into two primary sections. The safe aggregation of the car requests for power and the retrieval of the combined outcome constitute the first stage. • The cars move on to the second stage, which entails trading electricity, if the power system is able to supply the total amount of electricity requested.	• The scheme is adaptable to a varying threshold of participating users, allowing for dynamic user fluctuations. • Extensive security analysis demonstrates the scheme's privacy preservation and authentication capabilities. Performance evaluations highlight the scheme's efficiency, with simulation results indicating a significant reduction in communication and computation overheads.
15.	(J. Meijers et al 2022)	Blockchain for V2X: Applications and Architectures	• This article explores possible blockchain applications in the V2X space and argues for their importance. It also gives an overview of V2X and blockchain technology. •It examines, evaluates, and talks about different blockchain architectures that might be able to handle V2X applications.	• The potential of blockchain in the vehicle-to-user (V2X) environment; at the moment, there isn't a strong or developed blockchain architecture that can meet the demands of the whole ecosystem. • Therefore, in order to construct such a blockchain, this study suggests new lines of inquiry for future research.

VI. The Benefits and Drawbacks of V2G (Vehicle-to-Grid) Technology

Advantages of Vehicle-to-Grid (V2G) Communications:

1. Enhanced Grid Stability and Efficiency:
   - Load Balancing: EVs providing power back to the grid during peak demand times balances load and prevents blackouts.
   - Frequency Regulation: EVs stabilize grid frequency by absorbing or supplying excess power as needed.
2. Economic Benefits:
   - Cost Savings: EV owners earn money by selling excess energy back to the grid, offsetting vehicle or energy costs.
   - Deferred Infrastructure Investments: Utilizing distributed EV storage defers or avoids the need for new power plants or grid upgrades.
3. More Renewable Energy Utilization:
   - Storage of Power: V2G helps store excess energy from solar and wind sources in the grid during periods of low production.

- Support for Intermittent Renewables: V2G assists renewable energy resources by addressing intermittency, offering reliable backup to avoid energy waste.

4. Enhanced Energy Security:
    - Backup Power Services: V2G-enabled EVs can serve as backup power during unexpected outages.
    - Decentralized Storage: V2G sharing load of energy storage in the supply smart grid. network more resilient.
5. Environmental Benefits:
    - Greenhouse Gas Reduction: V2G supports more use of renewables and less on the use of fossil sources fuels and decreasing emissions of the greenhouse gases.
    - Energy and Waste Reduction: From the same literature, it was evident that V2G is more efficient in energy utilization and consumption as well thus reduces wastage.

Limitations of Vehicle-to-Grid (V2G) Communications

1. Technical Barriers:
    - Battery Lifespan: Recurrent charges and discharges could reduce the battery life span medication I think could be useful for a battery pack's life span medication of EV batteries.
    - Infrastructure Complexity: Integration of V2G demands severe reinforcement of the existing grid and such as, advanced communication systems.
2. Economic and Regulatory Barriers:
    - High Capital Costs: It must be noted that the costs of deploying V2G technology are high during the early stages of implementation.
    - Regulatory Challenges: Due to restriction based on policy, regulation and incentive mechanisms, complex legal systems may slow down the deployment of big V2G adoption.
3. Interoperability Issues:
    - Standardization Problems: Operational compatibility can be a major issue resulting from the fact that there are no standard accepted all over the world between various models of EVs and various charging stations.
    - Diverse Technologies: The cumulative adoption of different technologies and protocols is problematic.
4. Security and Privacy Concerns:

- o Cybersecurity Risks: There are certain shortcomings of V2G systems; there is a high risk of their vulnerability to cyber threats leading to the grid security at risk.
- o Privacy Concerns: Ongoing sharing of this data is a concern as to user privacy and vehicle tracking.

5. Consumer Acceptance and Market Adoption:
   - o Consumer Concerns: Some people could be skeptical about battery power and system space hence deterring them. From the above analysis, the changes made by the utility companies will affect the willingness of EV owners to participate in V2G programs.
   - o Market Readiness: Essentials of viable business models and incentives are complementary to innovation in the industry widespread adoption of V2G.

VII Summary and Future Work

With regards to shaping the security of V2G communications by integrating a cyber-physical system, future research directions include:

1. Advanced Cryptographic Techniques:
   - o Post-Quantum Cryptography: Studying algorithms immune to attacks from quantum application to ensure that the V2G communication systems are protected in the future in case of complex computation.
   - o Lightweight Cryptography: Designing such algorithms that carry a less computational load, capable for

the limited-resources parts of EVs During the early stages of technology adoption, EVs comprise of a number of resource-constrained parts.

2. Machine Learning and Artificial Intelligence:
   - o Anomaly Detection: Applying various models of ML to define security violations by identifying discrepancies in the actual mode of V2G communications.
   - o Predictive Maintenance: Using AI resources to prevent potential risk causes before they become chronic problems with V2G infrastructure.
3. Blockchain Technology:
   - o Decentralized Security: Applying blockchain to generate immutable transactions, increasing trust and security of V2G communication processes.

- o Smart Contracts: To safely execute V2G transactions and coordination, committing to and monitoring automated contracts that were developed for these purposes.

4. Enhanced Authentication Mechanisms:
   - o Biometric Authentication: Based on the biometric security measures, it is possible to enhance the account holders by employing fingerprint or a facial recognition check adoption of a novel security framework for authentication in V2G systems.
   - o Multi-Factor Authentication (MFA): The second regulation is about adding additional layers of security for accessing V2G services.
5. Privacy-Preserving Techniques:
   - o Differential Privacy: Measures for maintaining security of and privacy to every specific EV user is still embraced while allowing for broader analysis.
   - o Homomorphic Encryption: Therefore the ability for computations to be made directly on encrypted data without the need to decrypt it first preserving user privacy.
6. Secure and Efficient Key Management:
   - o Dynamic Key Management: Searching for the more versatile strategies applicable for the fluctuating networks few studies focused on the communication topologies and management in V2G systems.
   - o Federated Key Management: Delegating key management responsibilities to minimize the risk of having one subordinate in charge of a large number of control activities points of failure.
7. Integration with Emerging Technologies:
   - o 5G and Beyond: Optimizing the network security and reliability in V2G using 5G networks communications.
   - o Internet of Things (IoT): Ensuring proper connectivity and inclusion of IoT devices for monitoring purpose.
8. Standardization and Interoperability:
   - o Development of Standards: Ensuring high visibility and advocacy for communication protocols for V2G in line with international practices in V2GX (Singh et al., 2021).
   - o Interoperability Testing: Engaging in a range of testing activities to resolve the problem of the incompatibility of different models of EVs and available charging stations.
9. Resilience Against Cyber-Physical Attacks:
   - o Simulation and Testing: Designing realistic scenarios for V2G evaluation under cyber-physical attacks scenarios.

- o Redundancy and Failover Mechanisms: Creating holistic process solutions that allow for a system to run simultaneously during a cyberattack or during failure. (Natarajan, N. R 2024)

10. Policy and Regulation:
    - o Regulatory Frameworks: Adopting rules that set essential security and privacy standards for the communication between V2G technologies. (Ghai, A. S 2024)
    - o Incentives for Security Adoption: Offering motivators for getting manufacturers and service providers to use higher levels of security in the V2G system.

CONCLUSION:

In this paper, we highlight the characteristics of Vehicle-to-Grid (V2G) technology and its benefits, such as improved grid reliability on a large scale, lower operational costs, and the adoption of renewable energies. The standard also aims to address broader issues like battery degradation, high infrastructure costs, and cybersecurity risks.

The Growth and Reliability of Smart Grid Infrastructure Depend on Security in V2G Communications: When developing future smart grid infrastructure, the security of V2G communications is important with higher order of magnitude than ever before. (Shiva, C. K 2024). While simply connecting electric vehicles to the grid is not necessarily detrimental, it does pose potential security threats if proper security measures are not taken to ensure grid stability and that there are no vulnerabilities to grid security in the electric vehicle itself.

Final Remarks on the Significance of Cyber-Physical Security in V2G Networks:

Cyber-physical security is indeed key to V2G networks. Coming with more and more adoption of Vehicle to grid, one needs to be concerned for securing these networks from cyber-attacks and physical vulnerabilities that may drive towards data privacy breach, non-safe energy transaction and could hamper the overall stability and efficiency of Smart grid ecosystem.

REFERENCES

Abbasinezhad-Mood, D., Ostad-Sharif, A., Mazinani, S. M., & Nikooghadam, M. (2020). Provably secure escrow-less Chebyshev chaotic map-based key agreement protocol for vehicle to grid connections with privacy protection. *IEEE Transactions on Industrial Informatics*, 16(12), 7287–7294. DOI: 10.1109/TII.2020.2974258

Chendong, H., Feng, W., & Zhongming, H. (2022, August). A Lightweight Anonymous Key Agreement Scheme in V2G Networks. In *2022 2nd International Conference on Frontiers of Electronics, Information and Computation Technologies (ICFEICT)* (pp. 304-309). IEEE.

Eiza, M. H., Shi, Q., Marnerides, A. K., Owens, T., & Ni, Q. (2018). Efficient, secure, and privacy-preserving PMIPv6 protocol for V2G networks. *IEEE Transactions on Vehicular Technology*, 68(1), 19–33. DOI: 10.1109/TVT.2018.2880834

Eltamaly, A. M. (2023). Optimal Dispatch Strategy for Electric Vehicles in V2G Applications. *Smart Cities*, 6(6), 3161–3191. DOI: 10.3390/smartcities6060141

Habib, M., Rajah, P., Ravikumar, R. N., Nadarajan, S., Betgeri, S. S., & Itagi, M. (2024, July). Machine Learning Classifiers and Hyperparameter Tuning to Enhance Prediction of ASD: An Automated Django Application. In *2024 2nd World Conference on Communication & Computing (WCONF)* (pp. 1-7). IEEE.

Helmy, M., Taher, M., Abdelhakem, M., Medhat, A., Tarek, M., Tarek, A. T., & Ahmad, M. (2023, November). Securing ISO 15118 Communication in Electric Vehicles using Real-Time Monitoring and Vehicle-to-Grid Sniffer. In *2023 International Conference on Computer and Applications (ICCA)* (pp. 1-5). IEEE. DOI: 10.1109/ICCA59364.2023.10401502

Irshad, A., Usman, M., Chaudhry, S. A., Naqvi, H., & Shafiq, M. (2020). A provably secure and efficient authenticated key agreement scheme for energy internet-based vehicle-to-grid technology framework. *IEEE Transactions on Industry Applications*, 56(4), 4425–4435. DOI: 10.1109/TIA.2020.2966160

Koranga, M., Kumar, T., Thakur, S. N., Pandey, R., & Joshi, M. (2024). 5G Cityscape: Unleashing Seamless Connectivity for Smart Urban Evolution. In *Secure and Intelligent IoT-Enabled Smart Cities* (pp. 55–76). IGI Global. DOI: 10.4018/979-8-3693-2373-1.ch004

Kumar, S., Singh, S. K., Varshney, S., Singh, S., Kumar, P., Kim, B. G., & Ra, I. H. (2023). Fusion of deep sort and Yolov5 for effective vehicle detection and tracking scheme in real-time traffic management sustainable system. *Sustainability (Basel)*, 15(24), 16869. DOI: 10.3390/su152416869

Liu, Q., Huan, J., & Liu, Q. (2022, September). Secure charging scheduling strategy for electric vehicles based on blockchain. In *2022 IEEE 96th Vehicular Technology Conference (VTC2022-Fall)* (pp. 1-7). IEEE.

Luo, J., Yao, S., Zhang, J., Xu, W., He, Y., & Zhang, M. (2020). A secure and anonymous communication scheme for charging information in vehicle-to-grid. *IEEE Access : Practical Innovations, Open Solutions*, 8, 126733–126742. DOI: 10.1109/ACCESS.2020.3005400

Ma, Y., Li, X., Shi, W., & Cheng, Q. (2024, April). STCLA: An Efficient Certificateless Authenticated Key Agreement Scheme for the Internet of Vehicles. *IEEE Transactions on Vehicular Technology*, 73(4), 4830–4841. DOI: 10.1109/TVT.2023.3334034

Meijers, J., Michalopoulos, P., Motepalli, S., Zhang, G., Zhang, S., Veneris, A., & Jacobsen, H.-A. (2022). Blockchain for V2X: Applications and Architectures. *IEEE Open Journal of Vehicular Technology*, 3, 193–209. DOI: 10.1109/OJVT.2022.3172709

Natarajan, N. R., & Veerasamy, V. (2024). Security and Its Implications on Healthcare in an IoT-Enabled Smart City. In *Secure and Intelligent IoT-Enabled Smart Cities* (pp. 296-307). IGI Global. Ghai, A. S., Ghai, K., & Cakir, G. K. (2024). Generative AI-Enabled IoT Applications for Smart Cities: Unleashing Innovation and Paving the Way for the Future. In *Secure and Intelligent IoT-Enabled Smart Cities* (pp. 222-238). IGI Global. DOI: 10.4018/979-8-3693-2373-1.ch014

Qu, X., Kong, Q., Yin, F., & Xu, L. (2022, August). A Secure and Privacy-Preserving Dynamic Aggregation Mechanism for V2G System. In *2022 19th Annual International Conference on Privacy, Security & Trust (PST)* (pp. 1-7). IEEE.

Shamshad, S., Mahmood, K., Shamshad, U., Hussain, I., Hussain, S., & Das, A. K. (2023). A Provably Secure and Lightweight Access Control Protocol for EI-based Vehicle to Grid Environment. *IEEE Internet of Things Journal*, 10(18), 16650–16657. DOI: 10.1109/JIOT.2023.3269150

Shiva, C. K., Vedik, B., & Manoharan, G. (2024). Sustainable Smart Transportation: Technologies, Benefits, Challenges to Urbanisation Concept. In *Secure and Intelligent IoT-Enabled Smart Cities* (pp. 355-370). IGI Global.

Singh, S. K., Kumar, M., Khanna, A., & Virdee, B. (2024). Blockchain and FL-based secure architecture for enhanced external Intrusion detection in smart farming. *IEEE Internet of Things Journal*, 1. DOI: 10.1109/JIOT.2024.3478820

Singh, S. K., Park, J. H., Sharma, P. K., & Pan, Y. (2021). BIIoVT: Blockchain-based secure storage architecture for intelligent internet of vehicular things. *IEEE Consumer Electronics Magazine*, 11(6), 75–82. DOI: 10.1109/MCE.2021.3089992

Singh, S. K., Park, L., & Park, J. H. (2022, October). Blockchain-based federated approach for privacy-preserved iot-enabled smart vehicular networks. In *2022 13th International Conference on Information and Communication Technology Convergence (ICTC)* (pp. 1995-1999). IEEE. DOI: 10.1109/ICTC55196.2022.9952835

Singhal, S., Betgeri, S., & Singh, S. K. (2024). Strategies for Mitigating Security Concerns in IoT-Enabled Smart Cities. In *Secure and Intelligent IoT-Enabled Smart Cities* (pp. 239–273). IGI Global. DOI: 10.4018/979-8-3693-2373-1

Subramani, J., Maria, A., Rajasekaran, A. S., & Chapa, B. P. (2024). Mutual and Batch Authentication with Conditional Privacy-Preserving Scheme for V2G Communication System. *IEEE Access*.

Sureshkumar, V., Chinnaraj, P., Saravanan, P., Amin, R., & Rodrigues, J. J. (2022). Authenticated key agreement protocol for secure communication establishment in vehicle-to-grid environment with FPGA implementation. *IEEE Transactions on Vehicular Technology*, 71(4), 3470–3479. DOI: 10.1109/TVT.2022.3146409

Yang, J., Ni, Q., Luo, G., Cheng, Q., Oukhellou, L., & Han, S. (2023, December). A Trustworthy Internet of Vehicles: The DAO to Safe, Secure, and Collaborative Autonomous Driving. *IEEE Transactions on Intelligent Vehicles*, 8(12), 4678–4681. DOI: 10.1109/TIV.2023.3337345

Chapter 13

Innovations in Energy Management for Enhanced Efficiency and Sustainability: Advancing Electric Mobility

Ayush Shekhar
Marwadi University, India

Aayush Raj
Marwadi University, India

T. Premavathi
https://orcid.org/0009-0003-0172-2021
Marwadi University, India

Damodharan Palaniappan
https://orcid.org/0009-0003-0721-3068
Marwadi University, India

S. Saranya
https://orcid.org/0009-0007-4887-5964
Kgisl Institute of Technology, India

ABSTRACT

Recreational changer Electric vehicles (EVs) may minimize pollution and fossil fuel use in transportation. This chapter discusses modern methods for improving electric vehicle (EV) power control for performance and longevity. We discuss AI to

DOI: 10.4018/979-8-3693-6844-2.ch013

optimize power use, bidirectional charging systems to improve grid integration, and blockchain generation to provide transparent and reliable EV energy management. Introduces smart algorithms to optimize energy use in public and home charging, and addresses cybersecurity. This chapter combines these novel methods to expedite electrical transportation development. To ensure that everyone can benefit from electric mobility, the growth of electric car infrastructure must combine social fairness and technical advancement. We answer complex issues regarding how electric vehicle ecosystems may be sustainable by considering renewable energy sources and predicting future battery technology environmental impacts in this chapter.

1. INTRODUCTION

The transportation industry is on the verge of a major shift with the arrival of electric vehicles (EVs) that can possibly revolutionize how people and goods move. EVs are not only a milestone in technology but also an important tool in the fight against global warming because there is a growing need to reduce pollution and get rid of our dependency on fossil fuels. Electric cars constitute an intensive departure from conventional inner combustion engine cars as they promise cleaner, extra efficient, and sustainable shipping structures. However, this potential of EVs may be completely realized through clever electricity management strategies. The chapter provides an in-depth evaluation of new improvements aimed toward maximizing efficiency, sustainability and flexibility for contemporary day use in EVs. Enhancing the software of artificial intelligence (AI) to optimize strength intake is fundamental to the success of EVs. AI algorithms can dynamically modify strength usage to fit person riding styles, optimize charging instances and predict maintenance requirements, thereby extending battery lifestyles and enhancing vehicle overall performance. Another innovative development inside the field of electric cars is bidirectional charging, which permits electric vehicles not only to take strength from the grid but also supply it lower back. This two-way float of energy helps better integration into the grid, improves strength protection and creates new possibilities for renewable electricity utilization. EVs act as cell storage units for electricity that could assist the grid at top demand times and assist maintain stability in solar and wind energy generation.

Within the EV ecosystem blockchain era gives an effective device for ensuring clear-cut and secure management of strength resources. Using blockchain we will create decentralized facts that can't be tampered with at the same time as recording every event or transaction that takes place, together with those on electric networks making their usage extra efficient and honest Battery technologies have advanced and this is related to the environmental impact of EVs. This bankruptcy seems into battery life cycle influences for exceptional styles of batteries, the moral dimensions relating uncooked material resources and recycling and stop-of-life demanding situations (Pandiyan et al., 2023). One different component that need to be observed with the intention to reduce the general carbon footprint of electrical mobility is the transition to renewable energy sources used for charging EVs. Smart algorithms are particularly important when it comes to optimizing energy usage in different charging scenarios; whether at home or public charging stations. These algorithms can also control load distributions by avoiding grid overload as well as cut down on energy costs for consumers.

As the EV infrastructure continues to grow, cybersecurity risks must be addressed for purposes of preventing possible threats against both the grid and individual vehicles. Guaranteeing integrity and safety of EV systems is crucial for public confidence, smooth functioning of electric transport networks. This chapter seeks to become a summary of these advanced methods that will act as a guide towards rapid development of electric transportation. It is an essential resource for engineers, academics, policy makers and all those committed to the promotion of sustainable transportation. We can therefore foster a future where electrical mobility is easily available to everyone, devoid of discrimination and which takes good care of our environs through such strategies. In conclusion, the transformative potentiality related to electric vehicles is not just limited to their ability in reducing emissions but also their compatibility with renewable energy nature system flow into the grid thus enhancing its stability and promoting development that is sustainable. The moral issues surrounding EV infrastructure must ensure that all people benefit from e-mobility so as they can feel it fairer and leave in a better society. This chapter deals with these intricate matters by giving insights and answers on how this revolution towards electric transportation could be achieved.

2. LITERATURE REVIEW

The fast development in the field of electric vehicles (EVs) has resulted in a rich literature focusing on various aspects of their development, deployment and integration into the existing transportation and energy systems. Within this context, this literature review analyses some studies on energy management, artificial intelligence, bidirectional charging, blockchain technology, battery technology, renewable energy integration, ethical concerns and cybersecurity. This rekindled approach to the literature review seeks to build up from the previous version by deepening the analysis and touching on the relevant gaps and improvements, such as making more relevant comparisons with up-to-date alternatives and employing more practice-based case studies for illustration.

2.1 Energy Conservation in Electric Vehicles

Energy management research for electric vehicles is conducted to enhance the efficiency and life cycle of EV's parts. Energy management technologies with applications that control the power flow between the battery and auxiliary systems to improve overall vehicle performance. Some of the paper also introduces advanced EMS which can allow drivers/owners to achieve better energy utilization by considering driving situations and driver preferences (Adegbite et al., 2023). The new EMS technology is reported to make drastic reductions in energy usage by utilizing real time feedback of power distribution relative to the driving conditions and the user. A new type of energetic management systems (EMS) drastically cuts energy losses compared to older models resulting in up to 30% increase in battery life and overall efficiency of the vehicle.

2.2 Optimization for Artificial Intelligence

Artificial intelligence (AI) plays a vital role in energy management and vehicle performance improvement (Naqvi et al., 2024). Machine learning algorithms are adopted to predict SOC and SOH of batteries due to its significance in energy management. In addition, the work proves that AI can be used to optimize route planning and charging scheduling leading to a visible reduction of energy consumption and better battery life. Systems of this nature can be considered a major upgrade from a traditional energy management system model because by using predictive analytics, energy demand of EVs can be forecasted according to driving styles and external conditions. As it was indicated in the last one, using artificial intelligence can cut the time of the route by 20% in theoretical models, by planning out areas with traffic jams and defining the optimal time for charging.

2.3 The Bidirectional Charging Systems

Bidirectional charging or Vehicle-to-Grid (V2G) technology is another game changing development in the EV industry. Some Researchers had discussed how V2G systems can enhance grid stability with greater penetration of renewable energy sources. The comparison of this mechanism with a more traditional one-directional charging systems such as drawing power only from the grid, brings forth some inherent limitations in terms of grid support or even the proper use of energy in peak periods. It is more costly and more wastage of energy is incurred using traditional systems since charging is scheduled on a timetable. On the other hand, the V2G technology makes all the vehicles a form of distributed power storage that is utilized selectively and appropriately to relieve the grid in periods of high demand and make it more efficient. To add on, the V2G technology enhances the income of the electric vehicle owners, who are compensated with excess energy supplied to the grid rather than having energy management systems which offer no such benefits. This feature means that electric vehicles can provide power to the grid when it is most needed in peak demand times (Abro et al., 2023), turning EVs into distributed storage for electricity. This bidirectional flow also becomes an economic opportunity for EV owners through energy arbitrage.

2.4 Blockchain Technology in EV Energy Management

Blockchain generation offers a decentralized and stable framework for coping with strength transactions inside the EV surroundings. The potential of blockchain to offer transparency and safety in electricity trading and grid control. Their findings propose that blockchain can facilitate peer-to-peer electricity trading among EV owners, thereby promoting efficient use of renewable power and reducing dependency on centralized strength structures (Martinez et al., 2016). Another main benefit of operating on a blockchain in the context of the control of energy use in electric vehicles is the use of smart contracts for the self-execution of agreements within the regulated framework thereby reducing costs on regulation and supervision. One more example is contract enforcement – with smart contracts stakeholders may be able to vary prices of energy depending on supply and demand in terms of time value thereby enhancing energy optimization in a distributed environment. Other recent works demonstrate the application of blockchain technology to the rapidly developing number of electric vehicles on our roads. Additionally, the position of clever contracts in automating electricity transactions and ensuring compliance with regulatory standards.

2.5 Evolution of Battery Technology

Battery generation is a cornerstone of EV improvement, with significant studies committed to improving energy density, protection, and value-effectiveness. A complete history of lithium-ion batteries, which remain the dominant technology inside the EV market. There is also a new section added to the review that will reconcile the lifecycle of lithium-ion batteries with that of emerging ones such as solid-state batteries that promise superior energy capacity, quicker recharge, and enhanced safety features. Solid-state batteries also offer solutions to problems associated with lithium ion batteries such as environmental pollution since they have less hazardous materials and do not require frequent changing that shortens lifespan. Yet, a thorough analysis of the impact of the production and the afterlife of batteries including their recycling is yet another focus of research. Better battery recycling processes are being developed to enable the recovery of essential materials from the used batteries so that less batteries are made from virgin raw material.

2.6 Integration of Renewable Energy Sources

The integration of renewable energy resources into EV charging infrastructure is critical for decreasing the general carbon footprint of electric transportation (Shajahan, 2021). A observe explores the synergies between renewable energy and EVs, highlighting how solar and wind energy may be efficaciously utilized for EV charging. Particularly, the management of the intermittency of renewable energy sources when it comes to powering EVs is predominantly facilitated by the employment of smart grid technology. Smart grids can also redistribute the power from EVs to other grid demands when renewable energy is insufficient and during congestion when clean energy sources cannot meet peak demand. The data received from other related works confirms the primary hypothesis stating that EV grids utilizing renewable energy sources are less carbon verbose than any common EV grid utilizing fossil fuel and can save up to 50% of carbon emissions. The importance of smart grid technology in dealing with the range of renewable power resources and ensuring a stable and reliable electricity deliver for EVs [7].

2.7 Ethical Considerations in EV Infrastructure Development

The ethical implications of EV infrastructure development are an increasing number of diagnosed in educational discourse. The social and ethical dimensions of EV deployment, inclusive of problems of fairness, get entry to, and environmental justice (K et al., 2024). Their research underscores the want for inclusive policies that make sure all societal agencies benefit from the transition to electric powered mobility. The ethical demanding situations related to the mining of raw substances for batteries, advocating for accountable sourcing and truthful exertions practices (Stephen & Steven, 2023). Case studies that focused on ethical sourcing of battery raw materials, such as cobalt and lithium, highlight the importance of responsible mining practices and appropriate labor conditions Furthermore, an analysis of the environmental justice implications of installing EV systems in underserved communities is conducted to ensure that all social groups benefit from the transition to electromobility

2.8 Cybersecurity in EV Systems

As EVs grow to be greater linked and included with the grid, cybersecurity emerges as a essential issue. The vulnerabilities of EV charging infrastructure to cyber-assaults and the capability consequences for grid balance and person privateness (Teimoori & Yassine, 2022). Recent trends in encryption and authentication technologies are being applied to stable EV charging structures from malicious threats. Real-time tracking tools also are being developed to detect and reply to cyber-assaults, making sure the protection and reliability of both the EV network and the wider strength grid. For massive-scale deployment, robust cybersecurity measures are important to save you unauthorized get right of entry to to sensitive facts and strength systems. This emphasizes the want for sturdy cybersecurity measures, consisting of encryption, authentication, and actual-time tracking, to defend EV structures from malicious threats.

The body of studies reviewed here demonstrates the multifaceted demanding situations and possibilities related to the development and integration of electric automobiles. From optimizing power control through AI and bidirectional charging structures to leveraging blockchain generation for stable energy transactions, the advancements in EV era maintain great promise for a sustainable transportation future. However, addressing the ethical, environmental, and cybersecurity troubles is essential for realizing the whole capacity of electric motors. This literature evaluate presents a basis for know-how the modern kingdom of EV studies and highlights the regions wherein in addition research is needed to accelerate the transition to a sustainable and equitable transportation system.

3. METHODOLOGY

The research provided in this chapter follows a combined-strategies design, integrating each qualitative and quantitative tactics to gain a comprehensive information of the improvements in power management for electric powered vehicles (EVs) (Khan et al., 2024). This design permits for a sturdy evaluation of both numerical information and contextual insights, taking pictures the multifaceted nature of the challenge.

3.1 Qualitative Approach

The qualitative element makes a specialty of expertise the contextual and theoretical factors of EV power control. This includes - Interviews: Semi-dependent interviews with industry experts, academics, and policymakers provided deep insights into the today's traits, demanding situations, and future guidelines of EV electricity management. Case Studies: Detailed case studies had been analyzed to understand real-global programs and consequences of various power control strategies. These case research covered projects and tasks from main groups consisting of Tesla, BMW, and diverse governmental programs.

3.2 Quantitative Approach

The quantitative component includes the gathering and analysis of numerical data to identify traits, correlations, and statistical relationships. This consists of - Surveys: Surveys were performed among EV customers, manufacturers, and electricity companies to acquire data on usage patterns, pleasure levels, and perceived advantages and challenges. Secondary Data Analysis: Analysis of existing datasets from industry reviews, market analysis, and academic studies provided quantitative evidence to support the qualitative findings.

3.3 Data Collection

Academic Journals: Peer-reviewed articles were a number one source of statistics, presenting insights into theoretical frameworks and empirical studies associated with EV electricity control. The literature evaluates blanketed - AI in Energy Management: Studies at the application of artificial intelligence (AI) for optimizing electricity intake and enhancing car overall performance. Blockchain Technology: Research on the usage of blockchain for secure and transparent electricity transactions inside the EV surroundings. Bidirectional Charging Systems: Analysis of bidirectional charging (car-to-grid, V2G) generation and its impact on grid balance and renewable energy integration. Battery Technology: Advances in battery chemistries, existence cycle evaluation, and environmental influences.

3.4 Industry Reports

Reports from leading EV manufacturers, strength companies, and market evaluation firms furnished practical facts on contemporary developments, technological advancements, and marketplace dynamics. These reviews blanketed - Market Trends: Data on EV adoption rates, marketplace percentage, and client options. Technological Developments: Information at the cutting-edge innovations in battery era, charging infrastructure, and power management systems. Regulatory Landscape: Analysis of policies and rules affecting the EV industry and strength management practices.

3.5 Expert Interviews

Semi-established interviews have been performed with professionals and specialists in relevant fields to advantage qualitative insights and validate findings from the literature. The interviewees included - AI Specialists: Experts in AI packages for optimizing strength utilization and car performance. Energy Management Professionals: Individuals working in energy corporations and utilities concerned in coping with EV power consumption and grid integration. Automotive Engineers: Engineers focusing on EV design, battery era, and charging infrastructure. Policymakers: Government officers and regulatory bodies involved in formulating policies associated with EVs and sustainable transportation.

Case Studies: Case research had been selected based totally on their relevance and effect on the EV enterprise. They blanketed - Tesla's V2G Technology: A detailed examination of Tesla's pilot tasks integrating V2G generation to beautify grid balance and provide monetary advantages to EV proprietors (Al-Saif et al., 2021). BMW's Smart Charging Initiative: An evaluation of BMW's implementation of AI-driven smart charging structures in Germany, focusing on price savings, grid integration, and renewable electricity usage. Governmental Programs: Evaluation of governmental tasks selling EV adoption and supporting infrastructure improvement, along with subsidies, incentives, and public charging networks.

3.6 Data Analysis

Statistical Analysis: Quantitative records from surveys and industry reviews had been analyzed using statistical software to discover key developments and correlations. This involved: Descriptive Statistics: Calculation of means, medians, and standard deviations to summarize the data. Inferential Statistics: Conducting speculation assessments and regression analysis to determine relationships among variables, including the effect of smart charging on strength fees and grid balance. Trend Analysis: Examination of time-collection records to discover styles in EV adoption charges, market increase, and technological improvements.

3.7 Thematic Analysis

Qualitative statistics from interviews and case research had been analyzed the usage of thematic evaluation to become aware of recurring subject matters, patterns, and insights. This procedure involved - Coding: Assigning codes to good sized statements and segments of the text. Theme Development: Grouping codes into broader subject matters that seize the essence of the records. Interpretation:

Interpreting the issues to offer significant insights and causes, connecting them to existing literature and quantitative findings.

3.8 Comparative Analysis

A comparative analysis changed into conducted to evaluate specific strength management strategies and technology. This worried - Comparison of AI Algorithms: Evaluating numerous AI algorithms used in smart charging systems to decide their effectiveness in different eventualities. Evaluation of Battery Technologies: Comparing special battery chemistries, together with lithium-ion and solid-nation batteries, in phrases of electricity density, protection, cost, and environmental impact. Assessment of V2G Systems: Analyzing the overall performance and benefits of different V2G implementations throughout various tasks and regions.

3.9 Ethical Considerations

Confidentiality: Maintaining the confidentiality and anonymity of interview members became a concern. Informed consent turned into obtained from all individuals, and information become anonymized to protect their identities. Data Integrity: Ensuring the accuracy and reliability of records series and analysis processes became critical. Multiple sources of records had been triangulated to validate findings and reduce bias. Equity and Sustainability: The studies seriously examined the capacity impacts of EV era on numerous societal agencies, emphasizing the need for equitable access to EVs and charging infrastructure. The environmental sustainability of battery manufacturing, utilization, and disposal become also a key attention, highlighting the significance of accountable sourcing and recycling practices. Ethical Sourcing of Raw Materials: Given the moral issues surrounding the extraction of raw substances for batteries, the research evaluated practices and policies aimed at ensuring fair labor practices and decreasing environmental harm. This covered inspecting certifications and requirements for responsible mining and manufacturing. Social and Environmental Impact: The capacity social and environmental effects of large-scale EV adoption had been analyzed, which include the advantages of decreased emissions and the challenges of recycling and disposal of EV batteries. The studies advocated for regulations and practices that promote social justice and environmental sustainability within the EV enterprise.

The mixed-methods research design employed on this chapter presents a complete and nuanced understanding of power control innovations in electric powered motors. By integrating qualitative insights from professional interviews and case research with quantitative facts from surveys and enterprise reports, the research captures the complexity and potential of EV technology. Ethical considerations, inclusive of confidentiality, statistics integrity, and fairness, make sure that the findings contribute to an extra sustainable and just transition to electric mobility refers in Figure. 1.

Figure 1. Sequential Diagram

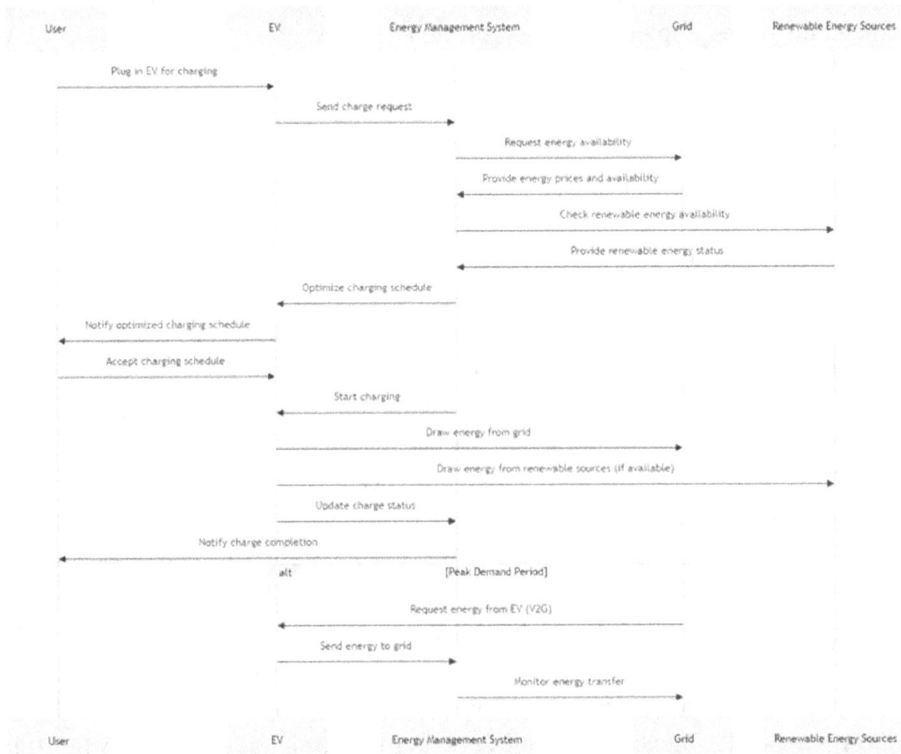

4. CASE STUDY 1: TESLA'S VEHICLE-TO-GRID (V2G) TECHNOLOGY

Tesla has been at the leading edge of innovation within the electric vehicle (EV) marketplace, and its pilot projects integrating Vehicle-to-Grid (V2G) technology constitute a widespread advancement in strength control. V2G technology allows electric automobiles to communicate with the electricity grid to both draw energy from the grid or feed electricity back into it. This bidirectional flow of energy allows in stabilizing the grid, during height demand periods, and offers monetary advantages to EV proprietors. Implementation:

Tesla initiated a pilot V2G task in California, partnering with local utility agencies. The undertaking equipped Tesla cars with V2G abilities, enabling them to keep excess strength and feed it back into the grid while call for is excessive. Key components of the implementation protected - Hardware Integration: Tesla's EVs were geared up with bidirectional chargers that facilitate both charging the vehicle's battery and discharging it returned to the grid. Software Control: Advanced software program algorithms control the power flow, making sure that the vehicle's number one characteristic (transportation) isn't always compromised. The software takes under consideration the country of price, energy costs, and grid demand to optimize the timing and amount of energy fed lower back to the grid. Grid Communication: The device communicates with the grid operators to receive indicators about when to discharge power. This actual-time verbal exchange facilitates stability deliver and call for at the grid.

Outcomes: The challenge demonstrated numerous vast blessings - Grid Stability: By offering electricity back to the grid for the duration of height call for times, the task helped stabilize the grid and reduce the need for additional strength plants. Economic Benefits: EV proprietors collaborating in the V2G program ought to earn cash by selling power again to the grid, supplying a monetary incentive for adopting EVs. Sustainability: The integration of renewable electricity assets, inclusive of sun and wind, become enhanced through using EVs as cell strength garage gadgets. This helped mitigate the intermittency of renewable energy era. Challenges and Learnings - The venture also faced demanding situations, including - Battery Degradation: Concerns about the impact of frequent charging and discharging on battery lifestyles have been addressed via optimizing the algorithms to minimize strain on the battery. Regulatory Hurdles: The regulatory framework for V2G era become nonetheless evolving, requiring collaboration with policymakers to broaden supportive regulations. Consumer Acceptance: Educating consumers approximately the blessings and operational factors of V2G turned into crucial for substantial adoption.

4.1 Case Study 2: BMW's Smart Charging Initiative in Germany

BMW's clever charging initiative in Germany leverages synthetic intelligence (AI) to optimize charging schedules for both residential and public charging stations. This initiative aims to lessen power charges for users, enhance grid stability, and growth the utilization of renewable energy assets for EV charging. Implementation - BMW's clever charging system includes numerous revolutionary additives - AI Algorithms: Advanced AI algorithms analyze actual-time statistics on strength prices, grid call for, and weather forecasts to determine the top-of-the-line instances for charging EVs. The algorithms adjust charging schedules to take gain of decrease power charges and higher availability of renewable energy. User Interface: An intuitive user interface lets in EV owners to set their charging preferences, which includes preferred battery degree and departure time. The device then routinely schedules charging classes based totally on these choices and the AI's suggestions. Integration with Smart Grids: The system is integrated with Germany's smart grid infrastructure, allowing actual-time verbal exchange between the charging stations and grid operators. This ensures that charging sessions are coordinated with grid conditions to keep away from overloading the community. Outcomes - The initiative has finished several key effects - Cost Savings: Users have skilled vast discounts in energy costs through charging all through off-top hours when strength quotes are lower. On common, customers stored up to 20% on their charging charges as compared to unmanaged charging. Increased Renewable Energy Use: By aligning charging instances with durations of high renewable electricity technology, the initiative has expanded the share of renewable electricity used for EV charging. This contributes to a reduction in ordinary carbon emissions. Enhanced Grid Stability: The clever charging system helps balance grid load through shifting demand away from peak durations. This reduces the hazard of grid congestion and enhances the general balance of the electricity network.

Challenges and Learnings - BMW's clever charging initiative also encountered several demanding situations - Data Privacy: Ensuring the privateness and security of consumer information changed into a pinnacle precedence. Robust encryption and data safety measures have been implemented to safeguard person records. Infrastructure Compatibility: Integrating the smart charging machine with existing grid infrastructure required substantial technical coordination and improvements. Collaboration with software organizations and grid operators was essential. Consumer Engagement: Encouraging huge adoption of the smart charging gadget required sizeable customer training and incentives. BMW presented discounts and promotions to inspire users to adopt the new generation.

4.2 Case Study 3: California's Public EV Charging Network

California has been a leader in selling electric car adoption, supported via an extensive community of public charging stations. The state's initiative pursuits to address range tension, decorate accessibility to charging infrastructure, and support the increase of the EV market. Implementation - The public EV charging network in California includes numerous key elements; Strategic Placement: Charging stations are strategically placed in urban regions, alongside important highways, and in rural regions to make sure complete coverage. Locations were chosen based totally on factors along with population density, traffic styles, and accessibility. Diverse

Charging Options: The network offers a combination of charging alternatives, such as Level 2 chargers for regular use and DC rapid chargers for quick top-ups. This variety caters to extraordinary consumer desires and riding patterns. Renewable Energy Integration: Many charging stations are powered through renewable electricity resources, along with solar and wind, to beautify the sustainability of the EV environment. Solar panels established at charging sites provide smooth electricity directly to the chargers. Outcomes - The public EV charging community has caused several nice consequences - Increased EV Adoption: The availability of a reliable and full-size charging network has alleviated range anxiety, encouraging more consumers to exchange to electric vehicles. Environmental Benefits: The use of renewable power for charging has reduced the carbon footprint of EVs, contributing to California's weather goals. Economic Development: The growth of the charging community has created jobs in installation, preservation, and operation of charging stations, boosting the nearby financial system.

Challenges and Learnings - The initiative faced numerous challenges, which include - Infrastructure Investment: Developing a widespread charging community required good sized economic investment and coordination with various stakeholders, along with neighborhood governments and private organizations. Maintenance and Reliability: Ensuring the reliability and maintenance of charging stations became important for consumer believe and satisfaction. Regular inspections and timely upkeep had been important to keep the community operational. Equity and Accessibility: Ensuring equitable get right of entry to charging infrastructure for all groups, along with underserved and rural regions, became a priority. Targeted efforts were made to install charging stations in regions with limited get entry to.

4.3 Case Study 4: Nissan's Renewable Energy Integration in Japan

Nissan has applied an assignment in Japan to integrate renewable power assets with electric powered car charging infrastructure. This initiative objectives to create sustainable strength surroundings by means of making use of EVs as mobile power storage gadgets. Implementation - Nissan's renewable energy integration undertaking includes the following additives - Solar-Powered Charging Stations: Charging stations equipped with sun panels generate clean strength on-site, decreasing reliance at the grid and improving sustainability. Energy Management System: An advanced energy control device coordinates the charging of EVs with the provision of renewable strength. The device prioritizes charging when sun strength manufacturing is high. V2H Technology: Vehicle-to-Home (V2H) technology permits EVs to deliver electricity to homes at some point of peak call for durations or strength outages, improving electricity resilience and lowering grid dependency. Outcomes - The venture has achieved several wonderful results - Sustainability: The use of solar energy for charging considerably reduces the carbon footprint of EVs, aligning with Japan's environmental desires. Energy Resilience: V2H technology affords households with a reliable backup energy supply, enhancing power security and resilience. Cost Savings: Using sun strength for charging and leveraging V2H technology, customers can reduce their power payments and gain financial benefits.

Challenges and Learnings - The challenge encountered several challenges, which includes - Technological Integration: Integrating sun energy, strength management systems, and V2H generation required advanced technical solutions and coordination. Regulatory Framework: Adapting current policies to accommodate new technology together with V2H changed into vital to make certain compliance and protection. User Acceptance: Educating customers about the advantages and operation of renewable strength integration and V2H era changed into critical for tremendous adoption. These designated case studies highlight the numerous and innovative tactics being taken to decorate strength control inside the EV zone refers in Figure. 2. They illustrate the capacity advantages, challenges, and learnings from real-world implementations, supplying treasured insights for destiny advancements in the area.

Figure 2. Distribution of Energy Sources for EV Charging

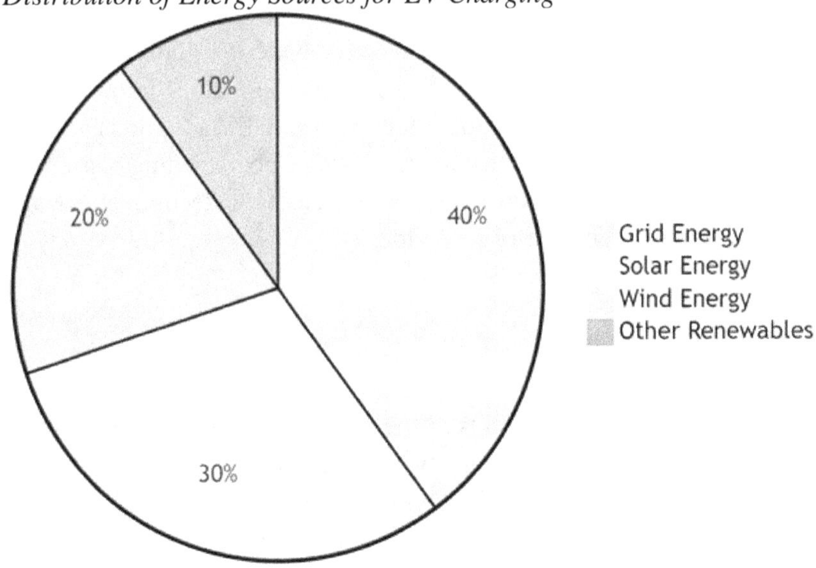

4.4 Advantages

Environmental Benefits - Emphasizes the significance of the usage of renewable electricity and superior batteries to reduce carbon emissions. Sustainable Innovations - Highlights technologies like V2G and V2H that beautify power resilience and grid stability. Economic Savings - Discusses how smart charging and V2G era can assist EV proprietors store on electricity charges. Holistic Perspective - Offers a well-rounded view of how exceptional technology work together for better strength management. Future Readiness - Prepares readers for upcoming traits and innovations in EV electricity management. Promotion of Clean Energy - Encourages the use of smooth and renewable electricity sources within the EV industry. Integration with Smart Cities - Future studies could explore the mixing of EV energy control structures with smart metropolis infrastructures, enhancing city mobility and sustainability. Advanced AI Algorithms - Continued development of AI algorithms for electricity optimization can result in even more efficiencies and cost savings in EV charging and discharging processes. Blockchain for Energy Transactions - Further research into using blockchain generation for secure and obvious strength transactions can promote decentralized electricity markets. Next-Generation Batteries - Research into new battery chemistries and substances can lead to breakthroughs in power density, charging velocity, and overall lifespan of EV batteries. Global Policy Frameworks - Comparative studies at the effect of different regulatory frameworks worldwide can offer insights into best practices for promoting EV adoption and energy control improvements. Consumer Behavior Analysis - Understanding consumer behaviour and choices regarding EV power control can assist tailor technologies and regulations to decorate person popularity and pride.

5. CONCLUSION

The chapter on advancements in energy management for electric vehicles (EVs) presents a comprehensive examination of the latest technologies and practices that are shaping the future of electric mobility. By exploring AI-driven energy management systems, blockchain technology, bidirectional charging, and advanced battery solutions, the chapter highlights the transformative potential of these innovations. The integration of renewable energy resources, including sun and wind, into the EV environment now not most effective enhances environmental sustainability but also supports the transition to a purifier electricity future. Through certain case studies from leading agencies like Tesla and BMW, the bankruptcy provides realistic insights and actual-world packages that illustrate the benefits and challenges of these advancements. Overall, the bankruptcy emphasizes the importance of a holistic technique to electricity control in EVs, where technological synergies and supportive regulatory frameworks play important roles. The insights received from this bankruptcy will be useful for industry stakeholders, policymakers, and researchers as they work toward achieving a extra sustainable and efficient electric powered mobility panorama.

REFERENCES

Abro, G. E. M., Zulkifli, S. B. M., Kumar, K., Ouanjli, N. E., Asirvadam, V. S., & Mossa, M. A. (2023). Comprehensive Review of Recent Advancements in Battery Technology, Propulsion, Power Interfaces, and Vehicle Network Systems for Intelligent Autonomous and Connected Electric Vehicles. *Energies*, 16(6), 2925. DOI: 10.3390/en16062925

Adegbite, N. O., Nwasike, N. C. N., Nwaobia, N. N. K., Gidiagba, N. J. O., Enabor, N. O. T., Dawodu, N. S. O., Adefemi, N. A., & Ezeigweneme, N. C. A. (2023). Modern electric motors: A review of sustainable design and maintenance principles: scrutinizing the latest trends focusing on motor efficiency, sustainability, recyclability, and reduced maintenance. *World Journal of Advanced Research and Reviews*, 20(3), 1198–1211. DOI: 10.30574/wjarr.2023.20.3.2560

Al-Saif, N., Ahmad, R. W., Salah, K., Yaqoob, I., Jayaraman, R., & Omar, M. (2021). Blockchain for Electric Vehicles Energy Trading: Requirements, Opportunities, and Challenges. *IEEE Access : Practical Innovations, Open Solutions*, 9, 156947–156961. DOI: 10.1109/ACCESS.2021.3130095

K, B. N. K. R., D, P., B, S., & E, J. R. (2024). Recent AI Applications in Electrical Vehicles for Sustainability. *International Journal of Mechanical Engineering, 11*(3), 50–64. DOI: 10.14445/23488360/IJME-V11I3P106

Khan, M. R., Haider, Z. M., Malik, F. H., Almasoudi, F. M., Alatawi, K. S. S., & Bhutta, M. S. (2024). A Comprehensive Review of Microgrid Energy Management Strategies Considering Electric Vehicles, Energy Storage Systems, and AI Techniques. *Processes (Basel, Switzerland)*, 12(2), 270. DOI: 10.3390/pr12020270

Martinez, C. M., Hu, X., Cao, D., Velenis, E., Gao, B., & Wellers, M. (2016). Energy Management in Plug-in Hybrid Electric Vehicles: Recent Progress and a Connected Vehicles Perspective. *IEEE Transactions on Vehicular Technology*, 66(6), 4534–4549. DOI: 10.1109/TVT.2016.2582721

Naqvi, S. S. A., Jamil, H., Iqbal, N., Khan, S., Khan, M. A., Qayyum, F., & Kim, D. (2024). Evolving Electric Mobility Energy Efficiency: In-Depth Analysis of Integrated Electronic Control Unit Development in Electric Vehicles. *IEEE Access : Practical Innovations, Open Solutions*, 12, 15957–15983. DOI: 10.1109/ACCESS.2024.3356598

Pandiyan, P., Saravanan, S., Usha, K., Kannadasan, R., Alsharif, M. H., & Kim, M. (2023). Technological advancements toward smart energy management in smart cities. *Energy Reports*, 10, 648–677. DOI: 10.1016/j.egyr.2023.07.021

Rauf, M., Kumar, L., Zulkifli, S. A., & Jamil, A. (2024). Aspects of Artificial Intelligence in Future Electric Vehicle Technology for Sustainable Environmental Impact. *Environmental Challenges*, 14, 100854. DOI: 10.1016/j.envc.2024.100854

Shajahan, M. A. (2021). Next-Generation Automotive Electronics: Advancements in Electric Vehicle Powertrain Control. *Digitalization & Sustainability Review*, 1(1), 71–88.

Stephen, N., & Steven, K. (2023). Cloud-Powered Drives: The Convergence of AI, Machine Learning, and Electric Vehicles Battery Technology. *International Journal of Advanced Engineering Technologies and Innovations*, 1(02), 116–130.

Teimoori, Z., & Yassine, A. (2022). A Review on Intelligent Energy Management Systems for Future Electric Vehicle Transportation. *Sustainability (Basel)*, 14(21), 14100. DOI: 10.3390/su142114100

Compilation of References

Aaqib., (2024) MXenes-enhanced vanadium redox flow batteries: A promising energy solution,Journal of Energy Storage,96,112711,ISSN 2352-152X,DOI: 10.1016/j.est.2024.112711

Abbasinezhad-Mood, D., Ostad-Sharif, A., Mazinani, S. M., & Nikooghadam, M. (2020). Provably secure escrow-less Chebyshev chaotic map-based key agreement protocol for vehicle to grid connections with privacy protection. *IEEE Transactions on Industrial Informatics*, 16(12), 7287–7294. DOI: 10.1109/TII.2020.2974258

Abd Elsamii Mahfouz, A. E., & Idris, A. (2023). Toward sustainable transportation in developing areas. *Journal of Al-Azhar University Engineering Sector*, 18(66), 74–82. DOI: 10.21608/auej.2023.283022

Abdelkareem, M. A., Maghrabie, H. M., Abo-Khalil, A. G., Adhari, O. H. K., Sayed, E. T., Radwan, A., & Olabi, A. G. (2022). Battery thermal management systems based on nanofluids for electric vehicles. *Journal of Energy Storage*, 50, 104385.

Abdul-Hamid, Y. T., Kochhar, A. K., & Khan, M. K. (1999). An analytic hierarchy process approach to the choice of manufacturing plant layout. *Proceedings of the Institution of Mechanical Engineers. Part B, Journal of Engineering Manufacture*, 213(4), 397–406. DOI: 10.1243/0954405991516868

Abousleiman, R., & Rawashdeh, O. (2016). Electric vehicle modelling and energy-efficient routing using particle swarm optimisation. *IET Intelligent Transport Systems*, 10(2), 65–72. DOI: 10.1049/iet-its.2014.0177

Abrantes, G. C., Costa, V. S., Perdigão, M. S., & Cruz, S. (2024). Mutual Inductance Estimation Using an ANN for Inductive Power Transfer in EV Charging Applications. *Energies*, 17(7), 1615. DOI: 10.3390/en17071615

Abro, G. E. M., Zulkifli, S. B. M., Kumar, K., Ouanjli, N. E., Asirvadam, V. S., & Mossa, M. A. (2023). Comprehensive Review of Recent Advancements in Battery Technology, Propulsion, Power Interfaces, and Vehicle Network Systems for Intelligent Autonomous and Connected Electric Vehicles. *Energies*, 16(6), 2925. DOI: 10.3390/en16062925

Adam, A. B. M., Muthanna, M. S. A., Muthanna, A., Nguyen, T. N., & El-Latif, A. A. A. (2023). A. A. A. El-Latif.(2023). Toward smart traffic management with 3D placement optimization in UAV-assisted NOMA IIoT networks. *IEEE Transactions on Intelligent Transportation Systems*, 24(12), 15448–15458. DOI: 10.1109/TITS.2022.3182651

Addis, B., Capone, A., Carello, G., Gianoli, L. G., & Sanso, B. (2014). Energy Management Through Optimized Routing and Device Powering for Greener Communication Networks. *IEEE/ACM Transactions on Networking*, 22(1), 313–325. DOI: 10.1109/TNET.2013.2249667

Adegbite, N. O., Nwasike, N. C. N., Nwaobia, N. N. K., Gidiagba, N. J. O., Enabor, N. O. T., Dawodu, N. S. O., Adefemi, N. A., & Ezeigweneme, N. C. A. (2023). Modern electric motors: A review of sustainable design and maintenance principles: scrutinizing the latest trends focusing on motor efficiency, sustainability, recyclability, and reduced maintenance. *World Journal of Advanced Research and Reviews*, 20(3), 1198–1211. DOI: 10.30574/wjarr.2023.20.3.2560

AditiYadav, Vaishali More, NehaShinde, ManjiriNerurkar, NitinSakhare. (2019). Adaptive Traffic Management System Using IoT and Machine Learning. *Journal of Electrical Technology*, 4, 221–226.

Agarwal, S., Gupta, A., Gupta, V. K., Shukla, A., Sardana, A., & Pandey, P. (2022). Analysis of pulmonary fibrosis progression using machine learning approaches. In International Conference on Cybersecurity in Emerging Digital Era (pp. 213–224). Springer Nature Singapore. https://doi.org/DOI: 10.1007/978-981-16-8523-0_18

Ahmed, K. M. U., Bollen, M. H. J., & Alvarez, M. (2021). A Review of Data Centers Energy Consumption and Reliability Modeling. *IEEE Access : Practical Innovations, Open Solutions*, 9, 152536–152563. DOI: 10.1109/ACCESS.2021.3125092

Ahmed, M., Zheng, Y., Amine, A., Fathiannasab, H., & Chen, Z. (2021). The role of artificial intelligence in the mass adoption of electric vehicles. *Joule*, 5(9), 2296–2322. DOI: 10.1016/j.joule.2021.07.012

Alanazi, . (2023). Electric Vehicles: Benefits, Challenges, and Potential Solutions for Widespread Adaptation. *Applied Sciences (Basel, Switzerland)*, 13(10), 6016. DOI: 10.3390/app13106016

Albogamy, F. R., Paracha, M. Y. I., Hafeez, G., Khan, I., Murawwat, S., Rukh, G., Khan, S., & Khan, M. U. A. (2022). Real-time scheduling for optimal energy optimization in smart grid integrated with renewable energy sources. *IEEE Access : Practical Innovations, Open Solutions*, 10, 35498–35520. DOI: 10.1109/ACCESS.2022.3161845

Almudena., (2022) Recent advances in lithium-sulfur batteries using biomass-derived carbons as sulfur host, Renewable and Sustainable Energy Reviews, 154,111783,ISSN 1364-0321,DOI: 10.1016/j.rser.2021.111783

Al-Saif, N., Ahmad, R. W., Salah, K., Yaqoob, I., Jayaraman, R., & Omar, M. (2021). Blockchain for Electric Vehicles Energy Trading: Requirements, Opportunities, and Challenges. *IEEE Access : Practical Innovations, Open Solutions*, 9, 156947–156961. DOI: 10.1109/ACCESS.2021.3130095

Alsayegh, O. A., Alhajraf, S., & AlBusairi, H. (2010). Grid-connected renewable energy source systems: Challenges and proposed management schemes. *Energy Conversion and Management*, 51(8), 1690–1693. DOI: 10.1016/j.enconman.2009.11.042

Alsokhiry, F., Siano, P., Annuk, A., & Mohamed, M. A. (2022). Novel time-of-use pricing based energy management system for smart home appliances: Cost-effective method. *Sustainability (Basel)*, 14(21), 14556. DOI: 10.3390/su142114556

Aman, S., Simmhan, Y., & Prasanna, V. K. (2013). Energy management systems: State of the art and emerging trends. *IEEE Communications Magazine*, 51(1), 114–119. DOI: 10.1109/MCOM.2013.6400447

AMTFT. (2024), Advanced machinery telematics and fuel monitoring by Technoton, available online, Retrieved October 2024, from https://jv-technoton.com/

An, X., Liu, Y., Li, J., Yu, H., & Hu, X. (2023). Travel Scenario Characteristics Effects on the Energy Consumption Deterioration of Electric Vehicles under Low Temperature. 2023 4th International Conference on Clean and Green Energy Engineering (CGEE), 128–133. DOI: 10.1109/CGEE59468.2023.10351963

Anadón Martínez, V., & Sumper, A. (2023). Planning and Operation Objectives of Public Electric Vehicle Charging Infrastructures: A Review. *Energies*, 16(14), 5431. DOI: 10.3390/en16145431

Andrae, A., & Edler, T. (2015). On Global Electricity Usage of Communication Technology: Trends to 2030. *Challenges*, 6(1), 117–157. DOI: 10.3390/challe6010117

Axsen, J., Mountain, D. C., & Jaccard, M. (2009). Combining stated and revealed choice research to simulate the neighbor effect: The case of hybrid-electric vehicles. *Resource and Energy Economics*, 31(3), 221–238. DOI: 10.1016/j.reseneeco.2009.02.001

Azim Mohseni, N., Bayati, N., & Ebel, T. (2024). Energy management strategies of hybrid electric vehicles: A comparative review. *IET Smart Grid*, 7(3), 191–220. DOI: 10.1049/stg2.12133

Azizighalehsari, , Venugopal, P., Pratap Singh, D., Batista Soeiro, T., & Rietveld, G. (2024). Empowering Electric Vehicles Batteries: A Comprehensive Look at the Application and Challenges of Second-Life Batteries. *Batteries*, 10(5), 161. DOI: 10.3390/batteries10050161

Badran, M. A., & Toha, S. F. (2024). Employment of Artificial Intelligence (AI) Techniques in Battery Management System (BMS) for Electric Vehicles (EV): Issues and Challenges. *Pertanika Journal of Science & Technology*, 32(2), 859–881. DOI: 10.47836/pjst.32.2.20

Badura, S., & Lieskovsky, A. (2010). Intelligent Traffic System: Cooperation of MANET and Image Processing. *1st International Conference on Integrated Intelligent Computing,* Bangalore, 119-123. DOI: 10.1109/ICIIC.2010.41

Baird, I. G., & Thorne, M. A. (2023). The downstream impacts of dams on the seasonally flooded riverine forests of the Mekong River in northeastern Cambodia. *South East Asia Research*, 31(4), 377–399. DOI: 10.1080/0967828X.2023.2243584

Barman, P., Dutta, L., Bordoloi, S., Kalita, A., Buragohain, P., Bharali, S., & Azzopardi, B. (2023). Renewable energy integration with electric vehicle technology: A review of the existing smart charging approaches. *Renewable & Sustainable Energy Reviews*, 183, 113518. DOI: 10.1016/j.rser.2023.113518

Barragán-Escandón, E. A., Zalamea-León, E. F., Terrados-Cepeda, J., & Vanegas-Peralta, P. F. (2020). Energy self-supply estimation in intermediate cities. *Renewable & Sustainable Energy Reviews*, 129, 109913. DOI: 10.1016/j.rser.2020.109913

Bayo, Y., Luo, Y., Zhang, W., Huang, M., Wang, L. Y., & Jiang, J. (2018). A Bi-Level Optimization Approach to Charging Load Regulation of Electric Vehicle Fast Charging Stations Based on a Battery Energy Storage System. *Energies*, 11(1), 229. DOI: 10.3390/en11010229

Bayram, I. S., Michailidis, G., Devetsikiotis, M., & Bhattacharya, S. (2011). Local energy storage sizing in plug-in hybrid electric vehicle charging stations under blocking probability constraints. In *Proceedings of the IEEE International Conference on Smart Grid Communications* (pp. 78–83). DOI: 10.1109/SmartGridComm.2011.6102396

Bayram, I. S., Michailidis, G., Devetsikiotis, M., Granelli, F., & Bhattacharya, S. (2012). Smart vehicles in the smart grid: Challenges, trends, and application to the design of charging stations. In Chakrabortty, A., & Ilic, M. D. (Eds.), *Control and Optimisation Methods for Electric Smart Grids* (Vol. 3, pp. 133–145). Springer. DOI: 10.1007/978-1-4614-1605-0_6

Becker, J., & Shukla, V. P. (2022, November). Electric Vehicle Battery Thermal Management Systems Current Trends and Future Directions. *SSRAML*, 5(2), 47–66.

Benato, A., & Stoppato, A. (2018). Pumped Thermal Electricity Storage: A technology overview. *Thermal Science and Engineering Progress*, 6, 301–315. DOI: 10.1016/j.tsep.2018.01.017

Bhagwat, R., & Sharma, M. K. (2007). Performance measurement of supply chain management using the analytical hierarchy process. *Production Planning and Control*, 18(8), 666–680. DOI: 10.1080/09537280701614407

Bishop, C. M. (2006). *Pattern recognition and machine learning*. Springer.

Bochkovskiy, A., Wang, C., & Liao, H. (2020). Yolov4: Optimal Speed and Accuracy of Object Detection. *arXiv*:2004.10934.

Bohlourihajar, K., & Harper, M. (2024). Catalyzing Urban Health and Sustainability: The Impact of Electric Bus Fleet Electrification in Justice 40 Communities of Salt Lake County. 2024 Forum for Innovative Sustainable Transportation Systems (FISTS), 1–6. DOI: 10.1109/FISTS60717.2024.10485534

Bohn, T., & Chaudhry, H. (2012). Overview of SAE standards for plug-in electric vehicles. In *Proceedings of IEEE PES Innovative Smart Grid Technologies* (pp. 1–7). DOI: 10.1109/ISGT.2012.6175597

Bracco, S., Bianco, G., Siri, S., Barbagelata, C., Casati, C., & Siri, E. (2021). Simulation Models for the Evaluation of Energy Consumptions of Electric Buses in Different Urban Traffic Scenarios. *2021 Sixteenth International Conference on Ecological Vehicles and Renewable Energies (EVER)*, 1–6. DOI: 10.1109/EVER52347.2021.9456661

Canco, I., Kruja, D., & Iancu, T. (2021). AHP, a reliable method for quality decision making: A case study in business. *Sustainability (Basel)*, 13(24), 13932. DOI: 10.3390/su132413932

Cao, K., Liu, Y., Meng, G., & Sun, Q. (2020). An overview on edge computing research. *IEEE Access : Practical Innovations, Open Solutions*, 8, 85714–85728. DOI: 10.1109/ACCESS.2020.2991734

Castelló-Sirvent, F., Meneses-Eraso, C., Alonso-Gómez, J., & Peris-Ortiz, M. (2022). Three decades of fuzzy AHP: A bibliometric analysis. *Axioms*, 11(10), 525. DOI: 10.3390/axioms11100525

Çelikbilek, Y., & Tüysüz, F. (2020). An in-depth review of theory of the TOPSIS method: An experimental analysis. *Journal of Management Analytics*, 7(2), 281–300. DOI: 10.1080/23270012.2020.1748528

Ceraolo, M., di Donato, A., & Franceschi, G. (2008). A General Approach to Energy Optimization of Hybrid Electric Vehicles. *IEEE Transactions on Vehicular Technology*, 57(3), 1433–1441. DOI: 10.1109/TVT.2007.909268

Chakraborty, C., Mishra, K., Majhi, S. K., & Bhuyan, H. (2023). Intelligent latency-aware tasks prioritization and offloading strategy in distributed fog-cloud of things. *IEEE Transactions on Industrial Informatics*, 19(2), 2099–2106. DOI: 10.1109/TII.2022.3173899

Chaudhary, P., Goel, S., Jain, P., Singh, M., & Aggarwal, P. K., Anupam.(2021)."The Astounding Relationship: Middleware, Frameworks, and API. 9th *International Conference on Reliability, Infocom Technologies and Optimization (Trends and Future Directions)* (ICRITO), Noida, India, 1-4, https://doi: DOI: 10.1109/ICRITO51393.2021.9596088

Chen, Y., Wu, G., Sun, R., Dubey, A., Laszka, A., & Pugliese, P. (2020). A review and outlook of energy consumption estimation models for electric vehicles. *arXiv preprint arXiv:2003.12873.*

Chen, C. F., de Rubens, G. Z., Noel, L., Kester, J., & Sovacool, B. K. (2020). Assessing the socio-demographic, technical, economic and behavioral factors of Nordic electric vehicle adoption and the influence of vehicle-to-grid preferences. *Renewable & Sustainable Energy Reviews*, 121, 109692. DOI: 10.1016/j.rser.2019.109692

Chendong, H., Feng, W., & Zhongming, H. (2022, August). A Lightweight Anonymous Key Agreement Scheme in V2G Networks. In *2022 2nd International Conference on Frontiers of Electronics, Information and Computation Technologies (ICFEICT)* (pp. 304-309). IEEE.

Chen, L. M., Liao, Z. L., & Ma, X. J. (2021). Hierarchicalcontrol-basedreal-timeenergy management strategy for hybrid electric vehicles. *ActaArmamentar*, II(8), 1580–1591. DOI: 10.3969/j.issn.1000-1093.2021.08.002

Chen, Z. Y., Xiong, R., Wang, K. Y., & Jiao, B. (2015). Optimalenergymanagementstrategyofaplug-inhybridelectricvehiclebasedonaparticleswarmoptimizationalgorithm. *Energies*, 8(08), 3661–3678. DOI: 10.3390/en8053661

Chen, Z., Xia, B., You, C., & Mi, C. C. (2015). A novel energy management method for series plug-in hybrid electric vehicles. *Applied Energy*, 145, 172–179. DOI: 10.1016/j.apenergy.2015.02.004

Choudhury, N., Matam, R., Mukherjee, M., Lloret, J., & Kalaimannan, E. (2021). NCHR: A non-threshold- based cluster-head rotation scheme for IEEE 802.15.4 cluster-tree networks. *IEEE Internet of Things Journal*, 8(1), 168–178. DOI: 10.1109/JIOT.2020.3003320

Choudhury, N., Matam, R., Mukherjee, M., & Shu, L. (2018). Beacon synchronization and duty-cycling in IEEE 802.15.4 cluster-tree networks: A review. *IEEE Internet of Things Journal*, 5(3), 1765–1788. DOI: 10.1109/JIOT.2018.2827946

Chui. K.T, Gupta. B.B., Liu. R.W, Zhang. X, Vasant. P, Thomas.(2021). J.J. Extended-Range Prediction Model Using NSGA-III Optimized RNN-GRU-LSTM for Driver Stress and Drowsiness. *Sensors*, 6412.

Chung, Y., & Kim, M. S. (2019, September). Thermal analysis and pack level design of battery thermal management system with liquid cooling for electric vehicles," *. *Energy Conversion and Management*, 196, 105–116. DOI: 10.1016/j.enconman.2019.05.083

Collotta, M., Pau, G., Scatà, G., & Campisi, T. (2014). A Dynamic Traffic Light Management System Based on Wireless Sensor Networks for the Reduction of the Red-light Running Phenomenon. *Journal of Transport and Telecommunications*, 15(1), 1–11. DOI: 10.2478/ttj-2014-0001

Colombo, C. G., Longo, M., Foiadelli, F., Garcia, M. D., & Yaici, W. (2023). Green Mobility Application in Malaga: Analysis of the Transition from ICE to EV. IEEE Int. Conference on Environment and Electrical Engineering (EEEIC / I&CPS Europe), 1–5. DOI: 10.1109/EEEIC/ICPSEurope57605.2023.10194877

Colombo, C. G., Miraftabzade, S., Aimar, M., Zaninelli, D., Longo, M., & Yaici, W. (2022). A Comprehensive Study on Electrification of Old Bus Fleets: A Real Case Study in Ottawa. *2022 International Symposium on Power Electronics, Electrical Drives, Automation and Motion (SPEEDAM)*, 844–849. DOI: 10.1109/SPEEDAM53979.2022.9842033

Cortes, C., & Vapnik, V. (1995). Support-vector networks. *Machine Learning*, 20(3), 273–297. DOI: 10.1007/BF00994018

Das, , Bhat, M. Y., & Sajith, S. (2024). Life cycle assessment of electric vehicles: A systematic review of literature. *Environmental Science and Pollution Research International*, 31(1), 73–89. DOI: 10.1007/s11356-023-30999-3 PMID: 38038907

Das, H., Rahman, M. M., Li, S., & Tan, C. H. (2020). Electric vehicles standards, charging infrastructure, and impact on grid integration: A technological review. *Renewable & Sustainable Energy Reviews*, 120, 109618. DOI: 10.1016/j.rser.2019.109618

Ding, H., Hu, Z., & Song, Y. (2015). Value of the energy storage system in an electric bus fast charging station. *Applied Energy*, 157, 630–639. DOI: 10.1016/j.apenergy.2015.01.058

Ding, T., Wang, J., Matas, J., Guerrero, J. M., Qiao, R., & Wang, C. (2023, June). Multi-Time Scale Energy Management Strategy based on MPC for 5G Base Stations Considering Backup Energy Storage and Air Conditioning. In *2023 IEEE International Conference on Predictive Control of Electrical Drives and Power Electronics (PRECEDE)* (pp. 1-6). IEEE. DOI: 10.1109/PRECEDE57319.2023.10174449

Dixon, J., & Bell, K. (2020). Electric vehicles: Battery capacity, charger power, access to charging and the impacts on distribution networks. *ETransportation*, 4, 100059. DOI: 10.1016/j.etran.2020.100059

Dubal, D. P., Ayyad, O., Ruiz, V., & Gómez-Romero, P. (2014). Hybrid energystorage: The merging of battery and supercapacitor chemistries. *Chemical Society Reviews*, 44(7), 1777–1790. DOI: 10.1039/C4CS00266K PMID: 25623995

Ecer, F. (2021). A consolidated MCDM framework for performance assessment of battery electric vehicles based on ranking strategies. *Renewable & Sustainable Energy Reviews*, 143, 110916. DOI: 10.1016/j.rser.2021.110916

Egbue, O., & Long, S. (2012). Barriers to widespread adoption of electric vehicles: An analysis of consumer attitudes and perceptions. *Energy Policy*, 48, 717–729. DOI: 10.1016/j.enpol.2012.06.009

Ehsan, A., & Yang, Q. (2020). Active Distribution System Reinforcement Planning With EV Charging Stations—Part I: Uncertainty Modeling and Problem Formulation. *IEEE Transactions on Sustainable Energy*, 11(2), 970–978. DOI: 10.1109/TSTE.2019.2915338

Eiza, M. H., Shi, Q., Marnerides, A. K., Owens, T., & Ni, Q. (2018). Efficient, secure, and privacy-preserving PMIPv6 protocol for V2G networks. *IEEE Transactions on Vehicular Technology*, 68(1), 19–33. DOI: 10.1109/TVT.2018.2880834

Elagouz, N., Onat, N., & Kucukvar, M. (2021). Environmental, Economic, and Social Life Cycle Impacts of Alternative Fuel Buses: The Case for Qatar. 2021 10th International Conference on Power Science and Engineering (ICPSE), 92–97. DOI: 10.1109/ICPSE53473.2021.9656863

Elbert, P., Ebbesen, S., & Guzzella, L. (2012). Implementation of dynamic programming for $ n $-dimensional optimal control problems with final state constraints. *IEEE Transactions on Control Systems Technology*, 21(3), 924–931. https://ieeexplore.ieee.org/document/6178777. DOI: 10.1109/TCST.2012.2190935

Electric Power Research Institute. (2019). *Interoperability of public electric vehicle charging infrastructure.* Retrieved June 20, 2021, from https://www.eei.org

ElGhanam, E. A., Hassan, M. S., & Osman, A. H. (2020). Deployment optimization of dynamic wireless electric vehicle charging systems: A review. In *Proceedings of the 2020 IEEE International IoT, Electronics, and Mechatronics Conference (IEMTRONICS)* (pp. 1–7). DOI: 10.1109/IEMTRONICS51293.2020.9216415

ElGhanam, E., Hassan, M., & Osman, A. (2021). Design of a high-power, LCC-compensated, dynamic, wireless electric vehicle charging system with improved misalignment tolerance. *Energies*, 14(4), 885. DOI: 10.3390/en14040885

Eltamaly, A. M. (2023). Optimal Dispatch Strategy for Electric Vehicles in V2G Applications. *Smart Cities*, 6(6), 3161–3191. DOI: 10.3390/smartcities6060141

Esmaeili, M., Anvari-Moghaddam, A., Muyeen, S. M., & Perić, V. S. (2022). On the role of renewable energy policies and electric vehicle deployment incentives for a greener sector coupling. *IEEE Access : Practical Innovations, Open Solutions*, 10, 53873–53893. DOI: 10.1109/ACCESS.2022.3176012

Etezadi-Amoli, M., Choma, K., & Stefani, J. (2010). Rapid-Charge Electric-Vehicle Stations. *IEEE Transactions on Power Delivery*, 25(3), 1883–1887. DOI: 10.1109/TPWRD.2010.2047874

Fayaz, H., Afzal, A., Samee, A. D. M., Soudagar, M. E. M., Akram, N., Mujtaba, M. A., Jilte, R. D., Islam, M. T., Ağbulut, Ü., & Saleel, C. A. (2022). Optimization of Thermal and Structural Design in Lithium-Ion Batteries to Obtain Energy Efficient Battery Thermal Management System (BTMS): A Critical Review. *Archives of Computational Methods in Engineering*, 29(1), 129–194. DOI: 10.1007/s11831-021-09571-0 PMID: 33935484

Fiore, P., Sicignano, E., & Donnarumma, G. (2020). An AHP-based methodology for the evaluation and choice of integrated interventions on historic buildings. *Sustainability (Basel)*, 12(14), 5795. DOI: 10.3390/su12145795

Fiori, C., Ahn, K., & Rakha, H. A. (2016). Power-based electric vehicle energy consumption model: Model development and validation. *Applied Energy*, 168, 257–268. DOI: 10.1016/j.apenergy.2016.01.097

Fisher, W., Suchara, M., & Rexford, J. (2010). Greening backbone networks. In Proceedings of the first*ACM SIGCOMM Workshop on Green Networking*; ACM: New York, NY, USA, 29–34.

Foles, A., Fialho, L., Horta, P., &Collares-Pereira, M. (2023).Economic and energetic assessment of a hybrid vanadium redox flow and lithium-ion batteries, considering power sharing strategies impact.Journal of Energy Storage, 71, (pp.108167). .DOI: 10.1016/j.est.2023.108167

Frieske, B., Kloetzke, M., & Mauser, F. (2013, November). *Trends in vehicle concept and key technology development for hybrid and battery electric vehicles. In 2013 world electric vehicle symposium and exhibition (EVS27)*. IEEE.

Frith, , Lacey, M. J., & Ulissi, U. (2023). A non-academic perspective on the future of lithium-based batteries. *Nature Communications*, 14(1), 420. DOI: 10.1038/s41467-023-35933-2 PMID: 36702830

Gaertner, E., Rinker, J., Sethuraman, L., Zahle, F., Anderson, B., Barter, G. E., & Viselli, A. (2020). *IEA wind TCP task 37: Definition of the IEA 15-megawatt offshore reference wind turbine (No. NREL/TP-5000-75698). National Renewable Energy Lab*. NREL.

Ganesh, S., Nivas, K. V., Kavya, A., Arunteja, S., & Guruvulunaidu, P., & KUMAR, B. (. (2023). DESIGNING OF ELECTRIC VEHICLE WITH REGENERATIVE BRAKING SYSTEM. *Journal on Electrical Engineering*, 16(3). Advance online publication. DOI: 10.26634/jee.16.3.19419

Gan, L., Topcu, U., & Low, S. (2011). Optimal decentralized protocol for electric vehicle charging. In *Proceedings of the IEEE Decision and Control and European Control Conference* (pp. 5798–5804). DOI: 10.1109/CDC.2011.6161220

García-Magariño, M. M. Nasralla,J. Lloret. (2021). 'A repository of method fragments for agent-oriented development of learning-based edge computing systems. *IEEE Network.*, vol.

García-Magariño, I., Nasralla, M. M., & Nazir, S. (2020). Real-time analysis of online sources for supporting business intelligence illustrated with Bitcoin investments and IoT smart-meter sensors in smart cities. *Electronics (Basel)*, 9(7), 1101. DOI: 10.3390/electronics9071101

Garg, H., Singh, M., Sharma, V., & Agarwal, M. (2022). Decentralized application (DAPP) to enable e-voting system using blockchain technology. In *2022 Second International Conference on Computer Science, Engineering and Applications (ICCSEA)* (pp. 1–6). https://doi.org/DOI: 10.1109/ICCSEA54677.2022.9936413

Garud, K. S., Tai, L. D., Hwang, S.-G., Nguyen, N.-H., & Lee, M.-Y. (2023). A Review of Advanced Cooling Strategies for Battery Thermal Management Systems in Electric Vehicles. *Symmetry*, 15(7), 1322. DOI: 10.3390/sym15071322

George, A. M., George, V. I., & George, M. A. (2018, March). IoT based smart traffic light control system. In *2018 International conference on control, power, communication and computing technologies (ICCPCCT)* (pp. 148-151). IEEE.

Ghadikolaei, M. A., Wong, P. K., Cheung, C. S., Zhao, J., Ning, Z., Yung, K. F., Wong, H. C., & Gali, N. K. (2021). Why is the world not yet ready to use alternative fuel vehicles? *Heliyon*, 7(7), e07527. DOI: 10.1016/j.heliyon.2021.e07527 PMID: 34337177

Ghazal, B., ElKhatib, K., Chahine, K., & Kherfan, M. (2016). Smart Traffic Light Control System. Proceedings of the *3rd International Conference on Electrical, Electronics, Computer Engineering and their Applications, Beirut,* 140-145.

Giansoldati, M., Rotaris, L., Scorrano, M., & Danielis, R. (2020). Does electric car knowledge influence car choice? Evidence from a hybrid choice model. *Research in Transportation Economics*, 80, 100826. DOI: 10.1016/j.retrec.2020.100826

Gicha, , Tufa, L. T., Nwaji, N., Hu, X., & Lee, J. (2024). Advances in All-Solid-State Lithium–Sulfur Batteries for Commercialization. *Nano-Micro Letters*, 16(1), 172. DOI: 10.1007/s40820-024-01385-6 PMID: 38619762

Goel, A., Singh, M., Gupta, J., & Mangla, N. (2023). Med Card: an innovative way to keep your medical records handy and safe. In Advances in *intelligent systems and computing*,51–60. https://doi.org/DOI: 10.1007/978-981-99-0550-8_4

Goodfellow, I., Bengio, Y., & Courville, A. (2016). *Deep learning*. MIT Press.

Gowri, K., Pratt, R. G., Tuffner, F. K., & Kintner-Meyer, M. C. W. (2011). *Vehicle-to-grid communication standards development, testing, and validation: Status report.* Pacific Northwest National Laboratory. DOI: 10.2172/1029430

Günther, H.-O., Kannegiesser, M., & Autenrieb, N. (2015, March). The role of electric vehicles for supply chain sustainability in the automotive industry," *. *Journal of Cleaner Production*, 90, 220–233. DOI: 10.1016/j.jclepro.2014.11.058

Gupta, M., Singh, M., Sharma, A., Sukhija, N., & Aggarwal, P. K. (2023). Unification of machine learning and blockchain technology in healthcare industry. *Innovations in Healthcare Informatics*, 185–206. DOI: 10.1049/PBHE041E_ch6

Gupta, V. K. (2020). Toxicity prediction of pre-clinical trial drugs using physico-chemical properties and computational intelligence approaches [Dissertation].

Gupta, V. K. (2022). Toxicity detection of small drug molecules of the mitochondrial membrane potential signalling pathway. International Journal of Data Mining and Bio-informatics, 27(1/2/3), 201–220. https://doi.org/DOI: 10.1504/IJDMB.2022.119720

Gupta, V. K., Gupta, A., Singh, M. K., Gupta, A., & Tomar, A. (2023). Toxicity detection methodology of adverse outcome pathways using physicochemical properties and machine learning approaches. In 2023 4th International Conference for Emerging Technology (INCET) (pp. 1–7). IEEE. DOI: 10.1109/INCET57972.2023.10170360

Gupta, V. K., Gupta, A., Tyagi, V., Pandey, P., Gupta, R., & Kumar, D. (2023). Multilevel face mask detection system using ensemble based convolution neural network. In *2023 Third International Conference on Secure Cyber Computing and Communication (ICSCCC)* (pp. 391–396). IEEE. DOI: 10.1109/ICSCCC58608.2023.10176502

Gür, T. M. (2018). Review of electrical energy storage technologies, materials and systems: Challenges and prospects for large-scale grid storage. *Energy & Environmental Science*, 11(10), 2696–2767. DOI: 10.1039/C8EE01419A

Habib, M., Rajah, P., Ravikumar, R. N., Nadarajan, S., Betgeri, S. S., & Itagi, M. (2024, July). Machine Learning Classifiers and Hyperparameter Tuning to Enhance Prediction of ASD: An Automated Django Application. In *2024 2nd World Conference on Communication & Computing (WCONF)* (pp. 1-7). IEEE.

Habib, S., Khan, M. M., Abbas, F., Sang, L., Shahid, M. U., & Tang, H. (n.d.). A comprehensive study of implemented international standards, technical challenges, impacts, and prospects for electric vehicles.

Habib, A. A., Hasan, M. K., Islam, S., Sharma, R., Hassan, R., Nafi, N., & Alotaibi, S. D. (2022). Energy-efficient system and charge balancing topology for electric vehicle application. *Sustainable Energy Technologies and Assessments*, 53, 102516. DOI: 10.1016/j.seta.2022.102516

Haleema, K., Akhilesh, R., Shashank, G., Baig, M. A., & Krishnamoorthy, M. (2024, May).A Comprehensive Analysis of Load Management Strategies and Hybrid Renewable Solutions for Sustainable Rural Electrification System.In *2024 5th International Conference for Emerging Technology (INCET)* (pp. 1-6).IEEE. DOI: 10.1109/INCET61516.2024.10593192

Hamada, A. T., &Orhan, M. F. (2022).An overview of regenerative braking systems. Journal of Energy Storage, 52, (pp.105033). DOI: 10.1016/j.est.2022.105033

Hamzaoui, M. A., & Julien, N. (2024). A Generic Deployment Methodology for Digital Twins–First Building Blocks. In *Handbook of Digital Twins* (pp. 102–121). CRC Press. DOI: 10.1201/9781003425724-10

Harakawa, T., & Tujimoto, T. (2002). Efficient solar power equipment for electric vehicles: improvement of energy conversion efficiency for charging electric vehicles. https://doi.org/DOI: 10.1109/IVEC.2001.961718

Hardman, S., Jenn, A., Tal, G., Axsen, J., Beard, G., Daina, N., Figenbaum, E., Jakobsson, N., Jochem, P., Kinnear, N., Plötz, P., Pontes, J., Refa, N., Sprei, F., Turrentine, T., & Witkamp, B. (2018). A review of consumer preferences and interactions with electric vehicle charging infrastructure. *Transportation Research Part D, Transport and Environment*, 62, 508–523. DOI: 10.1016/j.trd.2018.04.002

Hardman, S., Shiu, E., & Steinberger-Wilckens, R. (2016). Comparing high-end and low-end early adopters of battery electric vehicles. *Transportation Research Part A, Policy and Practice*, 88, 40–57. DOI: 10.1016/j.tra.2016.03.010

Hawkins, T. R., Gausen, O. M., & Strømman, A. H. (2012). Environmental impacts of hybrid and electric vehicles—A review. *The International Journal of Life Cycle Assessment*, 17(8), 997–1014. DOI: 10.1007/s11367-012-0440-9

Helmy, M., Taher, M., Abdelhakem, M., Medhat, A., Tarek, M., Tarek, A. T., & Ahmad, M. (2023, November). Securing ISO 15118 Communication in Electric Vehicles using Real-Time Monitoring and Vehicle-to-Grid Sniffer. In *2023 International Conference on Computer and Applications (ICCA)* (pp. 1-5). IEEE. DOI: 10.1109/ICCA59364.2023.10401502

He, Y., Venkatesh, B., & Guan, L. (2012). Optimal scheduling for charging and discharging of electric vehicles. *IEEE Transactions on Smart Grid*, 3(3), 1095–1105. DOI: 10.1109/TSG.2011.2173507

He, Z. C., Yang, G., & Lu, L. G. (2015). BatteryDCinternalresistancetestmethodbased on the constant current external characteristic and SOC. [SciTechnol]. *J Tsinghua Univ*, (5), 9–18.

Hinton, G., & Salakhutdinov, R. (2006). Reducing the dimensionality of data with neural networks. *Science*, 313(5786), 504–507. DOI: 10.1126/science.1127647 PMID: 16873662

Holland, S. P., Mansur, E. T., Muller, N. Z., & Yates, A. J. (2021). The environmental benefits of transportation electrification: Urban buses. *Energy Policy*, 148, 111921. DOI: 10.1016/j.enpol.2020.111921

Hong, S. H., Jang, D. S., Park, S., Yun, S., & Kim, Y. (2020, June). Thermal performance of direct two-phase refrigerant cooling for lithium-ion batteries in electric vehicles," *. *Applied Thermal Engineering*, 173, 115213. DOI: 10.1016/j.applthermaleng.2020.115213

Hoque, M. M., Hannan, M. A., Mohamed, A., & Ayob, A. (2017). Battery charge equalization controller in electric vehicle applications: A review. *Renewable & Sustainable Energy Reviews*, 75, 1363–1385. DOI: 10.1016/j.rser.2016.11.126

Hossain, E., Han, Z., & Poor, H. V. (2012). *Smart grid communications and networking*. Cambridge University Press. DOI: 10.1017/CBO9781139013468

Hou. Y, Zheng. X, Han. C, Wei. W. Scherer, R. Połap.(2022). Deep Learning Methods in Short-Term Traffic Prediction.*A Survey. Inf. Technol. Control, 51*, 139–157.

Huang, S. C., & Wu, C. F. (2018). Energy Commodity Price Forecasting with Deep Multiple Kernel Learning. *Energies*, 2018(11), 3029. DOI: 10.3390/en11113029

Huan, N., Yao, E. J., & Yang, Y.. (2019). Stochastic dynamic user equilibriumassignment model considering penetration of electric vehicle. *Journal of Traffic and Transportation Engineering*, (10), 150–160. DOI: 10.19818/j.cnki.1671-1637.2019.05.015

Hu, J. J., Zheng, Y., & Hu, Z. H.. (2018). Parameter matching and controlstrategiesofhybridenergystoragesystemforpureelectricvehicle.*ChinaJ.Highways and Transportation*, (3), 142–150. DOI: 10.19721/j.cnki.1001-7372.2018.03.016

Hussain, A., Kim, H., Baek, J., & Kim, H. (2019). Stationary Energy Storage System for Fast EV Charging Stations: Simultaneous Sizing of Battery and Converter. *Energies*, 12(23), 4516. DOI: 10.3390/en12234516

Hussain, M. Z., Anbalagan, R., Jayabalakrishnan, D., Muruga, D. N., Prabhahar, M., Bhaskar, K., & Sendilvelan, S. (2021). Charging of car battery in electric vehicle by using wind energy. *Materials Today: Proceedings*, 45, 5873–5877. DOI: 10.1016/j.matpr.2020.08.341

Hutchinson, L. (2023). *Design and simulation of photovoltaic system for reduced demand from electric bus charging*. University of Delaware.

Hwang, F. S., Confrey, T., Reidy, C., Picovici, D., Callaghan, D., Culliton, D., & Nolan, C. (2024). Review of battery thermal management systems in electric vehicles. *Renewable & Sustainable Energy Reviews*, 192, 114171. Advance online publication. DOI: 10.1016/j.rser.2023.114171

Ianzino, A. P., Chaudet, C., Rossi, D., & Rougier, J. L. (2012). A Survey of Green Networking Research. *IEEE Communications Surveys and Tutorials*, 14(1), 3–20. DOI: 10.1109/SURV.2011.113010.00106

Ibrahim, H., Younes, R., Ilinca, A., Dimitrova, M., & Perron, J. (2010). Study and design of a hybrid wind–diesel-compressed air energy storage system for remote areas. *Applied Energy*, 87(5), 1749–1762. DOI: 10.1016/j.apenergy.2009.10.017

Iris, Ç., & Lam, J. S. L. (2019). A review of energy efficiency in ports: Operational strategies, technologies and energy management systems. *Renewable & Sustainable Energy Reviews*, 112, 170–182. DOI: 10.1016/j.rser.2019.04.069

Irshad, A., Usman, M., Chaudhry, S. A., Naqvi, H., & Shafiq, M. (2020). A provably secure and efficient authenticated key agreement scheme for energy internet-based vehicle-to-grid technology framework. *IEEE Transactions on Industry Applications*, 56(4), 4425–4435. DOI: 10.1109/TIA.2020.2966160

Islam, M., Shahid, N., Ul Karim, D., Al Mamun, A., & Rhaman, M. (2016). An Efficient Algorithm for Detecting Traffic Congestion and a Framework or Smart Traffic Control System. Proceedings of the *18th International Conference on Advanced Communication Technology, PyeongChang,* p802-807.

Itani, , & De Bernardinis, A. (2023). A. Review on New-Generation Batteries Technologies: Trends and Future Directions. *Energies*, 16(22), 7530. DOI: 10.3390/en16227530

Jain, P., Sharma, S., Arora, S., Shokeen, V., Aggarwal, P. K., & Singh, M. (2024). Edge Computing-Based design for IoT security. In *Chapman and Hall/CRC eBooks* (pp. 298–309). DOI: 10.1201/9781003405535-22

Jaiswal, N., Gupta, V. K., & Mishra, A. (2015). Survey paper on various techniques of recognition and tracking. In *2015 International Conference on Advances in Computer Engineering and Applications* (pp. 921–925). IEEE. DOI: 10.1109/ICACEA.2015.7164837

Jang, Y. J. (2018). Survey of the operation and system study on wireless charging electric vehicle systems. *Transportation Research Part C, Emerging Technologies*, 95, 844–866. DOI: 10.1016/j.trc.2018.04.006

Jang, Y. J., Suh, E. S., & Kim, J. W. (2016). System architecture and mathematical models of electric transit bus system utilizing wireless power transfer technology. *IEEE Systems Journal*, 10(2), 495–506. DOI: 10.1109/JSYST.2014.2369485

Javaid, S., Sufian, A., Pervaiz, S., & Tanveer, M. (2018).Smart Traffic Management System Using Internet of Things. in Proceedings of the 20th *International Conference on Advanced Communication Technology, Chuncheon,* 393-398.

Javaid, N., Hafeez, G., Iqbal, S., Alrajeh, N., Alabed, M. S., & Guizani, M. (2018). Energy-efficient integration of renewable energy sources in the smart grid for demand-side management. *IEEE Access : Practical Innovations, Open Solutions*, 6, 77077–77096. DOI: 10.1109/ACCESS.2018.2866461

Javed, S., Raza, S., & Riaz, M. (2023). An energy management approach for isolated power system using low load diesel. *Sukkur IBA Journal of Emerging Technologies*, 6(1), 1–10. DOI: 10.30537/sjet.v6i1.1268

Jena, R. (2020). An empirical case study on Indian consumers' sentiment towards electric vehicles: A big data analytics approach. *Industrial Marketing Management*, 90, 605–616. DOI: 10.1016/j.indmarman.2019.12.012

Jeong, S., Jang, Y. J., & Kum, D. (2015). Economic analysis of the dynamic charging electric vehicle. *IEEE Transactions on Power Electronics*, 30(7), 6368–6377. DOI: 10.1109/TPEL.2015.2424712

Jiang, J., Bao, B., Meng, F., Ma, Y., Zhang, H., Jin, Y., Pan, F., & Yuan, X. (2022). Probabilistic Energy Consumption Estimation for Electric Buses. 2022 4th International Conference on Smart Power & Internet Energy Systems (SPIES), 1767–1771. DOI: 10.1109/SPIES55999.2022.10082616

Jiang, J., Jiang, Q., Chen, J., Zhou, X., Zhu, S., & Chen, T. (2021).Advanced power management and control for hybrid electric vehicles: a survey.Wireless Communications and Mobile Computing, 2021(1), (pp.6652038). .DOI: 10.1155/2021/6652038

Jian, L., Xue, H., Xu, G., Zhu, X., Zhao, D., & Shao, Z. Y. (2013). Regulated charging of plug-in hybrid electric vehicles for minimizing load variance in household smart microgrid. *IEEE Transactions on Industrial Electronics*, 60(8), 3218–3226. DOI: 10.1109/TIE.2012.2198037

Ji, C. W., Pan, S., & Wang, S.. (2020). Experimental study on effect factors ofagingrateforpowerlithium-ionbatteries.*J.BeijingUniv.Technol.*, (10), 1272–1282. DOI: 10.11936/bjutxb2019040020

Ji, J., Bie, Y., Zeng, Z., & Wang, L. (2022). Trip energy consumption estimation for electric buses. *Communications in Transportation Research*, 2, 100069. DOI: 10.1016/j.commtr.2022.100069

John, S., & G., V. (2021). PV fed Electric Vehicle Charging Station With Power Backup. 2021 13th IEEE PES Asia Pacific Power & Energy Engineering Conference (APPEEC), 1–6. DOI: 10.1109/APPEEC50844.2021.9687739

Junior, R. R. R., Vieira, M. A. M., Vieira, L. F. M., & Loureiro, A. A. F. (2022). Loureiro, A.A.F. Intra and inter-flow link aggregation in SDN. *Telecommunication Systems*, 79(1), 95–107. DOI: 10.1007/s11235-021-00841-7

K, B. N. K. R., D, P., B, S., & E, J. R. (2024). Recent AI Applications in Electrical Vehicles for Sustainability. *International Journal of Mechanical Engineering, 11*(3), 50–64. DOI: 10.14445/23488360/IJME-V11I3P106

Kaleem, S., Sohail, A., Tariq, M. U., & Asim, M. (2023). An improved big data analytics architecture using federated learning for Io Tenabled urban intelligent transportation systems. *IEEE Access : Practical Innovations, Open Solutions.*

Kanungo, A., Sharma, A., & Singla, C. (2014). Smart Traffic Lights Switching and Traffic Density Calculation Using Video Processing. in Proceedings of the *Recent Advances in Engineering and Computational Sciences*, Chadigarh, 1-6.

Kaur, N., & Bindal, R. K. (2023). Modeling and Simulation Analysis of Solar Charging Station for Electric Vehicle. 2023 5th International Conference on Smart Systems and Inventive Technology (ICSSIT), 344–349. DOI: 10.1109/ICSSIT55814.2023.10061147

Keerthi. R, HariharaGopalan. S.(2016).A Survey on Various Traffic Management Schemes for Traffic Clearance, Stolen Vehicle and Emergency Vehicle. *International Journal of Emerging Technology in Computer Science and Electronics,* vol. 19, no. 2, 46-50.

Kempton, W. (2016). Electric vehicles: Driving range. *Nature Energy*, 1(9), 16131. Advance online publication. DOI: 10.1038/nenergy.2016.131

Khalid, M., Alam, M. K., Sarwar, A., & Asghar, M. I. (2019). A Comprehensive Review of electric vehicles charging infrastructures and their impacts on power-quality of the utility grid. *eTransportation*, 1, 100006. DOI: 10.1016/j.etran.2019.100006

Khan, F., Ali, Y., & Khan, A. U. (2020). Sustainable hybrid electric vehicle selection in the context of a developing country. *Air Quality, Atmosphere & Health*, 13(4), 489–499. DOI: 10.1007/s11869-020-00812-y

Khan, F., Ullah, F., Kaleem, Z., Ur Rahman, S., Anwar, H., & Cho, Y.-Z. (2018). EVP-STC: Emergency vehicle priority and self-organising traffic control at intersections using Internet-of-Things platform. *IEEE Access : Practical Innovations, Open Solutions*, 6, 68242–68254. DOI: 10.1109/ACCESS.2018.2879644

Khan, M. R., Haider, Z. M., Malik, F. H., Almasoudi, F. M., Alatawi, K. S. S., & Bhutta, M. S. (2024). A Comprehensive Review of Microgrid Energy Management Strategies Considering Electric Vehicles, Energy Storage Systems, and AI Techniques. *Processes (Basel, Switzerland)*, 12(2), 270. DOI: 10.3390/pr12020270

Khan, Z., Khan, S., Dey, K., & Chowdhury, M. (2019). Development and Evaluation of Recurrent Neural Network-Based Models for Hourly Traffic Volume and Annual Average Daily Traffic Prediction. *Transportation Research Record: Journal of the Transportation Research Board*, 2673(7), 489–503. DOI: 10.1177/0361198119849059

Khan, Z., Koubaa, A., Benjdira, B., & Boulila, W. (2023). A game theory approach for smart traffic management. *Computers & Electrical Engineering*, 110, 108825. DOI: 10.1016/j.compeleceng.2023.108825

Knez, M., Zevnik, G. K., & Obrecht, M. (2019). A review of available chargers for electric vehicles: United States of America, European Union, and Asia. *Renewable & Sustainable Energy Reviews*, 109, 284–293. DOI: 10.1016/j.rser.2019.04.013

Knezovic, K., Marinelli, M., Zecchino, A., Andersen, P., & Træholt, C. (2017). Supporting involvement of electric vehicles in distribution grids. *Energy*, 134, 458–468. DOI: 10.1016/j.energy.2017.06.075

Kobashi, T., Yoshida, T., Yamagata, Y., Naito, K., Pfenninger, S., Say, K., Takeda, Y., Ahl, A., Yarime, M., & Hara, K. (2020). On the potential of "Photovoltaics+ Electric vehicles" for deep decarbonisation of Kyoto's power systems: Techno-economic-social considerations. *Applied Energy*, 275, 115419. DOI: 10.1016/j.apenergy.2020.115419

Koohi-Kamali, S., Tyagi, V., Rahim, N. A., Panwar, N. L., & Mokhlis, H. (2013). Emergence of energy storage technologies as the solution for reliable operation of smart power systems. *Renewable & Sustainable Energy Reviews*, 25, 135–165. DOI: 10.1016/j.rser.2013.03.056

Koranga, M., Kumar, T., Thakur, S. N., Pandey, R., & Joshi, M. (2024). 5G Cityscape: Unleashing Seamless Connectivity for Smart Urban Evolution. In *Secure and Intelligent IoT-Enabled Smart Cities* (pp. 55–76). IGI Global. DOI: 10.4018/979-8-3693-2373-1.ch004

Kosmanos, D., Maglaras, L. A., Mavrovouniotis, M., Moschoyiannis, S., Argyriou, A., & Janicke, H. (2015). Information and communication technology research opportunities in dynamic charging for electric vehicles. In *Proceedings of the Euromicro Conference on Digital System Design* (pp. 297–300).

Kosmanos, D., Maglaras, L. A., Mavrovouniotis, M., Moschoyiannis, S., Argyriou, A., Maglaras, A., & Janicke, H. (2018). Route optimization of electric vehicles based on dynamic wireless charging. *IEEE Access : Practical Innovations, Open Solutions*, 6, 42551–42565. DOI: 10.1109/ACCESS.2018.2847765

Kumar, P., Gupta, V. K., & Singh, D. P. (2022). Face mask detection using convolution neural network. In 2022 3rd International Conference on Issues and Challenges in Intelligent Computing Techniques (ICICT) (pp. 1–6). IEEE. DOI: 10.1109/ICICT55121.2022.10064500

Kumar, B. A., Jyothi, B., Rathore, R. S., Singh, A. R., Kumar, B. H., & Bajaj, M. (2023). A novel framework for enhancing the power quality of electrical vehicle battery charging based on a modified Ferdowsi Converter. *Energy Reports*, 10, 2394–2416. DOI: 10.1016/j.egyr.2023.09.070

Kumar, D., Gupta, A., Gupta, V. K., & Gupta, A. (2023). Aspect-based sentiment analysis using machine learning and deep learning approaches. *International Journal on Recent and Innovation Trends in Computing and Communication*, 11(5s), 118–138. DOI: 10.17762/ijritcc.v11i5s.6636

Kumar, R. R., & Alok, K. (2020). Adoption of electric vehicle: A literature review and prospects for sustainability. *Journal of Cleaner Production*, 253, 119911. DOI: 10.1016/j.jclepro.2019.119911

Kumar, S., Singh, S. K., Varshney, S., Singh, S., Kumar, P., Kim, B. G., & Ra, I. H. (2023). Fusion of deep sort and Yolov5 for effective vehicle detection and tracking scheme in real-time traffic management sustainable system. *Sustainability (Basel)*, 15(24), 16869. DOI: 10.3390/su152416869

Kuo, T. (2017). A modified TOPSIS with a different ranking index. *European Journal of Operational Research*, 260(1), 152–160. DOI: 10.1016/j.ejor.2016.11.052

Kwasinski, A., & Kwasinski, A. (2012). Signal processing in the electrification of vehicular transportation: Techniques for electric and plug-in hybrid electric vehicles on the smart grid. *IEEE Signal Processing Magazine*, 29(5), 14–23. DOI: 10.1109/MSP.2012.2186645

Lach, D., & Macioszek, E. (2018). Analysis of Selected ITS Services Used in Public Collective Transport. *WUT Journal of Transportation Engineering*, 121(1), 211–222. DOI: 10.5604/01.3001.0014.4596

Laimon, M., & Yusaf, T. (2024). Towards energy freedom: Exploring sustainable solutions for energy independence and self-sufficiency using integrated renewable energy-driven hydrogen system. *Renewable Energy*, 222, 119948. DOI: 10.1016/j.renene.2024.119948

Larsson, V., Johannesson, L., & Egardt, B. (2014). Analytic solutions to the dynamic programming subproblem in hybrid vehicle energy management. *IEEE Transactions on Vehicular Technology*, 64(4), 1458–1467. DOI: 10.1109/TVT.2014.2329864

Le Fouest, S., & Mulleners, K. (2024). Optimal blade pitch control for enhanced vertical-axis wind turbine performance. *Nature Communications*, 15(1), 2770. DOI: 10.1038/s41467-024-46988-0 PMID: 38553502

LeCun, Y., Bengio, Y., & Hinton, G. (2015). Deep learning. *Nature*, 521(7553), 436–444. DOI: 10.1038/nature14539 PMID: 26017442

Leung, L. C., & Cao, D. (2000). On consistency and ranking of alternatives in fuzzy AHP. *European Journal of Operational Research*, 124(1), 102–113. DOI: 10.1016/S0377-2217(99)00118-6

Lin, C., Liang, S., Chen, J., & Gao, X. (2019). A multi-objective optimal torque distribution strategy for four in-wheel-motor drive electric vehicles. *IEEE Access : Practical Innovations, Open Solutions*, 7, 64627–64640. DOI: 10.1109/ACCESS.2019.2917313

Lin, Y., Zhang, K., Shen, Z.-J. M., Ye, B., & Miao, L. (2019). Multistage large-scale charging station planning for electric buses considering transportation network and power grid. *Transportation Research Part C, Emerging Technologies*, 107, 423–443. DOI: 10.1016/j.trc.2019.08.009

Li, T., Zou, Y., & Liu, D. (2015). EnergymanagementstrategyforHEVsincludingbatterylifeoptimization. *IEEE Transactions on Transportation Electrification*, 70(1), 98–112. DOI: 10.1109/TTE.2015.2471180

Liu et al, (2022) Tobias Placke, K.T. Chau,Overview of batteries and battery management for electric vehicles,Energy Reports,8,pp. 4058-4084,ISSN 2352-4847,DOI: 10.1016/j.egyr.2022.03.016

Liu, Q., Huan, J., & Liu, Q. (2022, September). Secure charging scheduling strategy for electric vehicles based on blockchain. In *2022 IEEE 96th Vehicular Technology Conference (VTC2022-Fall)* (pp. 1-7). IEEE.

Liu, X. D., Ma, J., Zhao, X., Zhang, Y., Zhang, K., and He, Y. (2019).Integrated-component optimization and energy management for plug-in hybrid electric buses. *Process*7(8),(pp.477).. https://doi.org/.DOI: 10.3390/pr7080477

Liu, Z., Lee, H., Ali, G., Pesch, D., & Xiao, P. (2019). A survey on resource allocation in vehicular networks. *arXiv, 1909.13587.*

Liu, L., Kong, F., Liu, X., Peng, Y., & Wang, Q. (2015). A review on electric vehicles interacting with renewable energy in smart grid. *Renewable & Sustainable Energy Reviews*, 51, 648–661. DOI: 10.1016/j.rser.2015.06.036

Liu, W., Placke, T., & Chau, K. T. (2022). Overview of batteries and battery management for electric vehicles. *Energy Reports*, 8, 4058–4084.

Liu, Z., Li, Y., Fan, G., Wu, D., Guo, J., Jin, G., & Yang, X. (2022). Co-optimization of a novel distributed energy system integrated with hybrid energy storage in different nearly zero energy community scenarios. *Energy*, 247, 123553. DOI: 10.1016/j.energy.2022.123553

Liu, Z., Zhu, Z., Gao, J., & Xu, C. (2021). Forecast Methods for Time Series Data: A Survey. *IEEE Access : Practical Innovations, Open Solutions*, 9, 91896–91912. DOI: 10.1109/ACCESS.2021.3091162

Li, Y., He, H., Peng, J., & Zhang, H. (2017). Power management for a plug-in hybrid electric vehicle based on reinforcement learning with continuous state and action spaces. *Energy Procedia*, 142, 2270–2275. DOI: 10.1016/j.egypro.2017.12.629

Logan, K. G., Nelson, J. D., Brand, C., & Hastings, A. (2021). Phasing in electric vehicles: Does policy focusing on operating emission achieve net zero emissions reduction objectives? *Transportation Research Part A, Policy and Practice*, 152, 100–114. DOI: 10.1016/j.tra.2021.08.001

Lu, J., Zheng, W., Yu, Z., Xu, Z., Jiang, H., & Zeng, M. (2024). *Optimizing Grid-Connected Multi-Microgrid Systems With Shared Energy Storage for Enhanced Local Energy Consumption*. IEEE., DOI: 10.1109/ACCESS.2024.3351855

Lujano-Rojas, J. M., Monteiro, C., Dufo-López, R., & Bernal-Agustín, J. L.Jatin Pal Singh. (2012). Optimum residential load management strategy for real time pricing (RTP) demand response programs. *Energy Policy*, 45, 671–679. DOI: 10.1016/j.enpol.2012.03.019

Luo, J., Yao, S., Zhang, J., Xu, W., He, Y., & Zhang, M. (2020). A secure and anonymous communication scheme for charging information in vehicle-to-grid. *IEEE Access : Practical Innovations, Open Solutions*, 8, 126733–126742. DOI: 10.1109/ACCESS.2020.3005400

Machín, , Morant, C., & Márquez, F. (2024). Advancements and Challenges in Solid-State Battery Technology: An In-Depth Review of Solid Electrolytes and Anode Innovations. *Batteries*, 10(1), 29. DOI: 10.3390/batteries10010029

Maglaras, L., Jiang, J., Maglaras, A., Topalis, F., & Moschoyiannis, S. (2015). Dynamic wireless charging of electric vehicles on the move with mobile energy disseminators. *International Journal of Advanced Computer Science and Applications*, 6(5), 239–251.

Mahadevan, P., Sharma, P., Banerjee, S., & Ranganathan, P. (2009). A power benchmarking framework for network devices. In Lecture Notes in Computer Science (including Subseries Lecture Notes in Artificial Intelligence and Lecture Notes in Bioinformatics). Proceedings of the *8th International IFIP-TC 6 Networking Conference, Aachen, Germany*, 795–808. Andreoletti, D. Troia, S., Musumeci. F, Giordano. S., Maier. G, Tornatore.M.(2019). Network Traffic Prediction based on Diffusion Convolutional recurrent neural networks. In Proceedings of the *IEEE INFOCOM 2019—IEEE Conference on Computer Communications Workshops (INFOCOM WKSHPS), Paris, France, IEEE: Piscataway*, NJ, USA, 246–251. DOI: 10.1007/978-3-642-01399-7_62

Mahmud, M., Rahman, K. S., Rokonuzzaman, M., Habib, A. K. M. A., Islam, M. R., Motakabber, S. M. A., Channumsin, S., & Chowdhury, S. (2023, December). Lithium-ion battery thermal management for electric vehicles using phase change material: A review," *. *Results in Engineering*, 20, 101424. DOI: 10.1016/j.rineng.2023.101424

Makar, K., Goel, S., Kaur, P., Singh, M., Jain, P., & Aggarwal, P. K. (2021). Reliability of Mobile Applications: A Review and Some Perspectives. *9th International Conference on Reliability, Infocom Technologies and Optimization (Trends and Future Directions)* (ICRITO), Noida, 1-4, https://doi: DOI: 10.1109/ICRITO51393.2021.9596350

Mangiaracina, R., Perego, A., Salvadori, G., & Tumino, A.BMangiaracina. (2017). A Comprehensive View of Intelligent Transport Systems for Urban Smart Mobility. *International Journal of Logistics*, 20(1), 39–52. DOI: 10.1080/13675567.2016.1241220

Mao, Q., Hu, F., & Hao, Q. (2018). Deep Learning for Intelligent Wireless Networks: A Comprehensive Survey. *IEEE Communications Surveys and Tutorials*, 20(4), 2595–2621. DOI: 10.1109/COMST.2018.2846401

Martel, F., Dube, Y., Kelouwani, S., Jaguemont, J., & Agbossou, K. (2016). Long-term assessment of economic plug-in hybrid electric vehicle battery lifetime degradationmanagement through near optimal fuel cell load sharing. *Journal of Power Sources*, 318(30), 270–282. DOI: 10.1016/j.jpowsour.2016.04.029

Martinez, C. M., Hu, X., Cao, D., Velenis, E., Gao, B., & Wellers, M. (2016). Energy Management in Plug-in Hybrid Electric Vehicles: Recent Progress and a Connected Vehicles Perspective. *IEEE Transactions on Vehicular Technology*, 66(6), 4534–4549. DOI: 10.1109/TVT.2016.2582721

Ma, S. C., Fan, Y., Guo, J. F., Xu, J. H., & Zhu, J. (2019). Analysing online behaviour to determine Chinese consumers' preferences for electric vehicles. *Journal of Cleaner Production*, 229, 244–255. DOI: 10.1016/j.jclepro.2019.04.374

Masoud, S., & Tare Yelmek, K. (2014). Multi-objective optimal design of hybridrenewable energy systems using PSO-simulation based approach. *Renewable Energy*, 68, 67–79. http://hdl.handle.net/1993/31040. DOI: 10.1016/j.renene.2014.01.011

Matheus, M. N., & Musti, K. S. (2023). Design and Simulation of a Floating Solar Power Plant for Goreagab Dam, Namibia. In Human Agro-Energy Optimization for Business and Industry (pp. 1-26). IGI Global. DOI: 10.4018/978-1-6684-4118-3.ch001

Ma, Y., Li, X., Shi, W., & Cheng, Q. (2024, April). STCLA: An Efficient Certificateless Authenticated Key Agreement Scheme for the Internet of Vehicles. *IEEE Transactions on Vehicular Technology*, 73(4), 4830–4841. DOI: 10.1109/TVT.2023.3334034

Ma, Z., Callaway, D. S., & Hiskens, I. A. (2010). Decentralized charging control for large populations of plug-in electric vehicles. In *Proceedings of the 49th IEEE Conference on Decision and Control (CDC)* (pp. 206–212). DOI: 10.1109/CDC.2010.5717547

Ma, Z., Callaway, D. S., & Hiskens, I. A. (2013). Decentralized charging control of large populations of plug-in electric vehicles. *IEEE Transactions on Control Systems Technology*, 21(1), 67–78. DOI: 10.1109/TCST.2011.2174059

Meegoda, , Charbel, G., & Watts, D. (2024). Second Life of Used Lithium-Ion Batteries from Electric Vehicles in the USA. *Environments (Basel, Switzerland)*, 11(5), 97. DOI: 10.3390/environments11050097

Meegoda, , Charbel, G., & Watts, D. (2024). Sustainable Management of Rechargeable Batteries Used in Electric Vehicles. *Batteries*, 10(5), 167. DOI: 10.3390/batteries10050167

Mehmet., ... (2024) A review on the lithium-ion battery problems used in electric vehicles,Next Sustainability,3,100036,ISSN 2949-8236,DOI: 10.1016/j.nxsust.2024.100036

Meijers, J., Michalopoulos, P., Motepalli, S., Zhang, G., Zhang, S., Veneris, A., & Jacobsen, H.-A. (2022). Blockchain for V2X: Applications and Architectures. *IEEE Open Journal of Vehicular Technology*, 3, 193–209. DOI: 10.1109/OJVT.2022.3172709

Meyers, J. C. (2021) "The Great Transition: A Cost-Benefit Analysis of Transitioning from Diesel Fuel Buses to Zero Emission Electric Buses for the NFTA in The Buffalo-Niagara Falls MSA". Applied Economics Theses. 44., available online. last accessed October 2024, https://digitalcommons.buffalostate.edu/economics_theses/44

Miao, R., Yang, J., Feng, X., Jia, H., Wang, J., & Nuli, Y. (2014). Novel dual-salts electrolyte solution for dendrite-free lithium-metal based rechargeable batteries with high cycle reversibility. *Journal of Power Sources*, 271, 291–297. DOI: 10.1016/j.jpowsour.2014.08.011

Mishra, S., Shukla, A., Arora, S., Kathuria, H., & Singh, M. (2020).Controlling Weather Dependent Tasks Using Random Forest Algorithm. Third *International Conference on Advances in Electronics, Computers and Communications* (ICAECC), Bengaluru, India, 2020, 1-8, https://doi: DOI: 10.1109/ICAECC50550.2020.9339508

Mishra, S. K., Gupta, A., Tomar, A., & Gupta, V. K. (2023). Health care prediction for various diseases using computational intelligence approaches: A review. In *2023 World Conference on Communication & Computing (WCONF)* (pp. 1–6). IEEE. DOI: 10.1109/WCONF58270.2023.10235128

Mohamed, A., Salehi, V., Ma, T., & Mohammed, O. (2013). Real-time energy management algorithm for plug-in hybrid electric vehicle charging parks involving sustainable energy. *IEEE Transactions on Sustainable Energy*, 5(2), 577–586. DOI: 10.1109/TSTE.2013.2278544

Mohamed, M. A., & Mohamed, F. A. (2020). Design and Simulate an Off-Grid PV System with a Battery Bank for EV Charging. *Universal Journal of Electrical and Electronic Engineering*, 7(5), 273–288. DOI: 10.13189/ujeee.2020.070502

Mohammad, A., Zamora, R., & Lie, T. T. (2020). Integration of Electric Vehicles in the Distribution Network. *Energies*, 13(17), 4541. DOI: 10.3390/en13174541

Mohanty, A., Ramasamy, A. K., Verayiah, R., & Mohanty, S. (2024). Neighborhood and community battery projects: A systematic analysis of their current state and future prospects. *Journal of Energy Storage*, 95, 112525. DOI: 10.1016/j.est.2024.112525

Mohanty, S., Panda, S., Parida, S. M., Rout, P. K., Sahu, B. K., Bajaj, M., Zawbaa, H. M., Kumar, N. M., & Kamel, S. (2022). Demand side management of electric vehicles in smart grids: A survey on strategies, challenges, modeling, and optimization. *Energy Reports*, 8, 12466–12490. DOI: 10.1016/j.egyr.2022.09.023

Mohsenzadeh, A., Pang, C., & Yang, L. (2021). Transmission congestion management considering EV parking lots and demand response programmes. *International Journal of Ambient Energy*, 42(15), 1727–1731. DOI: 10.1080/01430750.2019.1614985

Mori, . (2023). Cathode materials for lithium-sulfur battery: A review. *Journal of Solid State Electrochemistry*, 27(4), 813–839. DOI: 10.1007/s10008-023-05387-z

Mude, K. N., Dashora, H. K., Bertoluzzo, M., & Buja, G. (2014). From wired to in-moving charging of electric vehicles. In *Proceedings of the WSEAS International Conference on Development, Energy, Environment, Economics (DEEE)* (pp. 33–42).

Münderlein, J., Ipers, G., Steinhoff, M., Zurmühlen, S., & Sauer, D. U. (2020). Optimization of a hybrid storage system and evaluation of operation strategies. *International Journal of Electrical Power & Energy Systems*, 119, 105887. DOI: 10.1016/j.ijepes.2020.105887

Musti, K. S. S. (2023b). Energy Sector Growth and Sustainability: Role of Government. In Leadership and Governance for Sustainability (pp. 186-203). IGI Global. DOI: 10.4018/978-1-6684-9711-1.ch010

Musti, K. S. S. (2023a). Multicriteria Decision Analysis for Sustainable Green Financing in Energy Sector. In *Green Finance Instruments, FinTech, and Investment Strategies Sustainable Finance* (pp. 3–25). Springer International., DOI: 10.1007/978-3-031-29031-2_1

Mwasilu, F., Justo, J. J., Kim, E. K., Duc, T., & Jung, J. (2014). Electric vehicles and smart grid interaction. *Renewable & Sustainable Energy Reviews*, 34, 501–516. DOI: 10.1016/j.rser.2014.03.031

Naqvi, S. S. A., Jamil, H., Iqbal, N., Khan, S., Khan, M. A., Qayyum, F., & Kim, D. (2024). Evolving Electric Mobility Energy Efficiency: In-Depth Analysis of Integrated Electronic Control Unit Development in Electric Vehicles. *IEEE Access : Practical Innovations, Open Solutions*, 12, 15957–15983. DOI: 10.1109/ACCESS.2024.3356598

Natarajan, N. R., & Veerasamy, V. (2024). Security and Its Implications on Healthcare in an IoT-Enabled Smart City. In *Secure and Intelligent IoT-Enabled Smart Cities* (pp. 296-307). IGI Global. Ghai, A. S., Ghai, K., & Cakir, G. K. (2024). Generative AI-Enabled IoT Applications for Smart Cities: Unleashing Innovation and Paving the Way for the Future. In *Secure and Intelligent IoT-Enabled Smart Cities* (pp. 222-238). IGI Global. DOI: 10.4018/979-8-3693-2373-1.ch014

Natsuki Nakamura et al. (2022) Future potential for lithium-sulfur batteries,Journal of Power Sources,558,232566,ISSN 0378-7753,DOI: 10.1016/j.jpowsour.2022.232566

Nazari, F., Noruzoliaee, M., & Mohammadian, A. (2024). Electric vehicle adoption behavior and vehicle transaction decision: Estimating an integrated choice model with latent variables on a retrospective vehicle survey. *Transportation Research Record: Journal of the Transportation Research Board*, 2678(4), 378–397. DOI: 10.1177/03611981231184875

Nazari, F., Rahimi, E., & Mohammadian, A. K. (2019). Simultaneous estimation of battery electric vehicle adoption with endogenous willingness to pay. *ETransportation*, 1, 100008. DOI: 10.1016/j.etran.2019.100008

Neves, S. A., Marques, A. C., & Moutinho, V. (2020). Two-stage DEA model to evaluate technical efficiency on deployment of battery electric vehicles in the EU countries. *Transportation Research Part D, Transport and Environment*, 86, 102489. DOI: 10.1016/j.trd.2020.102489

Ningbo (2024), Ningbo Deye Inverter Technology Co., Ltd. | Solar Components | China. Available online, Retrieved October, 2024, from https://www.enfsolar.com/deye-inverter

Nordelöf, A., Messagie, M., Tillman, A. M., Ljunggren Söderman, M., & Van Mierlo, J. (2014). Environmental impacts of hybrid, plug-in hybrid, and battery electric vehicles—What can we learn from life cycle assessment? *The International Journal of Life Cycle Assessment*, 19(11), 1866–1890. DOI: 10.1007/s11367-014-0788-0

Ntombela, M., Musasa, K., & Moloi, K. (2023). A comprehensive review for battery electric vehicles (BEV) drive circuits technology, operations, and challenges. *World Electric Vehicle Journal*, 14(7), 195. DOI: 10.3390/wevj14070195

Olabi et al., (2022) Battery electric vehicles: Progress, power electronic converters, strength (S), weakness (W), opportunity (O), and threats (T),International Journal of Thermofluids,16,100212,ISSN 2666-2027,DOI: 10.1016/j.ijft.2022.100212

Olabi, , Allam, M. A., Abdelkareem, M. A., Deepa, T. D., Alami, A. H., Abbas, Q., Alkhalidi, A., & Sayed, E. T. (2023). Redox Flow Batteries: Recent Development in Main Components, Emerging Technologies, Diagnostic Techniques, Large-Scale Applications, and Challenges and Barriers. *Batteries*, 9(8), 409. DOI: 10.3390/batteries9080409

Ortiz, Y., Arevalo, W., Peña, D., & Jurado, F. (2024). Recent advances in thermal management strategies for lithium-ion batteries: A comprehensive review. *Batteries*, 10(3), 83. DOI: 10.3390/batteries10030083

Ouddah, N., Adouane, L., Abdrakhmanov, R., & Kamal, E. (2017, July). Optimal Energy Management Strategy of Plug-in Hybrid Electric Bus in Urban Conditions. In ICINCO (1) (pp. 304-311). DOI: 10.5220/0006436803040311

Panda, (2023) Comprehensive Review of Electric Vehicle Technology and Its Impacts: Detailed Investigation of Charging Infrastructure, Power Management, and Control Techniques. *Appl. Sci.13*, 8919. .DOI: 10.3390/app13158919

Panda, S., Mohanty, S., Rout, P. K., Sahu, B. K., Parida, S. M., Samanta, I. S., Bajaj, M., Piecha, M., Blazek, V., & Prokop, L. (2023). A comprehensive review on demand side management and market design for renewable energy support and integration. *Energy Reports*, 10, 2228–2250. DOI: 10.1016/j.egyr.2023.09.049

Panday, A., & Bansal, H. O. (2014). A review of optimal energy management-strategiesforhybridelectricvehicle. *International Journal of Vehicular Technology*, 2014(2), 1–19. DOI: 10.1155/2014/160510

Pandiyan, P., Saravanan, S., Usha, K., Kannadasan, R., Alsharif, M. H., & Kim, M. (2023). Technological advancements toward smart energy management in smart cities. *Energy Reports*, 10, 648–677. DOI: 10.1016/j.egyr.2023.07.021

Panta, U., Gairola, P., & Nezamuddin, N. (2024). Modelling benefit-to-cost ratio for initial phase electrification using battery electric bus. *Transport Policy*, 145, 137–149. DOI: 10.1016/j.tranpol.2023.10.016

Pashupatimath, S., & Madhavanavar, S. (2018). Different Techniques Used in Traffic Control System: An Introduction. *International Journal of Engineering Research & Technology (Ahmedabad)*, 6(3), 1–4.

Patil, D., McDonough, M. K., Miller, J. M., Fahimi, B., & Balsara, P. T. (2018). Wireless power transfer for vehicular applications: Overview and challenges. *IEEE Transactions on Transportation Electrification*, 4(1), 3–37. DOI: 10.1109/TTE.2017.2780627

Petavratzi, E., Sanchez-Lopez, D., Hughes, A., Stacey, J., Ford, J., & Butcher, A. (2022). The impacts of environmental, social and governance (ESG) issues in achieving sustainable lithium supply in the Lithium Triangle. *Mineral Economics*, 35(3-4), 673–699. DOI: 10.1007/s13563-022-00332-4

Potkány, M., Hlatká, M., Debnár, M., & Hanzl, J. (2018). Comparison of the Life-cycle Cost Structure of Electric and Diesel Buses. *Nase More (Dubrovnik)*, 65(4), 270–275. DOI: 10.17818/NM/2018/4SI.20

Pranshu, S., & Goyal, A. (2019). Study of computerized segmentation & classification techniques: An application to histopathological imagery. *Informatica (Vilnius)*, 43(4), 561–572.

Pranshu, S., & Goyal, A. (2022). Computer-assisted grading of follicular lymphoma: A classification based on SVM, machine learning, and transfer learning approaches. *Imaging Science Journal*, 70(1), 30–45. DOI: 10.1080/13682199.2022.2162663

Pranshu, S., & Goyal, A. (2022). Two-stage binary classification of follicular histology using support vector machine. *Zhonghua Yi Xue Yi Chuan Xue Za Zhi*, 31(3), 258–268.

Prasad et al., (2023) Environmental life cycle assessment of emerging solid-state batteries: A review, Chemical Engineering Journal Advances,13,100439,ISSN 2666-8211,DOI: 10.1016/j.ceja.2022.100439

Preetha Yesheswini, B., Jai Iswarya, S., Amani, B., Prakash, P., & Sindhu, M. R. (2020). Solar PV Charging Station for Electric Vehicles. *2020 International Conference for Emerging Technology (INCET)*, 1–7. DOI: 10.1109/INCET49848.2020.9154187

Protected, A. (2023). 4 Ways to Reduce Energy Grid Reliance! *Supply Chain Game ChangerTM*.https://supplychaingamechanger.com/4-ways-to-reduce-energy-grid-reliance/

Pundir, A., Singh, S., Kumar, M., Bafila, A., & Saxena, G. J. (2022). 'Cyberphysical systems enabled transport networks in smart cities: Challenges and enabling technologies of the new mobility era. *IEEE Access : Practical Innovations, Open Solutions*, 10, 16350–16364. DOI: 10.1109/ACCESS.2022.3147323

Qu, X., Kong, Q., Yin, F., & Xu, L. (2022, August). A Secure and Privacy-Preserving Dynamic Aggregation Mechanism for V2G System. In *2022 19th Annual International Conference on Privacy, Security & Trust (PST)* (pp. 1-7). IEEE.

Qu, J., Wang, H., Zou, L., Zhang, L., Zhang, T., Zhou, J., & Zhang, B. (2024). A Two-Layer Energy Management Strategy of Fuel Cell Hybrid System in Electric Ships. *IEEE Transactions on Industry Applications*, 1–11. Advance online publication. DOI: 10.1109/TIA.2024.3447628

Rahulkumar, J., Narayanamoorthi, R., Vishnuram, P., Bajaj, M., Blazek, V., Prokop, L., & Misak, S. (2022). An empirical survey on wireless inductive power pad and resonant magnetic field coupling for in-motion EV charging system. *IEEE Access : Practical Innovations, Open Solutions*, 11, 4660–4693. DOI: 10.1109/ACCESS.2022.3232852

Ramakrishnan, D., & Radhakrishnan, K. (2022). Applying Deep Convolutional Neural Network (DCNN) Algorithm in the Cloud Autonomous Vehicles Traffic Model. *The International Arab Journal of Information Technology*, 19(2), 186–194. DOI: 10.34028/iajit/19/2/5

Ratana, P. (2024). AI in Energy Development for Cambodia. .DOI: 10.1080/02671522.2014.9084081130

Rau, F., Soto, I., Adasme, P., Zabala-Blanco, D., & Azurdia-Meza, C. A. (2021). Network Traffic Prediction Using Online-Sequential Extreme Learning Machine. In Proceedings of the 2021 *Third South American Colloquium on Visible Light Communications (SACVLC), Toledo, Brazil, IEEE: Piscataway*, NJ, USA, 1–6. DOI: 10.1109/SACVLC53127.2021.9652247

Rauf, M., Kumar, L., Zulkifli, S. A., & Jamil, A. (2024). Aspects of Artificial Intelligence in Future Electric Vehicle Technology for Sustainable Environmental Impact. *Environmental Challenges*, 14, 100854. DOI: 10.1016/j.envc.2024.100854

Rehan, H. (2024). The Future of Electric Vehicles: Navigating the Intersection of AI, Cloud Technology, and Cybersecurity. Valley International Journal Digital Library, 1127-1143.

Robert, . (2022). The Regulatory Environment for Lithium-Ion Battery Recycling. *ACS Energy Letters*, 7(2), 736–740. DOI: 10.1021/acsenergylett.1c02724

Robin, C., Gérard, M., Quinaud, M., d'Arbigny, J., & Bultel, Y. (2016). Proton exchange membrane fuel cell model for aging predictions: Simulated equivalent active surface area loss and comparisons with durability tests. *Journal of Power Sources*, 326, 417–427. DOI: 10.1016/j.jpowsour.2016.07.018

Rodríguez, T., & García, N. (2010). An Adaptive, real-time, Traffic Monitoring System. *Machine Vision and Applications*, 21(4), 555–576. DOI: 10.1007/s00138-009-0185-z

Roga, S., Bardhan, S., Kumar, Y., & Dubey, S. K. (2022). Recent technology and challenges of wind energy generation: A review. *Sustainable Energy Technologies and Assessments*, 52, 102239. DOI: 10.1016/j.seta.2022.102239

Saeed, M. A., Ahmed, N., Hussain, M., & Jafar, A. (2016, January).A comparative study of controllers for optimal speed control of hybrid electric vehicle.In *2016 International Conference on Intelligent Systems Engineering (ICISE)* (pp. 1-4).IEEE. DOI: 10.1109/INTELSE.2016.7475142

Salama, A., Saleh, B., & Eassa, M. (2010).Intelligent Cross Road Traffic Management System (ICRTMS). Proceedings of the ICCTD 2nd *International Conference on Computer Technology and Development, Cairo,* 27-31, 2010.

Sanguesa, J. A., Torres-Sanz, V., Garrido, P., Martinez, F. J., & Marquez-Barja, J. M. (2021). A review on electric vehicles: Technologies and challenges. *Smart Cities*, 4(1), 372–404. DOI: 10.3390/smartcities4010022

Sankaran, S. (2022). Pattern Matching Based Vehicle Density Estimation Technique for Traffic Monitoring Systems. *The International Arab Journal of Information Technology*, 19(4), 575–581. DOI: 10.34028/iajit/19/4/1

Santucci, A., & Sorniotti, A. andLekakou,C.(2014).Powersplitstrategiesforhybridenergystoragesystemsforvehicularapplications.*J.PowerSources*258,(pp.395–407).. https://doi.org/DOI: 10.1016/j.jpowsour.2014.01.118

Sathya, G., & Fathima Shameema, S, Jyothi MolSebastian, Jemsya K S. (2010). Automatic Rescue System for Ambulance and Authoritative Vehicles. *IEEE Access : Practical Innovations, Open Solutions.*

Saxena, P., & Goyal, A. (2023). Accurate demarcation of a biased nucleus from H&E-stained follicular lymphoma tissue samples. *Imaging Science Journal*, 71(8), 715–727. DOI: 10.1080/13682199.2023.2192550

Saxena, P., Singh, S. K., & Agrawal, P. (2013). Texture classification of biased cytoplasmic tissue samples from histopathological imagery by Gabor application. *Journal of Network Innovation and Computation*, 1(1), 248–259.

Saxena, P., Singh, S. K., & Sinha, A. (2024). Computer-assisted interpretation, in-depth exploration, and single-cell type annotation of RNA sequence data using k-means clustering algorithm. *Computer Methods in Biomechanics and Biomedical Engineering*, 1–11. Advance online publication. DOI: 10.1080/10255842.2023.2300685 PMID: 38235728

Saxena, P., Sinha, A., & Singh, S. K. (2023). The outbreak of coronavirus (2019-nCoV) in India: A statistical case study. *Journal of International Science and Technology*, 11(4), 570.

Sayahpour, , Hirsh, H., Parab, S., Nguyen, L. H. B., Zhang, M., & Meng, Y. S. (2022). Perspective: Design of cathode materials for sustainable sodium-ion batteries. *MRS Energy & Sustainability : a Review Journal*, 9(2), 183–197. DOI: 10.1557/s43581-022-00029-9

Sbordone, D., Bertini, I., Di Pietra, B., Falvo, M. C., Genovese, A., & Martirano, L. (2015). EV fast charging stations and energy storage technologies: A real implementation in the smart microgrid paradigm. *Electric Power Systems Research*, 120, 96–108. DOI: 10.1016/j.epsr.2014.07.033

Schopper, D., Kübler, K., Rudolph, S., & Riedel, O. (2021). EIPPM—The Executable Integrative Product-Production Model. *Computers*, 2021(10), 72. DOI: 10.3390/computers10060072

Sean et al., (2023) Designing lithium-ion batteries for recycle: The role of adhesives,Next Energy,1,2,2023,100023,ISSN 2949-821X,DOI: 10.1016/j.nxener.2023.100023

Sebastian, R., & Alzola, R. P. (2012). Flywheel energy storage systems: Review and simulation for an isolated wind power system. *Renewable & Sustainable Energy Reviews*, 16(9), 6803–6813. DOI: 10.1016/j.rser.2012.08.008

Shajahan, M. A. (2021). Next-Generation Automotive Electronics: Advancements in Electric Vehicle Powertrain Control. *Digitalization & Sustainability Review*, 1(1), 71–88.

Shamshad, S., Mahmood, K., Shamshad, U., Hussain, I., Hussain, S., & Das, A. K. (2023). A Provably Secure and Lightweight Access Control Protocol for EI-based Vehicle to Grid Environment. *IEEE Internet of Things Journal*, 10(18), 16650–16657. DOI: 10.1109/JIOT.2023.3269150

Shanthamallu, U. S., Spanias, A., Tepedelenlioglu, C., & Stanley, M. (2017). 'A brief survey of machine learning methods and their sensor and IoT applications. *8th International Conference Inf., Intell., Syst. Appl*(IISA), 1–8. DOI: 10.1109/IISA.2017.8316459

Shan, Z., & Zhu, Q. (2015). Camera Location for RealTime Traffic State Estimation in Urban Road Network Using Big GPS Data. *Neurocomputing*, 169, 134–143. DOI: 10.1016/j.neucom.2014.11.093

Shariff, S. M., Alam, M. S., Ahmad, F., Rafat, Y., Asghar, M. S. J., & Khan, S. (2020). System Design and Realization of a Solar-Powered Electric Vehicle Charging Station. *IEEE Systems Journal*, 14(2), 2748–2758. DOI: 10.1109/JSYST.2019.2931880

Sharma, A., Singh, M., Gupta, M., Sukhija, N., & Aggarwal, P. K. (2022). IoT and blockchain technology in 5G smart healthcare. In *Blockchain Applications for Healthcare Informatics*, 137–161. https://doi.org/DOI: 10.1016/B978-0-323-90615-9.00004-9

Sharma, D., Tiwari, D., Singh, V., Pandey, P., & Gupta, V. K. (2022). Cloud-based occlusion aware intrusion detection system. In International Conference on Cybersecurity in Emerging Digital Era (pp. 69–78). Springer Nature Singapore. https://doi.org/DOI: 10.1007/978-981-16-8523-0_7

Sharmila, B., Srinivasan, K., Devasena, D., Suresh, M., Panchal, H., Ashokkumar, R., & Shah, R. R. (2022). Modelling and performance analysis of electric vehicle. *International Journal of Ambient Energy*, 43(1), 5034–5040. DOI: 10.1080/01430750.2021.1932587

Sha, S., Li, J., Zhang, K., Yang, Z., Wei, Z., Li, X., & Zhu, X. (2020). RNN-Based Subway Passenger Flow Rolling Prediction. *IEEE Access : Practical Innovations, Open Solutions*, 2020(8), 15232–15240. DOI: 10.1109/ACCESS.2020.2964680

Shetty, D. K., Shetty, S., Raj Rodrigues, L., Naik, N., Maddodi, C. B., Malarout, N., & Sooriyaperakasam, N. (2020). Barriers to widespread adoption of plug-in electric vehicles in emerging Asian markets: An analysis of consumer behavioral attitudes and perceptions. *Cogent Engineering*, 7(1), 1796198. DOI: 10.1080/23311916.2020.1796198

Shin, Y. B., Lee, S., Chun, S. G., & Chung, D. (2013). A critical review of popular multi-criteria decision making methodologies. *Issues in Information Systems*, 14(1), 358–365.

Shirunda O.S., Kanyimba K., Musti K.S.S (2023), Policy perspective for developing electric vehicle ecosystem, in Ecosystem Dynamics and Strategies for Startups Scalability, pp. 175 – 195, DOI: DOI: 10.4018/979-8-3693-0527-0.ch009

Shiva, C. K., Vedik, B., & Manoharan, G. (2024). Sustainable Smart Transportation: Technologies, Benefits, Challenges to Urbanisation Concept. In *Secure and Intelligent IoT-Enabled Smart Cities* (pp. 355-370). IGI Global.

Shwetha, R. J., Srinath, M., & Vibha, S. (2018). Intelligent traffic signal management system using cloud vision API and Machine learning. [IJERT]. *International Journal of Engineering Research & Technology (Ahmedabad)*.

Siddique, A. R. M., Mahmud, S., & Van Heyst, B. (2018, October). A comprehensive review on a passive (phase change materials) and an active (thermoelectric cooler) battery thermal management system and their limitations," *. *Journal of Power Sources*, 401, 224–237. DOI: 10.1016/j.jpowsour.2018.08.094

Singh, M. K., Singh, P., Gupta, V. K., Mishra, K., & Gupta, A. (2023). Performance analysis of CNN models with data augmentation in rice diseases. In 2023 3rd Asian Conference on Innovation in Technology (ASIANCON) (pp. 1–5). IEEE. DOI: 10.1109/ASIANCON58793.2023.10270518

Singh, M., & Malik, A. (2024). Multi-Hop routing protocol in SDN-Based wireless sensor network. In *CRC Press eBooks* (pp. 121–141). DOI: 10.1201/9781003432869-8

Singh, M., Sharma, R., Kumar, K., Singh, M., & Vajpayee, P. (2024). Securing lives and assets: IoT-based earthquake and fire detection for real-time monitoring and safety. In *Communications in Computer and Information Science* (pp. 15–25). https://doi.org/DOI: 10.1007/978-3-031-56703-2_2

Singh, M., Sukhija, N., Sharma, A., Gupta, M., & Aggarwal, P. K. (2021). Security and privacy requirements for IOMT-Based Smart Healthcare System. In *CRC Press eBooks* (pp. 17–37). https://doi.org/DOI: 10.1201/9781003032328-2

Singh, R., Sharma, R., Kumar, K., Singh, M., & Vajpayee, P. (2024). Securing lives and assets: IoT-Based earthquake and fire detection for Real-Time monitoring and safety. In *Communications in computer and information science*,15–25, https://doi .org/DOI: 10.1007/978-3-031-56703-22

Singh. M, Gupta. M, Sharma. A, Aggarwal. Jain.(2023). Role of deep learning in healthcare industry: Limitations, challenges and future scope, Deep learning for healthcare services. In *BENTHAM SCIENCE PUBLISHERS eBooks*. https://doi .org/.DOI: 10.2174/97898150802301230201

Singh. R, Kumar.H. Singla.(2015). An intrusion detection system using network traffic profiling and an online sequential extreme learning machine. *Expert Syst. Appl,* Vol 42, 8609–8624.

Singh, M., & Malik, A. (2024). Multi-hop routing protocol in SDN-based wireless sensor networks: A comprehensive survey. In *Software-Defined Network Frameworks* (pp. 121–141). CRC Press.

Singh, R., & Avikal, S. (2020). COVID-19: A decision-making approach for prioritization of preventive activities. *International Journal of Healthcare Management*, 13(3), 257–262. DOI: 10.1080/20479700.2020.1782661

Singh, S. K., Kumar, M., Khanna, A., & Virdee, B. (2024). Blockchain and FL-based secure architecture for enhanced external Intrusion detection in smart farming. *IEEE Internet of Things Journal*, 1. DOI: 10.1109/JIOT.2024.3478820

Singh, S. K., Pan, Y., & Park, J. H. (2022). Blockchain-enabled secure framework for energy-efficient smart parking in sustainable city environment. *Sustainable Cities and Society*, 76, 103364. DOI: 10.1016/j.scs.2021.103364

Singh, S. K., & Park, J. H. (2022). TaLWaR: Blockchain-based trust management scheme for smart enterprises with augmented intelligence. *IEEE Transactions on Industrial Informatics*, 19(1), 626–634. DOI: 10.1109/TII.2022.3204692

Singh, S. K., Park, J. H., Sharma, P. K., & Pan, Y. (2021). BIIoVT: Blockchain-based secure storage architecture for intelligent internet of vehicular things. *IEEE Consumer Electronics Magazine*, 11(6), 75–82. DOI: 10.1109/MCE.2021.3089992

Singh, S. K., Park, L., & Park, J. H. (2022). Blockchain-based federated approach for privacy-preserved IoT-enabled smart vehicular networks. In *The 13th International Conference of ICT Convergence (ICTC-2022)*, Jeju, Korea, Oct. 19-21, pp. 1-5. DOI: 10.1109/ICTC55196.2022.9952835

Singh, S. K., Rashid, M., Alshamrani, S. S., Alnfiai, M. M., Saxena, P., & Khamparia, A. (2024). Efficient transfer learning approach for acute lymphoblastic leukemia diagnosis: Classification of lymphocytes and lymphoblastic cells. *TS. Traitement du Signal*, 41(4), 1749–1761. DOI: 10.18280/ts.410409

Singh, S. K., Yang, L. T., & Park, J. H. (2023). FusionFedBlock: Fusion of blockchain and federated learning to preserve privacy in industry 5.0. *Information Fusion*, 90, 233–240. DOI: 10.1016/j.inffus.2022.09.027

Sisakat, S. T. andBarakati,S.M.(2016).Energymanagementusingfuzzycontroller-for hybrid electrical vehicles.*J. Intelligent Fuzzy Syst.* 30, (pp.1411–1420), DOI: 10.3233/IFS-152054

Skippon, S., & Garwood, M. (2011). Responses to battery electric vehicles: UK consumer attitudes and attributions of symbolic meaning following direct experience to reduce psychological distance. *Transportation Research Part D, Transport and Environment*, 16(7), 525–531. DOI: 10.1016/j.trd.2011.05.005

Śmieszek, M., & Mateichyk, V. (2021). Determining the fuel consumption of a public city bus in urban traffic. *IOP Conference Series. Materials Science and Engineering*, 1199(1), 012080. DOI: 10.1088/1757-899X/1199/1/012080

Smith, M., Zhao, L., Cordova, J., Jiang, X., & Ebrahimi, M. (2024, January).Machine Learning-Based Energy-efficient Workload Management for Data Centers. In *2024 IEEE 21st Consumer Communications & Networking Conference (CCNC)* (pp. 799-802). IEEE. DOI: 10.1109/CCNC51664.2024.10454842

Song, C. X., Zhou, F., & Xiao, F. (2017a). Energy management optimization ofhybrid energy storage system based on dynamic programming. *J. Jilin Univ. Eng. Technol.Ed.*, (1), 8–14. DOI: 10.13229/j.cnki.jdxbgxb201701002

Song, C. X., Zhou, F., & Xiao, F.. (2017b). Parameter matching of on-board hybridenergy storage system based on convex optimization method. *Jixie Gongcheng Xuebao*, 53(08), 44–51. DOI: 10.3901/JME.2017.16.044

Sørensen, Å. L., Jiang, S., Torsæter, B. N., & Völler, S. (2018). Smart EV charging systems for zero emission neighbourhoods.

Sortomme, E., Hindi, M. M., MacPherson, S. D. J., & Venkata, S. (2011). Coordinated charging of plug-in hybrid electric vehicles to minimize distribution system losses. *IEEE Transactions on Smart Grid*, 2(1), 198–205. DOI: 10.1109/TSG.2010.2090913

Speirs, J., Contestabile, M., Houari, Y., & Gross, R. (2014). The future of lithium availability for electric vehicle batteries. *Renewable & Sustainable Energy Reviews*, 35, 183–193.

Stephen, N., & Steven, K. (2023). Cloud-Powered Drives: The Convergence of AI, Machine Learning, and Electric Vehicles Battery Technology. *International Journal of Advanced Engineering Technologies and Innovations*, 1(02), 116–130.

Subramani, J., Maria, A., Rajasekaran, A. S., & Chapa, B. P. (2024). Mutual and Batch Authentication with Conditional Privacy-Preserving Scheme for V2G Communication System. *IEEE Access*.

Sun, L., Jin, Y., Pan, L., Shen, J., & Lee, K. Y. (2019). Efficiency analysis and control of a grid-connected PEM fuel cell in distributed generation. *Energy Conversion and Management*, 195, 587–596. DOI: 10.1016/j.enconman.2019.04.041

Sureshkumar, V., Chinnaraj, P., Saravanan, P., Amin, R., & Rodrigues, J. J. (2022). Authenticated key agreement protocol for secure communication establishment in vehicle-to-grid environment with FPGA implementation. *IEEE Transactions on Vehicular Technology*, 71(4), 3470–3479. DOI: 10.1109/TVT.2022.3146409

Surya, S., & Patil, V. (2019). Cuk Converter as an Efficient Driver for LED. 2019 4th International Conference on Electrical, Electronics, Communication, Computer Technologies and Optimization Techniques (ICEECCOT), 49–52. DOI: 10.1109/ICEECCOT46775.2019.9114648

Su, W., Eichi, H., Zeng, W., & Chow, M. (2012). A survey on the electrification of transportation in a smart grid environment. *IEEE Transactions on Industrial Informatics*, 8(1), 1–10. DOI: 10.1109/TII.2011.2172454

Swathi, K., Sivanagaraju, V., Manikanta, A., & Kumar, S. (2016). Traffic Density Control and Accident Indicator Using WSN. *International Journal for Modern Trends in Science and Technology*, 2(04), 18–21.

Syed, F. U., Kuang, M. L., Czubay, J., & Ying, H. (2006). Derivation and Experimental Validation of a Power-Split Hybrid Electric Vehicle Model. *IEEE Transactions on Vehicular Technology*, 55(6), 1731–1747. DOI: 10.1109/TVT.2006.878563

Szinai, J. K., Sheppard, C. J., Abhyankar, N., & Gopal, A. R. (2020). Reduced grid operating costs and renewable energy curtailment with electric vehicle charge management. *Energy Policy*, 136, 111051. DOI: 10.1016/j.enpol.2019.111051

Taghizad-Tavana, K., Alizadeh, A., Ghanbari-Ghalehjoughi, M., & Nojavan, S. (2023). A comprehensive review of electric vehicles in energy systems: Integration with renewable energy sources, charging levels, different types, and standards. *Energies*, 16(2), 630. DOI: 10.3390/en16020630

Teimoori, Z., & Yassine, A. (2022). A Review on Intelligent Energy Management Systems for Future Electric Vehicle Transportation. *Sustainability (Basel)*, 14(21), 14100. DOI: 10.3390/su142114100

Teoh, Y. K., Gill, S. S., & Parlikad, A. K. (2022). IoT and fog-computing-based predictive maintenance model for effective asset management in Industry 4.0 using machine learning. *IEEE Internet of Things Journal*, 10(3), 2087–2094. DOI: 10.1109/JIOT.2021.3050441

Tete, P. R., Gupta, M. M., & Joshi, S. S. (2021, March). Developments in battery thermal management systems for electric vehicles: A technical review," *. *Journal of Energy Storage*, 35, 102255. DOI: 10.1016/j.est.2021.102255

Thakur, A. K., Prabakaran, R., Elkadeem, M. R., Sharshir, S. W., Arıcı, M., Wang, C., Zhao, W., Hwang, J.-Y., & Saidur, R. (2020, December). A state of art review and future viewpoint on advance cooling techniques for Lithium–ion battery system of electric vehicles," *. *Journal of Energy Storage*, 32, 101771. DOI: 10.1016/j.est.2020.101771

Tian, M., Wei, Y., Huang, X., & Ding, Z. (2022). Integrated Charging Station and Mixed Fleet Planning for Electric Taxi and Electric Bus. 2022 IEEE/IAS Industrial and Commercial Power System Asia (I&CPS Asia), 687–692. DOI: 10.1109/ICPSAsia55496.2022.9949794

Valentine, K., Temple, W. G., & Zhang, K. M. (2011). Intelligent electric vehicle charging: Rethinking the valley-fill. *Journal of Power Sources*, 196(24), 10717–10726. DOI: 10.1016/j.jpowsour.2011.08.076

van Vliet, O., Brouwer, A. S., Kuramochi, T., van den Broek, M., & Faaij, A. (2011, February). Energy use, cost and CO2 emissions of electric cars. *Journal of Power Sources*, 196(4), 2298–2310. DOI: 10.1016/j.jpowsour.2010.09.119

Verbrugge, B., Hasan, M. M., Rasool, H., Geury, T., El Baghdadi, M., & Hegazy, O. (2021). Smart Integration of Electric Buses in Cities: A Technological Review. *Sustainability (Basel)*, 13(21), 12189. DOI: 10.3390/su132112189

Volvo 9700 | Volvo Buses. (2024). Available online, Retrieved October 2024, from https://www.volvobuses.com/en/coaches/coaches/volvo-9700.html

Von Jouanne, A., Husain, I., Wallace, A., & Yokochi, A. (2005). Gone with the wind: Innovative hydrogen/fuel cell electric vehicle infrastructure based on wind energy sources. *IEEE Industry Applications Magazine*, 11(4), 12–19. DOI: 10.1109/MIA.2005.1458270

Waag, W., Fleischer, C., & Sauer, D. U. (2014). A critical review of the methods for monitoring of lithium-ion batteries in electric and hybrid vehicles. *Journal of Power Sources*, 258, 321–339. DOI: 10.1016/j.jpowsour.2014.02.064

Wang, B. W. (2018). *Studyonparametermatchingandcontrolstrategyofthepowersystemofnewenergyvehicle*. HunanUniv., DOI: 10.1177/0954407020936781

Wang, W., Zhang, Q., Peng, Z., Shao, Z., & Li, X. (2020, May 1). An empirical evaluation of different usage pattern between car-sharing battery electric vehicles and private ones. *Transportation Research Part A, Policy and Practice*, 135, 115–129. DOI: 10.1016/j.tra.2020.03.014

Wang, Y. M., Luo, Y., & Hua, Z. (2008). On the extent analysis method for fuzzy AHP and its applications. *European Journal of Operational Research*, 186(2), 735–747. DOI: 10.1016/j.ejor.2007.01.050

Wang, Y., Sun, Z., & Chen, Z. (2019). Energy management strategy for battery/supercapacitor/fuel cell hybrid source vehicles based on finite state machine. *Applied Energy*, 254, 113707. DOI: 10.1016/j.apenergy.2019.113707

Wazeer, A., Das, A., Abeykoon, C., Sinha, A., & Karmakar, A. (2022, September). Phase change materials for battery thermal management of electric and hybrid vehicles: A review," *. *Energy Nexus*, 7, 100131. DOI: 10.1016/j.nexus.2022.100131

Wei, D., He, H., & Cao, J. (2020). Hybrid electric vehicle electric motors for optimum energy efficiency: A computationally efficient design. *Energy, 203*, 117779. Bhatti, G., Mohan, H., & Singh, R. R. (2021). Towards the future of smart electric vehicles: Digital twin technology. *Renewable & Sustainable Energy Reviews*, 141, 110801.

Wei, W., Cao, M., Jiang, Q., Ou, S. J., & Zou, H. (2020). What influences Chinese consumers' adoption of battery electric vehicles? A preliminary study based on factor analysis. *Energies*, 13(5), 1057. DOI: 10.3390/en13051057

Weldon, P., Morrissey, P., & O'Mahony, M. (2018). Long-term cost of ownership comparative analysis between electric vehicles and internal combustion engine vehicles. *Sustainable Cities and Society*, 39, 578–591. DOI: 10.1016/j.scs.2018.02.024

Wenz, K.-P., Serrano-Guerrero, X., Barragán-Escandón, A., González, L. G., & Clairand, J.-M. (2021). Route prioritization of urban public transportation from conventional to electric buses: A new methodology and a study of case in an intermediate city of Ecuador. *Renewable & Sustainable Energy Reviews*, 148, 111215. DOI: 10.1016/j.rser.2021.111215

Wong, L. T., Ramachandaramurthy, V. K., Taylor, P., Ekanayake, J., Walker, S. I., & Padmanaban, S. (2019). Review on the optimal placement, sizing, and control of an energy storage system in the distribution network. *Journal of Energy Storage*, 21, 489–504. DOI: 10.1016/j.est.2018.12.015

Wu, J., & He, Y. (2021). Prediction of GDP in Time Series Data Based on Neural Network Model. In Proceedings of the 2021 *IEEE International Conference on Artificial Intelligence and Industrial Design (AIID), Guangzhou, China, IEEE: Piscataway*, NJ, USA, 20–23. DOI: 10.1109/AIID51893.2021.9456509

Wu, C., Zhu, C., Sun, J., & Ge, Y. (2017). A synthesized diagnosis approach forlithium-ion battery in hybrid electric vehicle. *IEEE Transactions on Vehicular Technology*, 66(7), 5595–5603. DOI: 10.1109/TVT.2016.2634125

Wu, W., Wang, S., Wu, W., Chen, K., Hong, S., & Lai, Y. (2019, February). A critical review of battery thermal performance and liquid based battery thermal management," *. *Energy Conversion and Management*, 182, 262–281. DOI: 10.1016/j.enconman.2018.12.051

Xiao, C., Choi, E., & Sun, J. (2018). Opportunities and challenges in developing deep learning models using electronic health records data: A systematic review. *Journal of the American Medical Informatics Association : JAMIA*, 2018(25), 1419–1428. DOI: 10.1093/jamia/ocy068 PMID: 29893864

Xie, S., Hu, X., Qi, S., Kun, L., Zongke, X., & James, B.. (2019a). Modelpredictiveenergymanagementforplug-inhybridelectricvehiclesconsideringoptimalbatterydepthofdischarge. *Energy*, 173, 667–678. DOI: 10.1016/j.energy.2019.02.074

Xiong, R., Sun, W., Yu, Q., & Sun, F. (2020). Research progress, challenges and prospects of fault diagnosis on battery system of electric vehicles. *Applied Energy*, 279, 115855. DOI: 10.1016/j.apenergy.2020.115855

Xiong, X., Zheng, K., Lei, L., & Hou, L. (2020). Resource allocation based on deep reinforcement learning in IoT edge computing. *IEEE Journal on Selected Areas in Communications*, 38(6), 1133–1146. DOI: 10.1109/JSAC.2020.2986615

Xiqian et al., (2022) Battery Safety: From Lithium-Ion to Solid-State Batteries,Engineering, 21,pp.9-14,ISSN 2095-8099, https://doi.org/DOI: 10.1016/j.eng.2022.06.022

Xu, B., Li, P., & Chan, C. L. (2015). Application of phase change materials for thermal energy storage in concentrated solar thermal power plants: A review. *Applied Energy*, 160, 286–307. DOI: 10.1016/j.apenergy.2015.09.016

Xu, W., Chen, H., Zhao, H., & Ren, B. (2019). Torque optimization control for electric vehicles with four in-wheel motors equipped with regenerative braking system. *Mechatronics*, 57, 95–108. DOI: 10.1016/j.mechatronics.2018.11.006

Yang, J., Liu, B., Zhang, T., Hong, J., & Zhang, H. (2023). Multi-parameter controlled mechatronics-electro-hydraulic power coupling electric vehicle based on active energy regulation. *Energy*, 263, 125877. DOI: 10.1016/j.energy.2022.125877

Yang, J., Ni, Q., Luo, G., Cheng, Q., Oukhellou, L., & Han, S. (2023, December). A Trustworthy Internet of Vehicles: The DAO to Safe, Secure, and Collaborative Autonomous Driving. *IEEE Transactions on Intelligent Vehicles*, 8(12), 4678–4681. DOI: 10.1109/TIV.2023.3337345

Yang, Y., Pei, H., Hu, X., Liu, Y., Hou, C., & Cao, D. (2019). Fuel economy optimization of power split hybrid vehicles: A rapid dynamic programming approach. *Energy*, 166, 929–938. DOI: 10.1016/j.energy.2018.10.149

Yan, Y., Qian, Y., Sharif, H., & Tipper, D. (2012). A survey on smart grid communication infrastructures: Motivations, requirements, and challenges. *IEEE Communications Surveys and Tutorials*, 15(1), 1–16.

Yao, T. Z., Xie, C. J., & Zeng, T.. (2019). Multi-Fuzzy control based energymanagementstrategyofbattery/supercapacitorhybridenergysystemofelectricvehicles. *Automotive Engineer*, (6), 615–640. DOI: 10.19562/j.chinasae.qcgc.2019.06.002

Yiju., ... (2021) Material design and structure optimization for rechargeable lithium-sulfur batteries, Matter,4,4,2021,pp.1142-1188,ISSN 2590-2385,DOI: 10.1016/j.matt.2021.01.012

Zahraee, S. M., Assadi, M. K., & Saidur, R. (2016). Application of Artificial Intelligence Methods for Hybrid Energy System Optimization. *Renewable & Sustainable Energy Reviews*, 66, 617–630. DOI: 10.1016/j.rser.2016.08.028

Zanoletti, , Carena, E., Ferrara, C., & Bontempi, E. (2024). A Review of Lithium-Ion Battery Recycling: Technologies, Sustainability, and Open Issues. *Batteries*, 10(1), 38. DOI: 10.3390/batteries10010038

Zeng, C., Ma, C., Wang, K., & Cui, Z. (2022). Parking Occupancy Prediction Method Based on Multi Factors and Stacked GRU-LSTM. *IEEE Access : Practical Innovations, Open Solutions*, 10, 47361–47370. DOI: 10.1109/ACCESS.2022.3171330

Zhang,L.,Hu,X.S.,Wang,Z.B.,etal.(2017).Multi-objectiveoptimalsizingofhybridenergy storage system for electric vehicles.*IEEE Trans. Veh. Technol.*, 1.. https://doi .org/DOI: 10.19562/j.chinasae.qcgc.2017.12.003

Zhang, F., Wang, L., Coskun, S., Pang, H., Cui, Y., & Xi, J. (2020). Energy management strategies for hybrid electric vehicles: Review, classification, comparison, and outlook. *Energies*, 13(13), 3352. DOI: 10.3390/en13133352

Zhang, K., & Dai, J. (2022). A novel TOPSIS method with decision-theoretic rough fuzzy sets. *Information Sciences*, 608, 1221–1244. DOI: 10.1016/j.ins.2022.07.009

Zhang, P., Yan, F. W., & Du, C. Q. (2015). A comprehensive analysis of energymanagement strategies for hybrid electric vehicles based on bibliometrics. *Renewable & Sustainable Energy Reviews*, 48(48), 88–104. DOI: 10.1016/j.rser.2015.03.093

Zhang, Q., Deng, W., & Li, G. (2017). Stochastic control of predictive power management for battery/supercapacitor hybrid energy storage systems of electric vehicles. *IEEE Transactions on Industrial Informatics*, 14(7), 3023–3030. DOI: 10.1109/TII.2017.2766095

Zhang, S., Xiong, R., & Cao, J. (2016). Battery durability and longevity based power management for plug-in hybrid electric vehicle with hybrid energy storage system. *Applied Energy*, 179, 316–328. DOI: 10.1016/j.apenergy.2016.06.153

Zhang, X., Kuehnelt, H., & De Roeck, W. (2021). Traffic Noise Prediction Applying Multivariate Bi- Directional Recurrent Neural Network. *Applied Sciences (Basel, Switzerland)*, 2021(11), 2714. DOI: 10.3390/app11062714

Zhang, X., Li, Z., Luo, L., Fan, Y., & Du, Z. (2022, January). A review on thermal management of lithium-ion batteries for electric vehicles. *Energy*, 238, 121652. Advance online publication. DOI: 10.1016/j.energy.2021.121652

Zhang, Z., Guan, C., & Liu, Z. (2020). Real-time optimization energy management strategy for fuel cell hybrid ships considering power sources degradation. *IEEE Access : Practical Innovations, Open Solutions*, 8, 87046–87059. DOI: 10.1109/ACCESS.2020.2991519

Zhao et al, (2023) Engineering of Sodium-Ion Batteries: Opportunities and Challenges, Engineering, 24,pp. 172-183,ISSN 2095-8099,DOI: 10.1016/j.eng.2021.08.032

Zhao, D., Chu, L., Xu, N., Sun, C., & Xu, Y. (2018).Development of a cooperative braking system for front-wheel drive electric vehicles.Energies, 11(2), (pp.378). DOI: 10.3390/en11020378

Zhao,B.,Song,Q.,andLiu,W.(2012).Powercharacterizationofisolatedbidirectional dual-active-bridge DC–DC converter with dual-phase-shiftcontrol.*PowerElectron. IEEETransaction*27(9), (pp.4172–4176)..DOI: 10.1109/TPEL.2012.2189586

Zhao, J., Zhou, J., Wang, L., & Zhao, Y. (2023). An energy management strategy to reduce the comprehensive cost of hybrid energy storage systems in electric vehicles. *Frontiers in Energy Research*, 11, 1191565. DOI: 10.3389/fenrg.2023.1191565

Zhou et al., (2024) Life cycle assessment of recycling lithium-ion battery related mineral processing by-products: A review,Minerals Engineering,208,108600,ISSN 0892-6875,DOI: 10.1016/j.mineng.2024.108600

Zhou, F., Song, C. X., & Liang, T. W.. (2017). Parameter matching of on-boardhybrid energy storage system using NSGA-II algorithm. *J. Jilin Univ. Eng. Technol. Ed.*, (9), 1336–1343. DOI: 10.1109/TVT.2017.2762368

Zhou, M. L., Feng, J. F., & Zhang, Y. (2019). Composite energystorage system anditsenergycontrolstrategyforelectricvehicles. *Electr.Mach.Control*, (5), 51–59.

Zhou, W., Yang, L., Cai, Y., & Ying, T. (2018). Dynamic programming for new energy vehicles based on their work modes Part II: Fuel cell electric vehicles. *Journal of Power Sources*, 407, 92–104. DOI: 10.1016/j.jpowsour.2018.10.048

Zhou, Y., Liu, X. C., Wei, R., & Golub, A. (2021). Bi-Objective Optimization for Battery Electric Bus Deployment Considering Cost and Environmental Equity. *IEEE Transactions on Intelligent Transportation Systems*, 22(4), 2487–2497. DOI: 10.1109/TITS.2020.3043687

Zichichi, M., Ferretti, S., & D'angelo, G. (2020). A framework based on distributed ledger technologies for data management and services in intelligent transportation systems. *IEEE Access : Practical Innovations, Open Solutions*, 8, 100384–100402. DOI: 10.1109/ACCESS.2020.2998012

About the Contributors

Parita Jain completed her Ph.D. Degree from Amity University, Noida, and M. M.Tech. (IT) from CDAC, Noida. Having vast teaching experience of approx. 14 years. Currently, she is working as an associate professor and assistant dean (academics and autonomous) in the Department of Computer Science and Engineering of the KIET Group of Institutions, Ghaziabad. Her research interests include Software Engineering, Machine Learning, Data Mining, and Mobile Computing. She has published more than 40 research papers in various international conferences and journals.

Puneet Kumar Aggarwal completed his Ph.D. degree from Amity University, Noida, and his M.Tech. (IT) degree from CDAC, Noida. Having vast teaching experience of approx. 14 years. Currently, he is working as an associate professor in computer science and engineering-data science at ABES Engineering College, Ghaziabad. His research interests include Machine Learning, Internet of Things, Data Mining and Mobile Computing. He has published more than 45 research papers in various international conferences and journals.

Mandeep Singh is an accomplished academician and Assistant Professor in the School of Computer Science, Engineering, and Technology (SCSET) at Bennett University, Greater Noida, India. He holds a Master's degree with Distinction in Information Security from Ambedkar Institute of Advanced Communication Technologies and Research (AIACTR) (now NSUT East Campus). With over nine years of teaching experience, he is currently pursuing a Ph.D. in Computer Science & Engineering from Dr. B.R. Ambedkar National Institute of Technology (NIT) Jalandhar, where he completed his coursework with Distinction (8.00 CGPA). Mr. Singh's expertise spans Computer Networks, Mobile Ad Hoc Networks (MANETs), Machine Learning, and Deep Learning. His research contributions include book chapters, international journals, and conference proceedings, alongside a notable track record of patent publications. Some of his patents address cybersecurity in email communication, AI-powered agricultural robotics, and intelligent machine vision systems. Additionally, he actively reviews papers for international conferences and journals and has held special sessions at prestigious events, including the International Conference on Computing and Communication Networks (ICCCNet). His dedication to excellence has earned him awards and recognition from the Institute of Industrial Development (IID) and MSME, Government of India for contributions to research and entrepreneurship.

Sushil Kumar Singh is an Associate Professor in the Department of Computer Engineering at Marwadi University, Rajkot, India. He received Ph.D. degree from Seoul National University of Science and Technology, Seoul, South Korea. He received M.Tech. Degree in Computer Science and Engineering from Uttarakhand Technical University, Dehradun, India. He also received an M.E. degree in Information Technology from Karnataka State University, Mysore, India. He has also been the lab leader of the UCS Lab at the Department of Computer Science Engineering, Seoul National University of Science and Technology, Seoul, South Korea. He has received the Best Lab Leadership Award from UCS Lab for 2019-2021. He has more than 12 years of experience teaching in the field of computer science. He has published Four Books: Computer C Programming, Cyber Security, Big Data Analytics, and Mobile Computing. He has also published many high-quality papers (Q1, Top 10% JCR Rank) in international journals and conferences. He has already delivered international lectures in many countries. His research interests include Blockchain, Artificial Intelligence, Big Data, Internet of Things, Smart City Security, and Cyber-Physical Systems. He is an Associate Editor in the HCIS (Human-centric Computation and Information Sciences) Journal and Editor in IGI Global Publication. He is a reviewer of the IEEE Wireless Communication Magazine, IEEE SYSTEMS, IEEE Internet of Things, FGCS, TETT, EXSY, JISA, Computer Network, MDPI, CIE, HCIS, JIPS, Computing (COMP), Multimedia Tools & Applications, and SCIS Journal.

Shailendra Rathore is working as a Lecturer in Division of Cybersecurity at Abertay University, UK and has more than 2 years of working experience as a lecturer at universities based in UK. He is also an active member of the cyberQuarter team, which is a new £18m cybersecurity research and development center located at Abertay University. Prior to joining Abertay, he was a lecturer in cybersecurity at Plymouth University and was associated with the £3 million Research England-supported Cyber-SHIP Lab project. He received a Ph.D. degree from the Graduate School of Computer Science and Engineering at Seoul National University of Science and Technology, Seoul, South Korea. His broad research interests include Cybersecurity, Artificial Intelligence, Blockchain, and Machine

Learning. Dr. Rathore has published his research outcomes in various top-tier international journals, including IEEE Communication Magazine, IEEE Network, IEEE Consumer Electronics, FGCS, Information Sciences, and IEEE Access. He has also served as a chair, program committee member or organizing committee chair for many international conferences and workshops, including IEEE and ACM events.

***

Hilde Amushembe received her Diploma: Power Engineering in 2006 and Bachelor of Technology: Power Engineering in 2012 both from Polytechnic of Namibia (now known as Namibia University of Science and Technology). She studied further and obtained her MTech.: Electrical Engineering from Cape Peninsula University of Technology, South Africa, in 2015. She is currently pursuing her Ph.D. in Electrical Engineering at Namibia University of Science and Technology, in Namibia. Hilde has industrial experience in managing infrastructure capital projects of over 15 years. She is currently working as a lecturer in the Department of Mechanical, Industrial and Electrical Engineering at Namibia University of Technology. Her main research interests include areas of Energy Supply Industry (ESI), power quality studies, Quality of supply (QoS), Quality of Service Supply (QoSS) and customer engagement.

Vishan Kumar Gupta has received his Ph.D. degree in Computer Science and Engineering Department, Thapar Institute of Engineering and Technology, Patiala, Punjab, India. He has received his M.Tech degree in Information Technology from ABV-Indian Institute of Information Technology and Management, Gwalior, MP, India and Bachelor of Engineering in CSE from Rajiv Gandhi Technical University, Bhopal, MP, India, Currently, he is working as an Associate Professor in Graphic Era Deemed to be University, Dehradun, Uttarakhand, India. His areas of research are Computational Biology, Data Mining, and Machine learning.

Paras Jain is a Senior Assistant Professor in the School of Computing Science & Engineering at VIT Bhopal University, India. He was awarded with Ph.D from Jaypee University of Engineering & Technology (JUET), Guna in the year 2015. His Ph.D dissertation title is "Adaptive Edge-preserving Image Denoising in Wavelet Domain". Dr. Jain has more than 10 years of academic experience. He has published several research papers in international journals and conferences of repute. He won the Best Paper Award for a research paper with title "An Adaptive Edge-preserving Image Denoising using Block-based Singular Value Decomposition in Wavelet Domain" presented in ICICT 2015. He has actively participated in various conferences, workshops and FDPs. Dr. Jain presently supervising two Ph.D students.

Dr. Jain has also been a reviewer to various reputed International Conferences and Journals. His research interest includes Image Processing, Data Mining, Machine Learning, and Pattern recognition, with expertise in subjects like, Digital Image Processing, Theory of Computation, Compiler Design, Data Structures, Design and Analysis of Algorithms, Algorithm for Advanced Analytics, etc.

ilampoornan K, currently he is working as a Professor in Department of Electrical Engineering, Bharath Institute of Higher Education and Research. His area of interest is mainly focusing on High Voltage Engineering

Pawan Kumar has completed his bachelor, Master and doctorate degree in computer science and engineering. Dr Pawan have total 2I years of teaching Experience . He published and presented more than75 research papers in reupdated journals and conferences. Presently Dr Pawan holding position of Associate Professor in Department of Computer Science of Ajay Kumar Garg Engineering College, Ghaziabad, UP, India.

KS Sastry Musti is presently with Namibia University of Science and Technology, Windhoek, Namibia. He has than 30 years of professional experience. He obtained his BTech in Electrical Engineering from JNTU, Kakinada (A.P.) in 1990, MTech and PhD from NIT, Warangal in the years 1996 and 2002 respectively. Some of his earlier employments include University of the West Indies, St Augustine, Trinidad, ESIGELEC, Rouen, France and Sophia University, Tokyo, Japan. His research interests include Power and Energy Systems, Information Systems and Engineering Education.

Damodharan Palaniappan works as an Associate Professor & Head in the Department of Information Technology, Marwadi University, Rajkot, Gujarat. He completed his Ph.D. at Anna University, Chennai, M.E. in Computer Science and Engineering at C.I.E.T, Coimbatore and B.E. in Computer Science and Engineering at CSI College of Engineering, Ketty, Nilgiri. He has 19 years experience of teaching and research. Currently, he guides 6 Ph.D. research scholars. He has authored 8 book chapters, and he has published research papers in reputed journals. His interested areas of research are Data Mining, Image Processing, Software, Machine Learning and Deep Learning.

Suriya Ponnambalam, currently she is working as a Assistant Professor in Department of Electrical Engineering, Bharath Institute of Higher Education and Research. Her area of interest is mainly focusing on Renewable Energy and Electric vehicles.

Saranya S working as an Assistant professor in the department of Computer Science and Engineering at PPG institute of technology has about 9 years of Industry and Teaching experience. She received B.E degree in Computer Science from Avinashilingam University and M.E degree in Computer Science with distinction from Karpagam University. She has published papers in international conferences and journals. Her interested areas of academics are Theory of computation, Embedded system, Artificial intelligence, Compiler Design and Cryptography etc.

Pranshu Saxena received the B. E. in (2010), M. Tech. (2013), and PhD in 2023. He is currently an Assistant Professor in the School of Computer Science Engineering and Technology at the Bennett University, Greater Noida, India. He has published 24 research articles which include 6-SCI, 6-Scopus, and 12 peer-review journals/conferences. His research interests include medical image processing, automated image segmentation, Image texture analysis, intelligent system and Pattern Classification. He has been a designated reviewer (ISIEA 2012-2013, IJERT) and committee member (IJACR) for many international conferences and journals in image processing and medical imaging. He has published one book published by LAP Lambert Academic Publishing, USA.

Naman Sharma is currently pursuing a Bachelor of Technology in Computer Science and Engineering at KIET Group of Institutions, Ghaziabad. As a second-year student, Naman has developed a strong interest in research, particularly in the fields of technology and engineering. This interest has led to a focus on contributing to academic literature, with a special emphasis on topics that merge innovation and practicality. Naman is dedicated to exploring and advancing knowledge in areas such as Advanced Battery Cooling Systems, with the goal of enhancing efficiency and longevity in engineering applications.

Premavathi T is a working as Assistant Professor in the Department of Computer Engineering – Artificial Intelligence, Marwadi University, Rajkot, Gujarat. She did B.E in Computer Science and Engineering at Dhanalakshmi Srinivasan Engineering College, Perambalur, TamilNadu. She did her M.E in Software Engineering with Distinction in Anna University. She is pursuing her Ph.D. at the Marwadi University. She is having 14 years experiences in teaching. She authored 3 book chapters and she published 3 research papers in national and international seminars and conferences. Her interested areas of academics are Object Oriented Programming, Software Engineering, Database Management Systems, Design and Analysis of Algorithms, Python Programming, Machine Learning and Deep Learning.

Taimi Uutoni is currently with the Ministry of Works & Transport, Government of Namibia. She is an Electrical Power Engineer by qualification and training and has 10+ years of professional experience. Her research interests are in the extended areas of Sustainable Energy Systems, Decarbonization, Climate Change and Electric Vehicles.

Index

A

B

C

D

E

F

G

I

K

M

P

R

www.ingramcontent.com/pod-product-compliance
Lightning Source LLC
LaVergne TN
LVHW080342011026
849161LV00009B/2156